Adsorption Applications for Environmental Sustainability

Online at: https://doi.org/10.1088/978-0-7503-5598-8

Adsorption Applications for Environmental Sustainability

Edited by

Kingsley Eghonghon Ukhurebor

Department of Physics, Edo State University, Uzairue, Nigeria

Uyiosa Osagie Aigbe

Department of Mathematics and Physics, Cape Peninsula University of Technology, Cape Town, South Africa

Robert Birundu Onyancha

Department of Technical and Applied Physics, School of Physics and Earth Sciences Technology, Technical University of Kenya, Nairobi, Kenya

IOP Publishing, Bristol, UK

ISBN 978-0-7503-5598-8 (ebook)
ISBN 978-0-7503-5596-4 (print)
ISBN 978-0-7503-5599-5 (myPrint)
ISBN 978-0-7503-5597-1 (mobi)

DOI 10.1088/978-0-7503-5598-8

Version: 20230901

IOP ebooks

British Library Cataloguing-in-Publication Data: A catalogue record for this book is available from the British Library.

Published by IOP Publishing, wholly owned by The Institute of Physics, London

IOP Publishing, No.2 The Distillery, Glassfields, Avon Street, Bristol, BS2 0GR, UK

US Office: IOP Publishing, Inc., 190 North Independence Mall West, Suite 601, Philadelphia, PA 19106, USA

To our families: past, present, and future.

Contents

Preface

Given the intensifying environmental contamination of water bodies, which ensues mostly from both natural and anthropogenic activities in the absence of appropriate treatment, there is a strong requirement for innovative approaches that use biosorbents for the removal of these contaminants due to their exclusive features such as biocompatibility, biodegradability, and renewability. Hence, this book, entitled 'Adsorption Applications for Environmental Sustainability,' aims to present the development, utilization, and applications of biosorbents for the adsorption of contaminants (such as heavy metals and dyes) in support of environmental sustainability and safety.

The book is made up of 15 chapters that particularly focus on advanced research into the use of nanotechnology in the synthesis of novel biosorbents using facile synthesis methods, the use of analytical spectroscopic techniques for their characterization, and the extensive applications of synthesized biosorbents in the remediation and photocatalytic degradation of contaminants (such as noxious metal ions and dye molecules). It also highlights the types, characteristics, and management options (reusability, recyclability, and final disposal) of adsorbents commonly used for reasons of environmental sustainability. Also covered are the costs, environmental evaluations, and comparisons of commonly used biosorbents, as well as the future scopes of adsorption applications used in support of environmental sustainability. Hence, this book will serve as a textbook and will be of great assistance to students, professionals, practitioners, scientists, researchers, and academicians in various research domains, particularly those in materials and environmental science.

Acknowledgements

First and foremost, we are most grateful to Almighty God, our creator, who kept us alive and gave us all we needed to make this book a success. We owe Him everything. We have always been and will forever be grateful to Him.

Our profound gratitude goes to our respective institutions' managements, who have always been of great assistance in our research and academic developments. We are also grateful to all the chapter contributors as well as the authors and publishers whose publications were used as a basis for writing this book, without whom this book would not have been possible.

We thank our families, mentors, colleagues, friends, and all who contributed and are still contributing in one way or another to the success of our academic and research activities as well as our other endeavours. May God bless you all richly!

Editor biographies

Kingsley Eghonghon Ukhurebor

Kingsley Eghonghon Ukhurebor is a lecturer/researcher and the acting head of the Department of Physics at Edo State University, Uzairue, Nigeria. He is also a research fellow at the West African Science Service Center on Climate Change and Adapted Land Use (WASCAL), Competence Center, Ouagadougou, Burkina Faso, a climate institute sponsored by the Federal Ministry of Education and Research, Germany. He holds a PhD in Physics Electronics from the University of Benin, Benin City, Nigeria. He is a member of several learned academic organizations, such as the Nigerian Young Academy (NYA), etc. His research interests are in applied physics, climate physics, environmental physics, telecommunication physics, and materials science (nanotechnology). He serves as an editor and reviewer for several reputable journals and publishers, such as Springer Nature, Elsevier, the Royal Society of Chemistry (RSC), the Institute of Physics (IOP), Taylor and Francis, Wiley, the IEEE, Frontiers, Hindawi, etc. He has authored or co-authored several publications with these reputable journals and publishers. He is presently ranked among the top 50 authors in Nigeria by Scopus scholarly output.

Uyiosa Osagie Aigbe

Uyiosa Osagie Aigbe is a research fellow at the Department of Mathematics and Physics, Faculty of Applied Science, Cape Peninsula University of Technology, Cape Town, South Africa. He obtained his PhD degree in Physics from the prestigious University of South Africa, Pretoria, South Africa. He is currently a member of several learned academic organizations. His research interests are applied physics, nanotechnology, fluid dynamics, water purification processes, image processing, environmental physics, and materials science. He has also served as a reviewer for numerous highly regarded journals. He has authored and co-authored several research publications.

Robert Birundu Onyancha

Robert Birundu Onyancha is a lecturer and researcher working full-time at the department of Technical and Applied Physics, School of Physics and Earth Science at the Technical University of Kenya. He holds a PhD in Physics from the University of South Africa. His research interests are in materials science, waste water treatment technologies, superconductivity, and magnetism. He is a registered member of various research bodies and has authored and co-authored research papers and book chapters which have been published in reputable and accredited journals and publishers. Furthermore, he serves as an editor and reviewer of highly accredited and trustworthy journals.

List of contributors

Kenneth Kennedy Adama
Department of Chemical Engineering, Edo State University Uzairue, PMB 04 Auchi, Edo State, Nigeria

Charles Ifeanyi Aghanwa
Pure and Industrial Chemistry Department, Faculty of Physical Sciences, Nnamdi Azikiwe University, PMB 5025, Awka, Anambra State, Nigeria

Efosa Aigbe
Department of Electrical and Electronic Engineering, Nile University of Nigeria, Abuja, Nigeria

Uyiosa Osagie Aigbe
Department of Mathematics and Physics, Cape Peninsula University of Technology, PO Box 1906, Cape Town, South Africa

Bamikole Olaleye Akinsehinde
Department of Computer Science, Aberystwyth University, Penglais Campus, West Wales, UK

Abimbola Bankole Akinyele
Pure and Industrial Chemistry Department, Faculty of Physical Sciences, Nnamdi Azikiwe University, PMB 5025, Awka, Anambra State, Nigeria

Osikemekha Anthony Anani
Department of Department of Biological Science, Edo State University Uzairue, PMB 04 Auchi, Edo State, Nigeria

Abdullahi S Argungu
Department of Water Resources and Environmental Engineering, Ahmadu Bello University, Zaria, Nigeria

Harrison I Atagana
Institute for Nanotechnology and Water Sustainability (iNanoWS), College of Science, Engineering and Technology, University of South Africa

Michael O Atambo
Department of Technical and Applied Physics, School of Physics and Earth Sciences Technology, Technical University of Kenya, Nairobi, Kenya

Olusoji Anthony Ayeleso
Department of Environmental Resource Management, Nasarawa State University, Nasarawa State, Nigeria

Adeoye Victor Babalola
Research and Institution Area, Nile University of Nigeria, Jabi-Abuja, FCT, Nigeria

Fatima B Badiru
Department of Water Resources and Environmental Engineering, Ahmadu Bello University, Zaria, Nigeria

Stephen Chege
Department of Technical and Applied Physics, School of Physics and Earth Sciences Technology, Technical University of Kenya, Nairobi, Kenya

Stefan Csaki
Department of Physics, Constantine the Philosopher University in Nitra, 94901, Nitra, Slovakia

Megersa Olumana Dinka
Department of Civil Engineering Science, University of Johannesburg, PO Box 524, Auckland Park 2006, South Africa

Chioma Adaku Ejelonu
Department of Physics, University of Benin, PMB 1154, Benin City, Edo State, Nigeria

Tracy Egboduku
Department of Agronomy Faculty of Agriculture Delta State University Abraka Delta State, Abraka, Nigeria

Wisdom Oghenevwogaga Egboduku
Department of Botany, Delta State University, Abraka, Nigeria

Uche Eunice Ekpunobi
Pure and Industrial Chemistry Department, Faculty of Physical Sciences, Nnamdi Azikiwe University, PMB 5025, Awka, Anambra State, Nigeria

Dennis Eyetan Elete
Department of Physics, Federal University of Petroleum Resources, Effurun, Delta State, Nigeria

Ahmed El Nemr
Environment Division, National Institute of Oceanography and Fisheries (NIOF), Kayet Bey, Elanfoushy, Alexandria, Egypt

Joseph Onyeka Emegha
Department of Energy and Petroleum Studies, Novena University Ogume, Kwale, Delta State, Nigeria

Daniel Elaigwu Enenche
Pure and Industrial Chemistry Department, Faculty of Physical Sciences, Nnamdi Azikiwe University, PMB 5025, Awka, Anambra State, Nigeria

Juliana O Ijia
Department of Water Resources and Environmental Engineering, Ahmadu Bello University, Zaria, Nigeria

Atala Bihari Jena
Department of Neurosurgery, Brigham and Women's Hospital, Harvard Medical School, Boston, MA 02115, USA

Grace Jokthan
Africa Centre of Excellence on Technology Enhanced Learning (ACETEL), National Open University of Nigeria, Abuja, Nigeria

Salami A Kazeem
Multifunctional Materials Laboratory, Department of Mechanical Engineering, Ahmadu Bello University, Samaru, Zaria, Nigeria

Rout George Kerry
Department of Biotechnology, Utkal University, Vani Vihar, Bhubaneswar, Odisha 751004, India

Francis Jesmar P Montalbo
College of Informatics and Computing Sciences, Batangas State University, Batangas 4200, Philippines

Lunyanda Lunga Noto
Department of Physics, College of Science, Engineering and Technology, University of South Africa, South Africa

Joan Nyika
Department of Civil Engineering Science, University of Johannesburg, PO Box 524, Auckland Park 2006. South Africa

David O Obada
Multifunctional Materials Laboratory, Department of Mechanical Engineering, Ahmadu Bello University, Samaru, Zaria, Nigeria

Kingsley Obodo
HySA-Infrastructure, North-West University, Faculty of Engineering, Private Bag X6001, Potchefstroom 2520, South Africa and National Institute of Theoretical and Computational Sciences, Johannesburg 2000, South Africa

Cynthia U Odili
Department of Water Resources and Environmental Engineering, Ahmadu Bello University, Zaria, Nigeria

Philomina Nkeonye Okanigbuan
Department of Geology, Faculty of Science, Dennis Osadebay University, Anwai, Asaba, Delta State, Nigeria

Joel Okpoghono
Department of Chemical Sciences (Biochemistry), Novena University, Ogume, PMB 02, Kwale, Delta State

Benedict Okundaye
Centre for Urban Design, Architecture and Sustainability (CUDAS), Department of Architecture and 3D Design, School of Arts and Humanities, University of Huddersfield, UK

Timothy Imanobe Oliomogbe
Department of Energy and Petroleum Studies, Novena University, Ogume, PMB 02, Kwale, Delta State, Nigeria

Enaroseha O E Omamoke
Department of Physics, Delta State University, Abraka, Nigeria

Uzochukwu Abraham Onuigbo
Pure and Industrial Chemistry Department, Faculty of Physical Sciences, Nnamdi Azikiwe University, PMB 5025, Awka, Anambra State, Nigeria

Robert Birundu Onyancha
Department of Technical and Applied Physics, Technical University of Kenya, PO Box 52428-00200, Nairobi, Kenya

Ikenna Benedict Onyeachu
Department of Chemistry, Edo State University Uzairue, PMB 04 Auchi, Edo State, Nigeria

Adelaja Otolorin Osibote
Department of Mathematics and Physics, Faculty of Applied Sciences, Cape Peninsula University of Technology, Cape Town, South Africa

Ayodeji O Oyedeji
Multifunctional Materials Laboratory, Department of Mechanical Engineering, Ahmadu Bello University, Samaru, Zaria, Nigeria

Bill Clintone Oyomo
Department of Technical and Applied Physics, School of Physics and Earth Sciences Technology, Technical University of Kenya, Nairobi, Kenya

Austin Ozukwe
Pure and Industrial Chemistry Department, Faculty of Physical Sciences, Nnamdi Azikiwe University, PMB 5025, Awka, Anambra State, Nigeria

Kingsley Eghonghon Ukhurebor
Department of Physics, Edo State University Uzairue, PMB 04 Auchi, Edo State, Nigeria

Patrick Ugochukwu Umennadi
Pure and Industrial Chemistry Department, Faculty of Physical Sciences, Nnamdi Azikiwe University, PMB 5025, Awka, Anambra State, Nigeria

Maryann I Uzochukwu
Department of Polymer Technology, Nigerian Institute of Leather and Science Technology, Zaria, Nigeria

Aizebeoje Balogun Vincent
Department of Mechanical Engineering, Edo State University Uzairue, PMB 04 Auchi, Edo State, Nigeria

Chapter 1

Introduction to the state of the art and relevant aspects of the applications of adsorption for environmental safety and sustainability

Kingsley Eghonghon Ukhurebor, Uyiosa Osagie Aigbe, Robert Birundu Onyancha, Kenneth Kennedy Adama, Osikemekha Anthony Anani, Ikenna Benedict Onyeachu, Joseph Onyeka Emegha, Benedict Okundaye, Bamikole Olaleye Akinsehinde, Olusoji Anthony Ayeleso and Grace Jokthan

We are in the midst of a rising crisis ensuing from environmental contamination, which is mainly caused by industrial and domestic activities, as well as the significant amount of these contaminants released into the environment by both natural and anthropogenic activities, which disrupts atmospheric, aquatic, and terrestrial systems. Thus, there is an urgent necessity for innovative approaches based on biosorbents for the removal of these contaminants, owing to the unique features of biosorbents, such as biocompatibility, biodegradability, and renewability. Hence, this chapter aims to present the development, utilization, and applications of the adsorption of heavy metals and dyes for reasons of environmental sustainability and safety.

List of unusual acronyms used

Acronym	Definition
AC	Activated carbon
ASB	Adsorbates
AST	Adsorption
ESS	Environmental safety and sustainability
HMI	Heavy metal ions
HMs	Heavy metals
MT	Magnetic
MO	Metal oxide
TDS	Total dissolved solids

doi:10.1088/978-0-7503-5598-8ch1

1.1 Introduction

As a result of the world population's rapidly increasing numbers, natural events such as volcanic explosions/eruptions, forest fires, and particle suspension in the atmosphere as well as industrialization and urbanization processes such as mining, transportation, wastewater from agriculture, and emissions into the atmosphere have all made significant contributions to environmental pollution [1–5]. Toxic gases, heavy metals (HMs)/heavy metal ions (HMIs), dyes, residential and industrial effluents, and organophosphates are all released by these anthropogenic and natural processes, which pose a serious threat to the wellbeing of humans and the environment [6–13]. The main pollutants in the environment and their origins are shown in figure 1.1.

Consequently, one of the biggest problems facing society in the twenty-first century is sustainable wastewater treatment due to these pollutants in the environment [14–21]. It will continue to be of utmost importance to create ecologically benign and economically viable treatment alternatives, especially for emerging nations, as the dumping of hazardous industrial sludge into water bodies increases [5, 22–24]. One of the most studied and applied techniques for the removal of HMs, dyes, and other industrial effluents from wastewater is adsorption (AST) or biosorption [1, 2, 25–31].

AST or biosorption, which refers to the use of both living (biotic) and non-living (abiotic) biomass, has attained significant popularity during the last few decades as a result of certain encouraging findings [1, 2, 31]. Waste products from agriculture, such as fruit, crop, and other plant wastes (collectively referred to in this chapter as 'agricultural wastes'), have frequently been utilized to remove HMs, dyes, and other

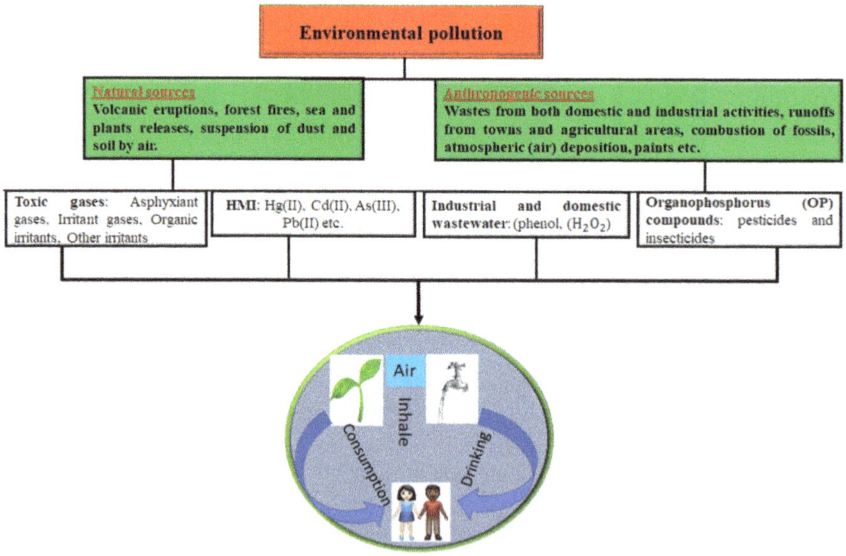

Figure 1.1. The main pollutants in the environment and their origins.

industrial effluents from wastewater via biosorption (or AST) processes [1, 2, 31]. Several of these potential uses of agricultural wastes as biosorbents for wastewater treatment have been investigated by a number of researchers, with encouraging findings [1, 2, 31]. These agricultural wastes predominantly absorb HMs, dyes, and other industrial effluents through the sorption of functional groups such as carboxyls and hydroxyls. The pH value and the co-solute content have been determined to be the key factors influencing the biosorption (or AST) of pollutants by agricultural wastes [1, 2, 31].

The quantity of solute that is adsorbed on the surface of the adsorbent per unit weight as a function of the equilibrium concentration at a fixed temperature is estimated using a representation known as an isotherm for AST [32, 33]. The Langmuir and Freundlich isotherms, which describe the AST (or biosorption) process, are the most frequently utilized [1, 34, 35]. The Temkin isotherm [34, 35], the Redlich–Peterson isotherm [36], the Koble–Corrigan isotherm [37], the Sips isotherm [38], the Toth isotherm [39], the Dubinin–Radushkevich isotherm [40, 41], Fritz–Schlunder isotherm [42], and the Radke and Prausnitz isotherm [43] models have also been utilized. The Langmuir isotherm model and pseudo second-order kinetic models are among the best models for explaining the agricultural waste-based biosorption of HMs, dyes, and other industrial effluents, according to a number of kinetic and isotherm investigations undertaken by various researchers and authors [1, 2, 31, 44, 45]. The outcomes of desorption investigations carried out by many researchers have further demonstrated the viability of agricultural wastes as reusable biosorbents [1, 2, 31]. The use of some agricultural wastes as biosorbents to remove HMs, dyes, and other industrial effluents from contaminated water, along with other significant features of this practice, has been extensively discussed in several publications [1, 2, 31]. Several reports have described the effects of a number of variables, including pH, initial solute concentration, biosorbent dose, co-solute concentration, activation, etc [1, 2, 31]. Hence, there is a substantial rationale for, and unique benefits to be obtained from investigating the variety of adsorbents and their effectiveness in removing HMs, dyes, and other industrial effluents from wastewater so as to sustain and preserve the environment (in support of environmental safety and sustainability). Nevertheless, a recent review study by Ukhurebor *et al* [31] placed a substantial emphasis on the numerous aspects of AST (or biosorption) for water treatment and some of the main issues, in addition to the potential of AST for purifying water. Figure 1.2, which was adapted from Ukhurebor *et al* [31], illustrates some of the processes involved in AST for water treatment and some of the treatment approaches for pollutants.

The underlying causes of pollutants and the idea of AST for removing emerging (newly identified) contaminants are addressed in this chapter. This chapter's major emphasis is to present an overview of what drives AST as an efficient technique for treating newly identified pollutants and the most recent developments in the technique of AST. Hence, this chapter introduces the theme of this book (which is

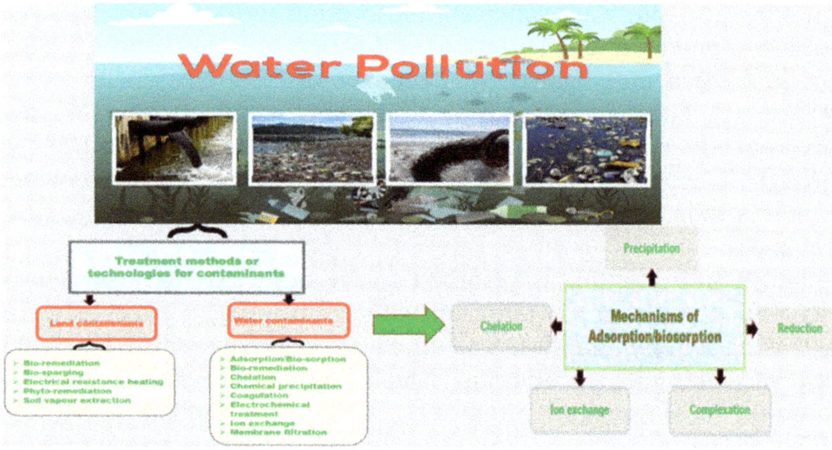

Figure 1.2. A schematic illustration of a variety of pollutant mitigation methods and their relationships to water purification AST (or biosorption) mechanisms. Reprinted by permission from Springer Nature [31], Copyright (2023).

centered on AST applications for environmental sustainability) and provides an introduction to the state of the art and relevant aspects of the applications of AST for environmental safety and sustainability (ESS). Furthermore, this book attempts to highlight the types, characteristics, and management options (reusability, recyclability, and final disposal) of adsorbents commonly used for environmental sustainability reasons in chapter 2. Chapter 3 addresses activated biosorbents for the removal of metals from aqueous solutions, while chapter 4 describes functionalized biosorbents for the sequestration of dyes from aqueous solutions, and chapter 5 discusses the sequestration of HMs from soil using functionalized biosorbents. The photocatalytic degradation of dyes and metal ions using functionalized biosorbents is discussed in chapter 6, and the effective sensing and sequestration of metal ions and dyes using functionalized biosorbents is discussed in chapter 7. Chapter 8 covers the use of microorganism-derived biosorbents in the removal of metal and dye ions, and chapter 9 describes biosorbents derived from invasive plants for environmental remediation. Chapter 10 deals with the use of microorganism-derived biosorbents in the sequestration of contaminants from the soil, followed by chapter 11, which deals with the modeling of AST processes that employ biosorbents for the removal of contaminants from water-soluble solutions using the kinetic and isotherm models. Green-derived biosorbents for the degradation of petroleum contaminants are discussed in chapter 12. A response surface methodology for the AST of HMs and dyes using biosorbents is discussed in chapter 13, and the cost, environmental evaluations, and comparisons of commonly used sorbents are discussed in chapter 14. The last chapter (chapter 15) deals with the challenges and perspectives of AST applications for environmental sustainability.

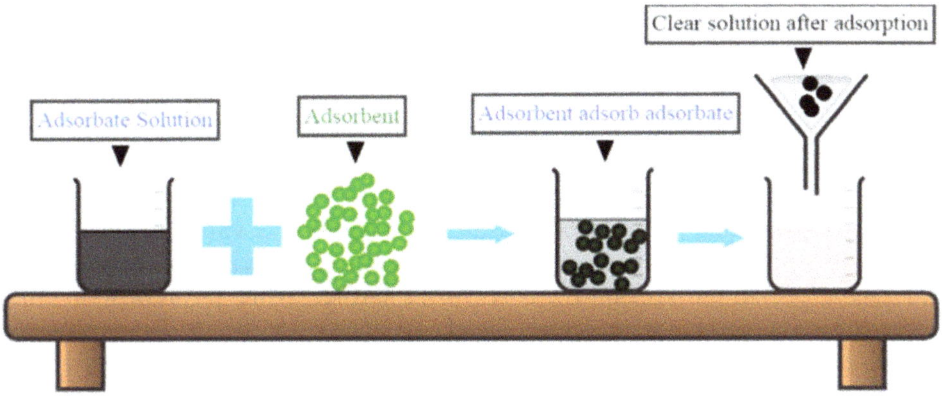

Figure 1.3. The procedure for AST. Reprinted from [1], Copyright (2023), with permission from Elsevier.

1.2 The development of AST

The process that occurs when ions or molecules from a gaseous or liquid bulk phase adhere to a solid surface is known as AST. Adsorbents are solids utilized for AST, whereas adsorbates (ASB) are molecules or ions [3]. Liquids are occasionally employed as adsorbents. ASBs should not penetrate the adsorbent's structure, since AST is a surface phenomenon that only affects the adsorbent's surface. The AST process is shown in figure 1.3 (adopted from Rathi and Kumar [1]). Desorption is the opposite process, in which a molecule is detached from an adsorbent surface [46].

1.2.1 The principle, concept, and mechanism of AST

The exchange of mass and the AST of carbon molecules from a liquid or gas (fluid) phase onto an adsorbent (solid surface) are the fundamental ideas behind carbon AST. This method of making activated carbon (AC) is used solely to produce very porous molecules of carbon with a sizable interior surface. The aforementioned porous structure captures and holds metals and organic and inorganic substances. According to Sagar *et al* [47], AST ensues when a contaminant has a restricted ability to dissolve in waste products, when it has a greater carbon affinity than its affinity for the waste materials, or both. AST is a result of the fact that the constituent parts (particles) on the surface of the adsorbent have unequal or residual attraction forces that are not in equilibrium with the larger adsorbent's molecules, in which all the forces are harmonized. The energy of the constituent parts at the surface of the adsorbent is considerably greater than that of the constituent parts within. According to Hessou *et al* [48], this extra energy per unit surface area is known as 'surface energy,' and it is necessary for the binding of the ASB to its adsorbent surface. The degree of AST tends to rise as the adsorbent's surface area per unit mass increases in a given setting. AST enthalpy and entropy are other significant factors that should be taken into account [49].

1.2.2 AST classification

Depending on how the ASB adheres to the surface of the adsorbent, AST may be categorized into two groups: physical AST and chemical AST [50]. Chemical AST is also referred to as 'chemisorption' because, in this case, the ASB is attached to the surface of the adsorbent through potent covalent bonds. Physical AST is also referred to as physisorption, in which the ASB attaches to the surface of the adsorbent owing to weaker forces such as electrostatic attraction and the van der Waals forces. Chemisorption is slower than physisorption and characteristically results in the development of a monolayer on the adsorbent's surface, while physisorption characteristically results in the development of multiple layers [35, 51]. At temperatures less than or close to the critical temperature of ASB, where it is very effective, physisorption is a process that can be reversed. Chemisorption, in contrast with physisorption, typically takes place at temperatures considerably greater than the critical temperature. Both types of AST might occur simultaneously or one may occur after the other under ideal conditions. Because physisorption is exothermic, i.e. it is defined by a decrease in free energy and entropy, it occurs relatively quickly at low temperatures and dramatically lessens with an increase in temperature. However, beyond a certain point, chemisorption increases with temperature and then starts to decline [52, 53].

1.3 Categories and sources of pollutants and emerging (newly identified) contaminants in water

The threat of emerging pollutants is rising. The environment contains an enormous variety of pollutants, each with unique physical, chemical, and toxicological properties. The pollutants come from industrial waste, household wastes, wastes from farm animals, agricultural waste, and radioactive waste. They include both organic and inorganic chemicals. A recent review by Rathi and Kumar [1] highlighted the various emerging (newly identified) pollutants and their sources.

1.3.1 Wastes from industry

Many companies produce industrial wastes which contain HMs and other pollutants [1, 2, 31]. Some of the main emerging (newly identified) pollutants from pharmaceuticals include sulfamethazine, diclofenac, carbamazepine, caffeine, ibuprofen, clofibric acid, ciprofloxacin, bisphenol A, atenolol, metronidazole, dimetridazole, datrizoate, metalaxyl, tricyclazole, fludioxonil, bentazon, carbofuran, iopamidol, etc [54–56], while the main emerging (newly identified) pollutants from the personal care industry include estrone, estradiol, progesterone, testosterone, estriol, methyl paraben, ethyl paraben, propyl paraben perflouro octane sulfonate, etc [57, 58]. The main emerging (newly identified) pollutants emitted from power plants are oxides of sulfur, nitrogen, and carbon, particulate matter (PM), formaldehyde, furans, dioxins, and HMs (such as mercury, cadmium, copper, chromium, lead, etc.) [1, 2, 31].

The next-largest industry that produces hazardous wastewater discharges is the manufacturing industry. Even though this sector includes a wide variety of production techniques, its distinctive pollutants are apparent in every manufacturing system. In addition, it is possible to identify the petrochemical and chemical industries, paper and pulp manufacturers, and electroplating businesses as sources of effluent contamination. Chemicals such as benzene, phenol, acetone, petroleum products, and nitrogen as well as other dangerous chemicals can be found in the wastes of petroleum refineries. HM ions from the paper and pulp industries, as well as zinc, calcium, magnesium, sodium, and potassium, can be found in the wastewater produced by electroplating facilities [59].

The manufacture of printed circuit boards is a significant source of HM pollution. Soldering plates made of zinc, lead, and nickel are the most frequently utilized prone overlays. The apparel, tannery, chemical, and textile-related sectors have all employed various colored pigments. Nickel is the main contaminant in the effluents from mining and smelting activities, while lead is the main contaminant in the effluent from the painting industry [1].

1.3.2 Wastes from domestic activities

Among the most common household wastes are significant sewage discharges. These are contaminants that might endanger people's health by harming wastewater treatment facilities, drainage systems, and the environment as well as by preventing the recycling of biosolids and wastewater. Caffeine, paraxanthine, and acetaminophen are a few of the emerging (newly identified) pollutants that are readily discovered in sewage systems [60]. The major pollutants in the drainage include total dissolved solids (TDS), colorants, cadmium, nickel, zinc, mercury, copper, lead, and arsenic, according to Metro Melbourne water utilities. Municipal wastewater is a mixture of the two main wastes produced: (a) gray water refers to the water from the kitchen sink, the washer, and the shower, and (b) black water is the water that bathrooms or restrooms discharge. The harmful substances found include boron, which is naturally present in glass, the majority of metals, steel, ceramic soap, and washing powder. Additionally, meals, cosmetics, and the flocculants used in water filtration frequently include iron. The natural deposition and movement of salt water involves sodium, which is also a component of water softening, cleaning, and soap-making chemicals. Ammonia is a by-product of the decomposition of biological (organic) materials, including urea from living organisms. It is frequently utilized in the manufacture of textiles, plastics, cleaning agents, and food items. These are some of the pollutants that can be found in household trash, according to some reports [1, 2, 31].

1.3.3 Wastes from animal farming

The residues from animal product contamination have an impact on the health of the population. Organic pollutants, elemental pollutants, and mycotoxins are three types of contaminants. Pesticides, animal pharmaceuticals, synthetic chemicals such as polybrominated fire retardants, and other pollutants such as dioxins are examples of natural substances that are a cause for concern. Dioxins and their related compounds, such as polychlorinated dibenzodioxins, polychlorinated dibenzofurans, and a small number of polychlorinated dioxin-like biphenyls, make up a significant compositional community. Animal health necessitates some chemical substances, among which feed is the foremost. Lead, zinc, and cadmium are others. Other pollutants include chemicals such as insecticides, pesticides, and disinfectants as well as inorganic and organic debris comprising endotoxins, bacteria, allergens, and fungi [61].

1.3.4 Wastes from agricultural activities

New agriculturally harmful compounds such as antibiotics, vaccines, growth boosters, and hormones have emerged during the last twenty years. These can enter water through runoff from the surface and leaching from fisheries and livestock farms, as well as through the fertilization of fields with sludge. Through the discharge of wastewater into aquatic environments, farming not only contributes to the presence of significant pollutants but also causes their resurgence [1]. Some of the emerging (newly identified) toxins from pesticides are caffeine, diclofenac, estriol, estrone, progesterone, testosterone, triclosan, bisphenol A, ametryn, atrazine, carbendazim, clomazone, difenoconazole, chlorpyrifos, etc [62]. Pesticides and fertilizers are used in large quantities in modern agricultural practice [1].

The greatest potential yields of food and fiber are the primary goals of such chemical alterations [1]. Nitrate poisoning of underground water typically occurs due to the overuse of nitrogen-bearing fertilizers, which cause leaching beneath the root region [1]. Fertilizers, which are typically mineral/inorganic compounds, are applied on a regular basis each year and include nitrogen, phosphorus, potassium (NPK), and metals to reduce the quantity of iron, zinc, copper, boron, and molybdenum, which are examples of micronutrients [1]. NH_4NO_3, $(NH_4)_2SO_4$, KNO_3, NO_3-, $NH_4PO_4^{3-}$, and other contaminants may behave as both pollutants and nutrients in waste from agriculture [63].

1.4 Various adsorbent types

Various chemical compositions and geometrical structural components can be present in solid adsorbent substances [1, 2, 31]. Better knowledge is required to support their use in commercial applications or laboratory procedures. Adsorbents are often divided into two categories: natural organic adsorbents and synthetic adsorbents, depending on what they are made from. Minerals, charcoal, ores, clays, and zeolites are examples of natural organic adsorbents [1, 2, 31], while synthetic adsorbents are made from waste materials such as waste sludge, agricultural waste, industrial waste, etc [1, 2, 31]. However, as rightly reported by Hoang *et al* [64], there is a third essential type of adsorbent, namely natural inorganic adsorbents (also known as commercial adsorbents).

1.4.1 Natural biological/organic adsorbents

Natural organic adsorbents are inexpensive, widely accessible, easy to use, and effective at absorbing or removing contaminants. According to Sharma *et al* [65], natural organic adsorbents are divided into four primary groups based on the material's source: (a) agricultural waste; (b) fruit waste; (c) plant waste; and (d) bio-adsorbents.

1.4.2 Synthetic adsorbents

The pharmaceutical industry and the food industry, as well as other industries, frequently employ synthetic adsorbents which are typically composed of polymers for the purification and separation of pollutants [66]. According to Weber and van Vliet [67], synthetic adsorbents are sphere-binding polymers with a substantial porosity and a particular surface area in the range of $5.00 \times 10^2 - 1.20 \times 10^3$ m^2 g^{-1} (dry).

1.4.3 Natural inorganic/mineral adsorbents (or commercial/industrial adsorbents)

Natural inorganic/mineral adsorbents (or commercial/industrial adsorbents) are commercialized adsorbents that are always readily accessible on the market. The commonest of these are activated carbon (AC), silica gel, activated alumina, zeolites, polymers, resins, and clay [1, 2, 31].

1.5 The application of the AST procedure

AST can occur in fluids (that is, in both the gaseous and liquid states). Recently, there has been a considerable increase in the number of AST technologies available in a variety of fields due to the increasing availability of different types of adsorbents. AST is used in a variety of processes, including enhanced wastewater treatment, the repossession and concentration of proteins, the removal of other emerging (newly identified) pollutants, the removal of dyes, the removal of HMs, the removal of phenolic compounds, the removal of odor and taste, etc [1, 68].

AST (or biosorption) methods are widely utilized for the removal of HMs, dye, and other industrial effluents from wastewater [1, 2, 31]. Activated carbon (AC) is the most popular adsorbent and provides the best removal of HMs, dye, and other industrial effluents [1, 2, 31]. AC is the most frequently utilized adsorbent. Although it produces the best outcomes, its expensive cost restricts its application. Its production and regeneration costs are significant. Freshwater resources are in short supply in the world today; therefore, it is necessary to find alternatives that reduce the strain on the available resources [1, 2, 31]. In addition, because HMs, dyes, and other industrial effluents are harmful even in tiny amounts, inexpensive adsorbents are necessary in order to develop environmentally acceptable techniques for their removal [1, 2, 31]. AST (or biosorption) is a technology that saves money and has become prevalent because it allows for the slightest amount of waste to be disposed of [1, 2, 31]. However, this brief study focuses on the different kinds of adsorbents that are now on the market as well as the AST process. It also includes inexpensive adsorbents, ranging from industrial waste to agricultural waste, and this helps to explain the performance of AST [1, 2, 31]. The criteria used to choose an adsorbent are cost-effectiveness, technological applicability, and ease of access to raw materials with little detrimental influence to the system [1, 2, 31]. However, this chapter only highlights the applications of AST in the removal of emerging (newly identified) pollutants, since other chapters describe further applications of AST in the removal of other pollutants such as dyes, HMs, etc.

1.5.1 The application of AST in the removal of emerging (newly identified) pollutants

The threat caused by emerging (newly identified) pollutants is rising [1]. Organic pollutants, such as those found in chemical fertilizers, pharmaceuticals, personal care products, hormonal substances, polymer compounds, food components, wood preservatives, products for cleaning, surfactants, antiseptics, and fire retardants, as well as other mineral/organic compounds, are the main emerging contaminants [1, 2, 31]. These substances are frequently found in the natural wastewater streams generated by human activities as well as in the industrial sector. The primary sources of emerging (newly identified) pollutants are illustrated in figure 1.4 (reprinted from Rathi and Kumar [1]). Even at modest concentrations, the majority of new pollutants can have harmful impacts on both people and aquatic species. Traditional basic and secondary treatment facilities can hardly remove or eliminate such dangerous poisons efficiently, necessitating cost-effective tertiary procedures.

The major roots of the emerging (newly identified) pollutants are shown in figure 1.5 (reprinted from Rathi and Kumar [1]). Some of the methods utilized to remove emerging (newly identified) pollutants are advanced oxidation procedures, membrane filtration, ion exchange, AST, coagulation, and sedimentation [1, 2, 31, 69]. AST is a successful strategy for eradicating new pollutants because it is highly efficient, has an affordable initial cost, and is simple to utilize. In research settings, activated carbons, modified biochars, nanoadsorbents, composite adsorbents, and other adsorbents have all been utilized to remove emerging (newly identified) pollutants from water and wastewater [1, 2, 31, 57, 69].

Figure 1.4. The primary sources of emerging (newly identified) pollutants. Reprinted from [1], Copyright (2021), with permission from Elsevier.

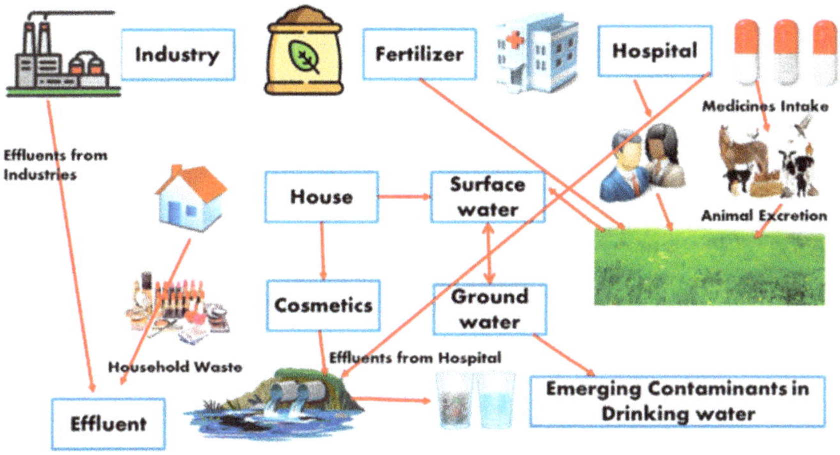

Figure 1.5. The major roots of emerging (newly identified) pollutants. Reprinted from [1], Copyright (2021), with permission from Elsevier.

Some of the emerging (newly identified) pollutants as reported by Rathi and Kumar [1] are bisphenol A, triclosan, ibuprofen, naproxen, ketoprofen, salicylic acid, diclofenac, acetaminophen, androstenedione, atrazine, caffeine, carbamazepine, Dilantin, estriol, estrone, ethynylestradiol, fluoxetine, hydrocodone, etc. The various adsorbents that have been utilized to remove the various emerging (newly identified) pollutants, along with their removal percentages, are described in a recent review work by Rathi and Kumar [1].

1.6 The state of the art and relevant advances in the applications of AST

Considering its accessibility, affordability, simplicity, and ecological friendliness, AST is a leading approach for the removal of emerging (newly identified) pollutants from wastewater [70, 71]. AST technology will continue to progress, overcoming its drawbacks and improving its ability to remove all pollutants, including emerging (newly identified) pollutants [72]. Only emerging (newly identified) pollutants from pharmaceuticals, insecticides, personal care items, and others are the main focus here. Since the AST process heavily depends on the surface area, and nanoparticle-based materials provide high surface areas, in turn boosting the AST performance, nanotechnology has recently become widespread in this industry. Nanomaterials can be developed into powerful adsorbents owing to their variable particle diameters [73, 74].

For the removal of mineral/organic pollutants or toxic chemicals, there are four distinct categories of nanoadsorbents: carbon nanotubes, silica-based nanoparticles, metal oxide (MO)-based nanomaterials, and chitosan-based nanoparticles [75, 76]. With the exception of chitosan-based nanoparticles, which are made from the shells of aquatic species and are biodegradable (also known as 'green adsorbents,' which are not harmful to the ecosystem), each form of nanoparticle has a shortcoming. It is also possible to use biomass-based nanoparticles for AST [77, 78].

Metal–organic framework adsorbents are thought to be among the latest developments in water treatment techniques [79]. In order to increase efficiency, hybrid technologies concentrate on techniques that may be combined with AST and chemical or biological treatments. The most recent methods of wastewater treatment may also be combined with physical techniques, such as the combination of ultrasound and gamma rays with AST using activated carbon [80]. The treatment and removal of contamination in wastewater may be done efficiently and precisely using magnetic (MT) carbon material. MT separation is an emerging form of water treatment because of its low cost and simplicity of application. In order to ease segregation, cleaning, and re-dispersion of the materials, separation is performed by introducing an external MT field. In actuality, the capability to recycle or reuse MT carbon substances or materials increases the chance that these products will be used in operations for the large-scale treatment and management of wastewater [81].

1.7 Conclusions and future insights

Contaminants are related to the emergence of dangerous substances emanating from humans and the industrial sector, which is alarmingly expanding daily. Pharmacological substances are a top target among emerging (newly identified) pollutants because of their ability to cause serious and extensive effects to the environment and human health. Numerous emerging (newly identified) contaminants present a serious risk to living things, even at extremely low concentrations.

In order to effectively remove pollutants, however, effective adsorbents such as nanoadsorbents, MT adsorbents, and MO adsorbents must be used together with hybrid AST techniques. The use of AST, its basic concepts and mechanisms, the sources and varieties of contaminants, the various adsorbents that may be utilized for treating ASB, and AST's role in removing emerging (newly identified) pollutants were all covered in this chapter.

AST-based methods for the treatment of hazardous and damaging wastewater are unquestionably a viable choice due to their affordability and accessibility. Finding the right adsorbent to take up the ASB of concern is a crucial step in the AST process.

A suitable, low-cost adsorbent must be chosen, and the best conditions must be established. This scenario must take into account variables including the starting ASB concentration, contact duration, pH, adsorbent, and operational conditions. The threat of emerging (newly identified) pollutants is rising.

The significance of AST methods for the elimination of emerging (newly identified) pollutants has been meticulously highlighted. AST has the ability to eliminate very close to 100% (about 95%) of emerging (newly identified) pollutants. For the removal of emerging (newly identified) pollutants, hybrid AST techniques, MO adsorbents, MT adsorbents, as well as nanoadsorbents can all be used to increase the effectiveness of AST. The AST procedure may successfully remove these pollutants from wastewater. By using a hybrid system consisting of AST, an

MO-based adsorbent, and a nanoscale green adsorbent, the effectiveness of the AST may be enhanced. A powerful alternative is offered by the development of contemporary adsorbents for waste and by-products that rely on organic materials, which offer the benefits of utilizing inexpensive adsorbents.

References

[1] Rathi B and Kumar P 2021 Application of adsorption process for effective removal of emerging contaminants from water and wastewater *Environ. Pollut.* **280** 116995

[2] Hussain A, Madan S and Madan R 2021 Removal of heavy metals from wastewater by adsorption *Heavy Metals—Their Environmental Impacts and Mitigation* (London: IntechOpen)

[3] Han D, iaqiang E J, Deng Y, Chen J, Leng E, Liao G, Zhao X, Feng C and Zhang F 2021 A review of studies using hydrocarbon adsorption material for reducing hydrocarbon emissions from cold start of gasoline engine *Renew. Sustain. Energy Rev.* **135** 110079

[4] Adetunji C *et al* 2021 Bionanomaterials for green bionanotechnology *Bionanomaterials: Fundamentals and Biomedical Applications* ed R Singh and K Singh (Bristol: IOP Publishing) 1–24

[5] Anani A, Adama K, Ukhurebor K, Habib A, Abanihi V and Pal K 2023 Application of nanofibrous protein for the purification of contaminated water as a next generational sorption technology: a review *Nanotechnology* **34** 1–18

[6] Ukhurebor K, Aigbe U, Onyancha R, Ndunagu J, Osibote O, Emegha J, Balogun V, Kusuma H and Darmokoesoemo H 2022 An overview on the emergence and challenges of land reclamation: issues and prospect *Appl. Environ. Soil Science* **5889823** 1–14

[7] Ukhurebor K, Onyancha R, Aigbe U, UK-Eghonghon G, Kerry R, Kusuma H, Darmokoesoemo H, Osibote O and Balogun V 2022 A methodical review on the applications and potentialities of using nanobiosensors for diseases diagnosis *BioMed Res. Int.* **1682502** 1–20

[8] Ukhurebor K, Aigbe U, Onyancha R, Nwankwo W, Osibote O, Paumo H, Ama O, Adetunji C and Siloko I 2021 Effect of hexavalent chromium on the environment and removal techniques: a review *J. Environ. Manage.* **280** 111809

[9] Onyancha R, Aigbe U, Ukhurebor K and Muchiri P 2021 Facile synthesis and applications of carbon nanotubes in heavy-metal remediation and biomedical fields: a comprehensive review *J. Mol. Struct.* **1238** 130462

[10] Onyancha R, Ukhurebor K, Aigbe U, Osibote O, Kusuma H, Darmokoesoemo H and Balogun V 2021 A systematic review on the detection and monitoring of toxic gases using carbon nanotube-based biosensors *Sens. Bio-Sens. Res.* **34** 100463

[11] Onyancha R, Ukhurebor K, Aigbe U, Osibote O, Kusuma H and Darmokoesoemo H 2022 A methodical review on carbon-based nanomaterials in energy-related applications *Adsorpt. Sci. Technol.* **4438286** 1–21

[12] Onyancha R, Aigbe U, Ukhurebor K, Osibote O, Balogun V and Kusuma H 2022 Current existing techniques for environmental monitoring *Nanobiosensors for Environmental Monitoring: Fundamental and Applications* ed R Singh, K Ukhurebor, J Singh, C Adetunji and K Singh (Berlin: Springer Nature) pp 239–62

[13] Kerry R *et al* 2021 A comprehensive review on the applications of nano-biosensor based approaches for non-communicable disease detection *Biomater. Sci.* **9** 3576–602

[14] Aigbe U, Onyancha R, Ukhurebor K and Obodo K 2020 Removal of fluoride ions using polypyrrole magnetic nanocomposite influenced by rotating magnetic field *RSC Adv.* **10** 595–609

[15] Aigbe U, Ukhurebor K, Onyancha R, Osibote O, Kusuma H and Darmokoeso H 2022 Measuring the velocity profile of spinning particles and its impact on Cr(IV) sequestration *Chem. Eng. Process.* **178** 1–15

[16] Aigbe U, Ukhurebor K, Onyancha R, Ama M, Osibote O, Kusuma H, Okanigbuan P, Azi S and Osifo P 2022 Dendrimers for environmental remediation *Nanotechnology for Environmental Remediation* ed S Thomas, M S Thomas and L A Pothen (New York: Wiley)

[17] Aigbe U, Onyancha R, Ukhurebor K, Osibote O, Atagana H, Ama O, Atagana H, Ogbemudia P and Akanji S 2022 Electrochemical detection of heavy metals *Modified Nanomaterials for Environmental Applications-Electrochemical Synthesis, Characterization and Properties* ed O Ama, S Ray and P Osifo (Berlin: Springer Nature) pp 25–63

[18] Aigbe U, Ukhurebor K, Onyancha R, Okundaye B, Pal K, Osibote O, Esiekpe E, Kusuma H and Darmokoesoemo H 2022 A facile review on the sorption of heavy metals and dyes using bionanocomposites *Adsorpt. Sci. Technol.* **8030175** 1–36

[19] Aigbe U, Ukhurebor K, Onyancha R, Ama M, Okundaye B, Esiekpe E, Osibote O, Kusuma H and Osifo P 2021 Utility of bionanocomposite for wastewater treatment *Bionanomaterials for Environmental and Agricultural Applications* ed R Singh and K Singh (Bristol: IOP Publishing) pp 8–25

[20] Aigbe U, Ukhurebor K, Onyancha R, Okundaye B and Osibote O 2023 Green nanomaterials for wastewater treatment analysis *Green Nanoarchitectonics* ed K Pal (Singapore: Jenny Stanford Publishing)

[21] Aigbe U, Ukhurebor K, Onyancha R, Osibote O, Darmokoesoemo H and Kusuma H 2021 Fly Ash-based adsorbent for adsorption of heavy metals and dyes from aqueous solution: a review *J. Mater. Res. Technol.* **14** 2751–74

[22] Sudarni D, Aigbe U, Ukhurebor K, Onyancha R, Kusuma H, Darmokoesoemo H, Osibote O, Balogun V and Widyaningrum B 2021 Malachite green removal by activated potassium hydroxide clove leaves agro-waste biosorbent: characterization, kinetics, isotherms and thermodynamics studies *Adsorpt. Sci. Technol.* **no. 1145312** 1–15

[23] Aidonojie P, Ukhurebor K, Oaihimire I, Ngonso B, Egielewa P, Akinsehinde B, Heri S and Darmokoesoemo H 2023 Bioenergy revamping and complimenting the global environmental legal framework on the reduction of waste materials: a facile *Heliyon* **9** E12860

[24] Aidonojie P, Ukhurebor K, Masajuwa F, Imoisi S, Edetalehn O and Nwazi J 2022 Legal implications of nanobiosensors concerning environmental monitoring *Nanobiosensors for Environmental Monitoring: Fundamentals and Applications* ed R Singh, K Ukhurebor, J Singh, C Adetunji and K Singh (Berlin: Springer Nature) pp 439–58

[25] El-Nemr M, Aigbe U, Ukhurebor K, Onyancha R, El Nemr A, Ragab S, Osibote O and Hassaan M 2022 Adsorption of Cr6+ ion using activated pisum sativum peels-triethylenetetramine *Environ. Sci. Pollut. Res.* **2022** 1–25

[26] Eldeeb T, Aigbe U, Ukhurebor K, Onyancha R, El-Nemr M, Hassaan M, Ragab S, Osibote O and El Nemr A 2022 Adsorption of methylene blue (MB) dye on ozone, purified and sonicated sawdust biochars *Biomass Convers. Biorefin.* **2022** 1–23

[27] Eldeeb T *et al* 2022 Biosorption of acid brown 14 dye to mandarin-biochar-CO-TETA derived from mandarin peels *Biomass Convers. Biorefin.* **2022** 1–21

[28] Eleryan A, Aigbe U, Ukhurebor K, Onyancha R, Hassaan M, Elkatory M, Ragab S, Osibote O, Kusuma H and El Nemr A 2023 Adsorption of direct blue 106 dye using zinc oxide nanoparticles prepared via green synthesis technique *Environ. Sci. Pollut. Res.* **2023** 1–17
[29] Eleryan A, Hassaan M, Aigbe U, Ukhurebor K, Onyancha R, El Nemr M, Ragab S, Hossain I and El Nemr A 2023 Kinetic and isotherm studies of acid orange 7 dye absorption using sulphonated mandarin biochar treated with TETA *Biomass Convers. Biorefin.* **2023** 1–12
[30] Kusuma H, Aigbe U, Ukhurebor K, Onyancha R, Okundaye B, Ama O, Darmokoesoemo H, Widyaningrum B, Osibote O and Balogun V 2023 Biosorption of methylene blue using clove leaves waste modified with sodium hydroxide *Results Chem.* **5** 1–15
[31] Ukhurebor K, Hossain I, Pal K, Jokthan G, Osang F, Ebrima F and Katal D 2023 Applications and contemporary issues with adsorption for water monitoring and remediation: a facile review *Top. Catal.* **2023**
[32] Foo K and Hameed B 2010 An overview of dye removal via activated carbon adsorption process *Desalin. Water Treatment* **19** 255–74
[33] Al-Ghouti M and Da'ana D 2020 Guidelines for the use and interpretation of adsorption isotherm models: a review *J. Hazard. Mater.* **393** 122383
[34] Ayawei N, Ebelegi A and Wankasi D 2017 Modelling and interpretation of adsorption isotherms *J. Chem.* 1–11
[35] Kecili R and Hussain C 2018 Mechanism of adsorption on nanomaterials *Nanomaterials in Chromatography* (Amsterdam: Elsevier) 89–115
[36] Redlich O and Peterson D 1959 A useful adsorption isotherm *J. Phys. Chem.* **63** 1024
[37] Koble R and Corrigan T 1952 Adsorption isotherms for pure hydrocarbons *Ind. Eng. Chem.* **44** 383–7
[38] Volesky B 2003 Biosorption process simulation tools *Hydrometallurgy* **71** 179–90
[39] Toth J 1971 State equation of the solid-gas interface layers *Acta Chim. Hung* **69** 311–28
[40] Dubinin M 1947 The equation of the characteristic curve of activated charcoal *InDokl. Akad. Nauk. SSSR* **55** 327–9
[41] Hu Q and Zhang Z 2019 Application of Dubinin–Radushkevich isotherm model at the solid/solution interface: a theoretical analysis *J. Mol. Liq.* **277** 646–8
[42] Fritz W and Schluender E 1974 Simultaneous adsorption equilibria of organic solutes in dilute aqueous solutions on activated carbon *Chem. Eng. Sci.* **29** 1279–82
[43] Jossens L, Prausnitz J, Fritz W, Schlünder E and Myers A 1978 Thermodynamics of multi-solute adsorption from dilute aqueous solutions *Chem. Eng. Sci.* **33** 1097–106
[44] Lima E, Hosseini-Bandegharaei A, Moreno-Pirajan J and Anastopoulos I 2019 A critical review of the estimation of the thermodynamic parameters on adsorption equilibria. Wrong use of equilibrium constant in the Van't Hoof equation for calculation of thermodynamics parameters of adsorption *J. Mol. Liq.* **273** 425–34
[45] Lima E, Gomes A and Tran H 2020 Comparison of the nonlinear and linear forms of the van't Hoff equation for calculation of adsorption thermodynamic parameters (D S and D H) *J. Mol. Liq.* **311** 113315
[46] Artioli Y 2008 *Adsorption: Encyclopaedia of Ecology* (Amsterdam: Elsevier) pp 60–5
[47] Sagar M, Belwalkar N S and Mane A A 2017 Adsorption and its isotherm e theory *Int. J. Eng. Res.* **6** 312–6

[48] Hessou E, Jabraoui H, Houngue M, Mensah J, Pastore M and Badawi M 2019 A first principle evaluation of the adsorption mechanism and stability of volatile organic compounds into NaY zeolite *Z. für Kristallogr.—Cryst. Mater.* **234** 469–82
[49] Tran N, Tieu A, Zhu H, Ta H, Ta T and Le H 2018 First-principles study of the adsorption and depolymerization mechanisms of sodium silicate on iron surfaces at high temperature *J. Phys. Chem.* C **122** 20827–40
[50] Kumar P, Joshiba G, Femina C, Varshini P, Priyadharshini S, Karthick M and Jothirani R 2019 A critical review on recent developments in the low-cost adsorption of dyes from wastewater *Desalin. Water Treat.* **172** 395–416
[51] Saha P and Chowdhury S 2011 Insight into adsorption thermodynamics *Thermodynamics* **16** 349–64
[52] Xia M, Chen Z, Li Y, Li C, Ahmad N, Cheema W and Zhu S 2019 Removal of Hg (II) in aqueous solutions through physical and chemical adsorption principles *RSC Adv.* **9** 20941–53
[53] Zhang K, Cheung W and Valix M 2005 Roles of physical and chemical properties of activated carbon in the adsorption of lead ions *Chemosphere* **60** 1129–40
[54] Lin Y, Lai W, Tung H and Lin A 2015 Occurrence of pharmaceuticals, hormones, and perfluorinated compounds in groundwater in Taiwan *Environ. Monit. Assess.* **187** 1–9
[55] Zenker A, Cicero M, Prestinaci F, Bottoni P and Carere M 2014 Bioaccumulation and biomagnification potential of pharmaceuticals with a focus to the aquatic environment *J. Environ. Manag.* **133** 378–87
[56] Tijani J, Fatoba O and Petrik L 2013 A review of pharmaceuticals and endocrine-disrupting compounds: sources, effects, removal, and detections *Water Air Soil Pollut.* **224** 1–29
[57] Lima E 2018 Removal of emerging contaminants from the environment by adsorption *Ecotoxicol. Environ. Saf.* **150** 1–17
[58] Caldas S, Bolzan C, Guilherme J, Silveira M, Escarrone A and Primel E 2013 Determination of pharmaceuticals, personal care products, and pesticides in surface and treated waters: method development and survey *Environ. Sci. Pollut. Control Ser.* **20** 5855–63
[59] Vysokomornaya O, Kurilenko E and Shcherbinina A 2015 Major contaminants in industrial and domestic wastewater *MATEC Web Conf.* **23** 01041
[60] Wells M, Morse A, Bell K, Pellegrin M and Fono L 2009 Emerging pollutants *Water Environ. Res.* **81** 2211–54
[61] MacLachlan D 2011 Estimating the transfer of contaminants in animal feedstuffs to livestock tissues, milk and eggs: a review *Anim. Prod. Sci.* **51** 1067
[62] Lopez-Doval J, Montagner C, de Alburquerque A, Moschini-Carlos V, Umbuzeiro G and Pompeo M 2017 Nutrients, emerging pollutants and pesticides in a tropical urban reservoir: spatial distributions and risk assessment *Sci. Total Environ.* **575** 1307–24
[63] Brusseau M and Artiola J 2019 Chemical contaminants *Environmental and Pollution Science* (Amsterdam: Elsevier) pp 175–90
[64] Hoang A, Le V, Al-Tawaha A, Nguyen D, Al-Tawaha A, Noor M and Pham V 2019 An absorption capacity investigation of new absorbent based on polyurethane foams and rice straw for oil spill clean-up *Petrol. Sci. Technol.* **36** 361–70
[65] Sharma P, Kaur H, Sharma M and Sahore V 2011 A review on applicability of naturally available adsorbents for the removal of hazardous dyes from aqueous waste *Environ. Monit. Assess.* **183** 151–95

[66] Singh N, Nagpal G and Agrawal S 2018 Water purification by using adsorbents: a review *Environ. Technol. Innov.* **11** 187–240
[67] Weber W and Van Vliet B 1981 Synthetic adsorbents and activated carbons for water treatment: overview and experimental comparisons *J. Am. Water Works Assoc* **73** 420–6
[68] Kaveeshwar A, Kumar P, Revellame E, Gang D, Zappi M and Subramaniam R 2018 Adsorption properties and mechanism of barium (II) and strontium (II) removal from fracking wastewater using pecan shell based activated carbon *J. Clean. Prod. 193* **193** 1–3
[69] Rathi B, Kumar P, Ponprasath R, Rohan K and Jahnavi N 2021 An effective separation of toxic arsenic from aquatic environment using electrochemical ion exchange process *J. Hazard. Mater.* **412** 125240
[70] Kumar P, Pavithra J, Suriya S, Ramesh M and Kumar K 2015 Sargassum wightii, a marine alga is the source for the production of algal oil, bio-oil, and application in the dye wastewater treatment *Desalin. Water Treatment* **55** 1342–58
[71] Ungureanu G, Santos S, Boaventura R and Botelho C 2015 Arsenic and antimony in water and wastewater: overview of removal techniques with special reference to latest advances in adsorption *J. Environ. Manag.* **151** 326–42
[72] Suganya S, Saravanan A and Ravikumar C 2017 Computation of adsorption parameters for the removal of dye from wastewater by microwave assisted sawdust: theoretical and experimental analysis *Environ. Toxicol. Pharmacol.* **50** 45–57
[73] Nithya K, Sathish A, Kumar P and Ramachandran T 2018 Fast kinetics and high adsorption capacity of green extract capped superparamagnetic iron oxide nanoparticles for the adsorption of Ni (II) ions *J. Ind. Eng. Chem.* **59** 230–41
[74] Prabu D, Parthiban R, Kumar P, Kumari N and Saikia P 2016 Adsorption of copper ions onto nano-scale zero-valent iron impregnated cashew nut shell *Desalin. Water Treat.* **57** 6487–502
[75] Gerard N, Krishnan R, Ponnusamy S, Cabana H and Vaidyanathan V 2016 Adsorptive potential of dispersible chitosan coated iron-oxide nanocomposites toward the elimination of arsenic from aqueous solution *Process Saf. Environ. Protect.* **104** 185–95
[76] Anitha T, Kumar P and Kumar K 2015 Binding of Zn (II) ions to chitosan-PVA blend in aqueous environment: adsorption kinetics and equilibrium studies *Environ. Prog. Sustain. Energy* **34** 15–22
[77] Thekkudan V, Vaidyanathan V, Ponnusamy S, Charles C, Sundar S, Vishnu D, Anbalagan S, Vaithyanathan V and Subramanian S 2016 Review on nanoadsorbents: a solution for heavy metal removal from wastewater *IET Nanobiotechnol.* **11** 213–24
[78] Saxena R, Saxena M and Lochab A 2020 Recent progress in nanomaterials for adsorptive removal of organic contaminants from wastewater *Chem. Select.* **5** 335
[79] Rojas S and Horcajada P 2020 Metaleorganic frameworks for the removal of emerging organic contaminants in water *Chem. Rev.* **120** 8378–415
[80] Ahmed M, Zhou J, Ngo H, Guo W, Thomaidis N and Xu J 2017 Progress in the biological and chemical treatment technologies for emerging contaminant removal from wastewater: a critical review *J. Hazard. Mater.* **323** 274–98
[81] Rocha L, Pereira D, Sousa E, Otero M, Esteves V and Calisto V 2020 Recent advances on the development and application of magnetic activated carbon and char for the removal of pharmaceutical compounds from waters: a review *Sci. Total Environ.* **718** 137272

Chapter 2

The types, characteristics, and management options (reusability/recyclability/final disposal) of commonly used adsorbents in environmental sustainability

Kingsley Obodo and Uyiosa Osagie Aigbe

The removal of contaminants from fluids and solids using adsorbents holds immense significance and plays a critical role in environmental remediation and sustainability. The effectiveness of contaminant removal greatly depends on the adsorbents' physicochemical properties, including their surface area, pore size, chemical affinity, reusability, etc. Adsorption techniques offer distinct advantages, since they utilize commonly employed adsorbents such as carbon-based materials, clay minerals, silica gel, zeolites, and polymeric materials for environmental remediation and sustainability. This chapter explores various adsorbent types and their characteristics and management options, with a specific emphasis on the reusability, recyclability, and final disposal of commonly utilized adsorbents to ensure environmental sustainability.

2.1 Introduction

The elimination of pollutants from liquids, gases, and solids using adsorbents is of the utmost importance and plays a crucial role in environmental remediation and sustainability. These materials are widely utilized in various industries, including water treatment, air purification, and waste management. The effectiveness of contaminant removal heavily relies on the adsorbents' physicochemical properties, such as surface area, pore size, and chemical affinity. Therefore, the success of sorption-based contaminant removal hinges on the specific characteristics and attributes of the adsorbent material. Adsorbents are materials that can attract and retain pollutants on their surfaces, allowing for their

doi:10.1088/978-0-7503-5598-8ch2

extraction from the environment. This process involves physical, chemical, or physicochemical (i.e. combined physical and chemical) mechanisms. Environmental remediation and sustainability require the optimization and utilization of adsorbents to efficiently remove pollutants while minimizing environmental impacts such as wastes [1].

Adsorbents can be derived from various materials, including carbon-based substances, natural and synthetic polymers, and inorganic materials such as clays and zeolites. Carbon-based adsorbents such as activated carbon (AC) and carbon nanotubes (CNTs) exhibit high efficiency in eliminating organic contaminants from water owing to their extensive surface areas and strong adsorption capabilities. Natural adsorbents such as biochar and chitosan are environmentally sustainable options, as they are renewable and biodegradable. To ensure environmental sustainability, the entire life cycle of adsorbents should be considered, from production to disposal. It is also important to adopt sustainable practices during adsorbent production in order to minimize energy consumption, waste generation, and greenhouse gas emissions. The utilization of adsorbents should prevent the release of noxious chemicals that could harm the ecosystem. Spent adsorbents should be discarded sustainably via reuse or safe disposal approaches that prevent the escape of pollutants into the environment [2].

Adsorbents can be characterized based on their reusability, recyclability, and alternatives for final disposal [3]. One such adsorbent is AC, which has an extremely porous structure and an enormous surface area, making it suitable for adsorbing volatile organic compounds (VOCs), pollutants, and impurities. AC can be recycled via reactivation, and its final disposal can be carefully carried out by incinerating it or sending it to a landfill [4]. Clay minerals, including kaolin, bentonite, and attapulgite, show an elevated cation exchange capacity, making them valuable for eliminating heavy metal ions (HMIs), dyes, and other contaminants. These minerals are plentiful, renewable, and biodegradable, making them an environmentally friendly choice for adsorption [5, 6]. Silica gel is a synthetic material generally exploited for moisture control and the adsorption of volatile organic compounds (VOCs), gases, and liquids; it is nontoxic, reclaimable, and recyclable. The disposal of silica gel can be carried out via regular waste management procedures [7, 8]. Zeolites, minerals with a well-ordered porous structure, are highly effective in removing pollutants such as HMIs, radioactive isotopes, and organic compounds. Zeolites can be regenerated for reuse, and their final disposal can be carried out using conventional waste management methods [9]. Polymeric adsorbents, such as polystyrene, polyvinyl alcohol, and polyvinyl chloride, are synthetic materials extensively employed for removing pollutants from liquids and gases. While these materials are not biodegradable and present disposal challenges, some polymeric adsorbents can be recycled and reused through appropriate treatment processes [10].

To ensure environmental sustainability, the entire life cycle of an adsorbent must be considered from production to disposal. Sustainable production practices should minimize energy consumption, waste generation, and greenhouse gas emissions. In addition, the application of adsorbents should avoid the use of toxic chemicals during the adsorption process to prevent harmful effects to the environment. Proper disposal methods, such as recycling or safe disposal techniques, should be employed to prevent the release of pollutants into the environment when dealing with used adsorbents. Thus, the selection of an appropriate adsorbent for environmental sustainability depends on factors such as the specific application and the type of contaminant to be removed [11]. It is essential to consider reusability, recyclability, and proper disposal options in order to ensure that the environmental impact of adsorbents is minimized.

Laskar *et al* [12] conducted a review study focused on the contamination of wastewater by dyes and the utilization of adsorption approaches to minimize the dye waste in the effluent released. Dyes are extensively used in various applications for industrial purposes, but their direct disposal into water bodies without adequate treatment poses a serious threat, as they act as pollutants. This review study provided a brief overview of current methodologies employed for the sorption mechanism, emphasizing the significance of sorption and the factors affecting the tested system. The study found that Langmuir (LNR) and Freundlich (FRH) isotherm models, along with a pseudo second-order (PSO) kinetic model, effectively represented the dye uptake. However, further investigation into theoretical dye adsorption modeling is necessary to gain a comprehensive consideration of the intricate molecular interactions occurring between the adsorbent and the adsorbate. The authors of this study explored the potential of discarded agronomic material, both in its original form and after modification, to serve as a cost-effective adsorbent for the removal of dyes and the production of AC. They also discussed the importance of kinetics, dynamic parameters, and isotherm models in comprehending the adsorption behavior and mechanism. Factors such as the initial concentration, pH, and temperature were thoroughly examined.

A review by Mehdi *et al* [13] presents an in-depth analysis of the sorptive confiscation process of harmful materials using carbon-based materials, including AC, CNTs, carbon nanofibers (CNFs), biochar, graphene, and carbon aerogels. This review highlights the fabrication techniques and characteristics of these carbon-based materials. Industrial wastewater containing dangerous compounds poses a significant threat to water supplies and the ecosystem. Therefore, adsorption is widely recognized as an effective and extensively used method among conventional technologies due to its simplicity and relatively low cost. This review comprehensively covers various aspects related to the physical and chemical properties of different carbon substances and their sorption capabilities, which are obtained by increasing the surface area or through alteration. It focuses on the adsorptive mechanisms used to eliminate heavy metals (HMs), pesticides, antibiotics, dyes, oils, phenolics, VOCs, and gas contaminants by leveraging the unique adsorptive removal properties of nanocarbon materials. Gisi *et al* [4] reviewed the utilization of low-cost by-products derived from the household, agricultural, and industrial sectors, which has gained recognition as an environmentally balanced approach to wastewater treatment. These by-products offer the dual benefits of removing pollutants from effluent and contributing to waste reduction, recovery, and reuse. The low-cost sorbents considered were categorized into five groups: domestic and agricultural wastes, marine materials, sludge, industrial by-products, and soil and ore materials, along with unique low-cost adsorbents. The study discusses the effectiveness of these sorbents in extracting various pollutants, their real wastewater applications, cost considerations, and the potential for their reuse after the adsorption process. Additionally, to better evaluate the sorbents' effectiveness in removing different types of pollutants such as dyes, HMs, biorecalcitrant compounds, nitrogen, and phosphate compounds, the study proposes and applies a simple methodological tool known as the 'adsorbent–pollutant' matrix. Such matrices aid in identifying potential adsorbents that can replace commercial AC(s). Shahadat *et al* [14] reviewed clay-based adsorbents in terms of their regeneration capacity as well as the efficiency of various adsorbents in the removal of industrial dyes. Azha *et al* [15] reviewed a clay-based flexible adsorbent coating intended for use in wastewater treatment, considering its prospects as well as its challenges, and demonstrated that this material can find significant application in industrial wastewater treatment. Other materials such as zeolites [16], carbon-based materials (such as coal [17], chitosan–calcite [18]), silica [19], nanoadsorbents [20], novel materials (such as MXenes [21, 22]), metal–organic frameworks [23, 24], etc. have been applied as adsorbent materials in the remediation, sustainability, and management of environmental pollution. Thus, the optimal conditions for adsorption vary depending on the material and material type. However, in most of the cases evaluated, the sorption mechanism appears to adhere to the LNR isotherm model or the PSO model [2, 25–28].

Thus, the mechanism of adsorption entails the transfer of mass in which substances amass at the interface between two phases, for example, gas–liquid, gas–solid, liquid–liquid, or liquid–solid interfaces [12]. The substance being accumulated or sorbed is called the sorbate, while the material accumulating the substance is known as the adsorbent. The properties of both the adsorbate and adsorbent are specific and dependent on their constituents. When the interaction occurring between the adsorbed molecules and the solid surface is primarily physical, it is referred to as physisorption. Physisorption is driven by weak van der Waals forces, making it reversible. It typically occurs at temperatures close to or below the critical temperature of the sorbed substance. Alternatively, if the adsorbed molecules establish strong attractive forces with the solid surface through chemical bonding, the phenomenon is referred to as chemisorption. Chemisorption always results in a monolayer, and the substances chemisorbed onto the solid surface are challenging to eliminate owing to the strong forces involved. Under suitable conditions, both chemisorption and physisorption can occur concurrently or interchangeably. Physical adsorption results in a decrease in free energy and an accompanying decrease in the entropy of the adsorption system, making it an exothermic process [13].

Adsorption serves as a widely employed technique for achieving environmental sustainability, and a multitude of adsorbents have been developed for applications in wastewater treatment, HM removal, and contaminated soil remediation [29, 30]. Appropriate adsorbent selection relies on the particular characteristics of the ecological matrix and the contaminants targeted for removal. Furthermore, the proper management of spent adsorbents is crucial for ensuring environmental sustainability. Continued research is necessary to advance the development of more efficient and sustainable adsorbents for environmental applications. Consequently, this chapter explores various types of adsorbents and their characteristics and management options, with a specific emphasis on the reusability, recyclability, and final disposal of commonly utilized adsorbents to ensure environmental sustainability.

2.2 Commonly used adsorbents

2.2.1 Carbon-based materials and activated carbon

2.2.1.1 Carbon-based materials.

Carbon-based materials, such as AC, graphene, CNTs, CNFs, biochar, lignite, and carbon aerogels, which are utilized for removing harmful substances from various media such as water and air, have immense potential for environmental remediation and sustainability [13, 31]. AC derived from coconut shells, coal, and wood is extensively used for adsorbing VOCs, pollutants, and impurities. Its high surface area and adsorption capacity make it a versatile adsorbent [32]. Graphene, CNTs, and CNFs are nanomaterials (NMs) with exceptional mechanical, electrical, and adsorptive properties. Their large surface areas and unique structures enable the effective adsorption of contaminants. Biochar, a carbonaceous material derived from biomass, lignite, low-grade coal, and carbon aerogels, is a porous and lightweight carbon material that exhibits promising adsorption capabilities for organic pollutants and HMs [33].

The method used to prepare these materials greatly influences their physicochemical properties, which also affects their efficacy in adsorbing pollutants. These materials are highly promising for various environmental applications due to their exceptional adsorption capacity as well as selectivity. They are very effective in the elimination of non-living and organic contaminants, which include but are not limited to antibiotics, dyes, HMs, pesticides, oils, phenolics, VOCs, and gas contaminants from various media [34]. Carbon NMs are superior adsorbents for separating pollutants due to their high adsorption capacity and robust selectivity. To improve the preparation process and optimize the performance of carbon-based adsorbents, more effort needs to be focused on the synthesis, functionalization, and/or decoration approaches that dominate their physicochemical adsorption properties. It is also essential to develop comprehensive and effective methods for the synthesis of carbon adsorbents that will lead to the reduction of secondary pollution. The behavior of carbon-based adsorbents in practical working environments needs to be understood [1, 35]. The use of various carbon-based adsorbent NMs (not only AC) will play a greater role in the adsorptive elimination of hazardous compounds as further research is carried out in this area.

2.2.1.2 Activated carbon (AC)

This is an extremely porous material that is commonly applied for the elimination of VOCs, contaminants, and impurities from gases and liquids. AC, also known as activated charcoal, is made from a wide range of materials, including coconut shells, coal, and wood, which undergo a process of carbonization and treatment with oxidizing agents to increase their surface area. The significant surface area of AC enables it to effectively sorb pollutants such as organic compounds and HMs. One of the key advantages of AC is its cost-effectiveness compared to other adsorbent materials [36]. In addition, it exhibits high selectivity for certain contaminants, particularly VOCs, making it an efficient solution for the treatment of contaminated liquids and gases [37, 38]. AC can be reused or recycled through reactivation processes, further contributing to its environmental sustainability [39]. Moreover, its final disposal can be carried out in an environmentally safe manner, including methods such as incinerating it or sending it to landfill [40].

AC finds extensive use in various industries, including water treatment, air purification, and waste management. In water treatment applications, AC is employed to eliminate dissolved organic compounds, HMs, and other impurities [41]. Similarly, in air purification, it effectively removes VOCs such as benzene, formaldehyde, and toluene, which are commonly present in both indoor and outdoor air. In waste management, AC is utilized to treat leachate, the liquid produced when waste materials are disposed of in landfills [4]. The choice of raw material from which to produce AC depends on the specific requirements of the application and the types of contaminants that need to be removed. Coconut shell-based AC is ideal for removing organic pollutants because of its extraordinary surface area as well as its affinity for organic substances. On the other hand, coal-based AC is regularly employed for the confiscation of inorganic pollutants, such as HMs, due to its elevated adsorption capacity. Wood-based AC serves as a low-cost complement to coconut shell AC and coal-based AC, and it finds applications in situations in which the confiscation of pollutants is not crucial [4].

The mechanism of action of AC includes the sorption of contaminants onto its surface. This adsorption process is controlled by a combination of physical and chemical forces, including hydrogen bonding, van der Waals forces, and ionic interactions. The enormous surface area of AC allows for a high adsorption capacity, making it an effective solution for pollutant removal [42].

AC is an extremely effective and efficient adsorbent material that is widely used in ecological sustainability efforts. Its substantial surface area, high selectivity, and cost-effectiveness make it the preferred choice for the confiscation of pollutants from gases and liquids. Furthermore, its ability to be reused or recycled, along with its ecologically safe final disposal procedures, contribute to its overall sustainability as a solution for the treatment of polluted liquids and gases [4, 43].

2.2.2 Clay minerals

Clay minerals, comprising attapulgite, bentonite, and kaolin are naturally occurring materials regularly utilized for the sequestration of dyes, HMIs, and other contaminants [18, 20, 44]. They are also referred to as phyllosilicates. They occur naturally and are extensively distributed in the Earth's crust. Owing to their adsorptive capabilities, clay minerals have been widely used in several applications, including ecological remediation. They offer various advantages in terms of environmental sustainability. They possess an elevated cation exchange capacity, which makes them beneficial for the elimination of pollutants from liquids and gases. They are abundant, renewable, and biodegradable, making them a green option for adsorption. The regenerative properties of clay minerals enable their repeated exploitation, and their final disposal can be managed via standard waste management systems. The utilization of clay minerals as sorbents is generally recognized as an economical, effective, and efficient technique for treating contaminated liquids and gases, especially in sectors such as agriculture and aquaculture [15, 45, 46]. Due to their extraordinary surface area, clay minerals demonstrate a substantial adsorption capacity, which contributes to their efficacy in eliminating pollutants [47]. Some key attributes of clay are: (i) it is readily available and abundant in numerous regions worldwide; (ii) its cost-effectiveness surpasses that of other adsorbent materials due to the ease of its extraction and minimal processing requirements, which eliminate the need for complex manufacturing processes; (iii) its extraordinary surface area provides a substantial adsorption capacity, making it effective in removing contaminants; (iv) disposal can be carried out in an environmentally safe manner, such as by sending it to landfill, or it can be repurposed for other applications, making it a sustainable solution for the treatment of contaminated liquids and gases [15].

A favorably promising adsorbent called attapulgite clay@carbon (ATP@C) nanocomposite (NC) was effectively produced via a suitable one-pot hydrothermal carbonization procedure by Chen *et al* [45]. This procedure utilizes two inexpensive and eco-friendly materials: attapulgite clay (ATP), a plentiful naturally occurring magnesium aluminum silicate and a green chemical derived from biomass, to which glucose was added. In contrast to carbon-based materials, the ATP@C NC displays an outstanding capacity for sorbing Cr(VI) and Pb(II), with the best possible adsorption capacities of 177.74 and 263.83 mg g^{-1}, respectively. These results convincingly establish the ATP@C NC as an exceptionally viable solution for the efficient, cost-effective, and sustainable removal of toxic ions from water. In water treatment, clay minerals are used to eliminate HMs, dissolved organic compounds, and other impurities [15]. In air purification, they serve as effective adsorbents for VOCs, such as benzene, formaldehyde, and toluene, which are commonly encountered in both indoor and outdoor environments [48, 49]. In waste management, clay minerals are utilized for the treatment of leachate, which denotes the liquid generated during the disposal of waste materials in landfills [47]. Clay minerals find widespread utilization in various industries, including water treatment, air purification, and waste management. Dyes and HMs have been removed from various aqueous solutions using natural clays, 1D clay nanotubes, and 2D clay nanosheets [50]. These clay nanotubes and nanosheets can easily be functionalized or modified to make multifunctional adsorbent clays.

The mechanism of action of clay minerals involves the adsorption of contaminants onto their surfaces. This adsorption process is governed by a combination of physical and chemical forces, including hydrogen bonding, van der Waals forces, and ionic interactions [46]. The unique properties of clay and clay-based minerals are attributed to their ability to donate and/or accept electrons in organic reactions, which is closely tied to their ion exchange and cation exchange capacity, as well as their swelling behavior. The swelling of clay minerals depends on various factors, including the type of clay mineral, the concentrations of electrolytes, and the characteristics of the cations present in the solution.

Clay minerals can be classified into two primary groups: kaolinite and montmorillonite. Kaolinite, a 1:1 layered silicate mineral, possesses a high surface area and exhibits a great adsorption capacity for organic contaminants. On the other hand, montmorillonite, a 2:1 layered silicate mineral, also has an extraordinary surface area and demonstrates an elevated adsorption capacity for HMs and other inorganic pollutants. The selection of the appropriate clay mineral depends on the particular pollutants to be eliminated and the conditions of the application [51–54].

2.2.3 Silica gel

Silica gel is an artificial material that is normally used for moisture control and the adsorption of VOCs, gases, and liquids. It has very high porosity with a significant surface area, making it an efficient adsorbent for the elimination of pollutants. It is non-lethal, reusable, and recyclable, and its definitive disposal can be achieved via standard waste management procedures. The use of silica gel as an adsorbent is extensively acknowledged in industries such as electronics, food packaging, and pharmaceuticals, in which moisture control is essential. Its ability to sorb moisture and VOCs makes it an effective and efficient way to remediate polluted liquids and gases. In water remediation, it is used to eliminate HMs, dissolved organic compounds, and other impurities. In air decontamination, it is employed to remove VOCs, such as benzene, formaldehyde, and toluene, which are normally found in indoor and outdoor air [19]. In waste management, it is employed to treat leachate, which is the liquid that is produced when waste materials are discarded in landfills. It is extensively utilized in several industries, including for environmental redress such as water treatment, air purification, and waste management, due to its extraordinary surface area and capability to sorb pollutants from liquids and gases [8, 55].

Silica gel has numerous benefits in terms of ecological sustainability. Primarily, as stated above, its porosity and significant surface area make it a beneficial solution for the elimination of pollutants. Furthermore, it is an artificial material which permits significant manipulation of its properties and the ability to alter it for specific purposes. Third, it is a non-lethal and chemically stable material, making it safe for ecological remediation. Finally, its ultimate disposal can be done in an ecologically safe way, such as by sending it to landfill, making it an ecological solution for the treatment of contaminated liquids and gases [2]. Overall, it possesses an outstanding ability to sorb considerable quantities of water, rendering it effective for both drying gases and preserving the dryness of various objects [56].

In addition, it is an extremely efficient and resourceful adsorbent material that is extensively applied in ecological sustainability. Its large surface area, its ability to be modified for specific purposes, and its nontoxicity lead to its selection for the management of polluted liquids and gases. Its definitive disposal can also be done in an ecologically safe way, making it a sustainable solution for the treatment of contaminated materials. The versatility and efficiency of silica gel make it an attractive alternative for several industries, including water treatment, air purification, and waste management [7, 8].

Farrukh *et al* [57] presented the successful use of surface-functionalized silica gel for sequestering water-soluble cationic dyes, attaining a sequestration rate of up to 98%. Due to the harmful and non-degradable properties of organic dyes, it is crucial to create and synthesize new adsorbents that can effectively eliminate them from the environment. The surface of the silica gel (SiO_2–SH) was decorated with thiol groups, which were subsequently oxidized to obtain sulfonic acid groups. As a result of this process, the silica gel's surface (SiO_2–SO_3H) had negatively charged components. X-ray photoelectron spectroscopy (XPS), transmission electron microscopy (TEM), and scanning electron microscopy (SEM) were used to characterize the morphology resulting from the developed adsorbents as well as the surface modifications.

To investigate the adsorbents' efficacy, the uptake of three cationic dyes (crystal violet, rhodamine 6G, and rhodamine B) by the SiO_2–SH and SiO_2–SO_3H was evaluated, considering factors such as the temperature, the pH of the solution versus the amount of dissolved adsorbent used, and the contact time. The interaction between the negatively charged species on the surface of SiO_2–SO_3H and the cationic dyes was more efficient due to electrostatic forces, leading to better remediation properties when compared to SiO_2–SH. In addition, the adsorbent's reusability was evaluated by examining the adsorption/desorption of dyes. The straightforward fabrication process offers an efficient approach to producing adsorbents that can effectively remove cationic dyes. Although silica gel lacks high selectivity towards CO_2, pressure swing adsorption (PSA) engineers compensate for this by adjusting the unit's operation. An advantage of using silica gel is its moderately steep adsorption isotherms for CO_2. In addition, it offers the convenience of performing both drying and upgrading steps in a single column. The evaluation of two silica gel samples, one with wide pores and the other with narrow pores, revealed that the material possessing the higher surface area and density holds greater potential for use in biogas upgrading [56].

In terms of its mechanism of action, silica gel works by sorbing pollutants on its surface. The pollutants are held on the silica gel surface by a combination of physical and chemical forces, such as hydrogen bonding, van der Waals forces, and ionic interactions. The extraordinary surface area of silica gel provides a better adsorption capacity, making it an efficient solution for the elimination of pollutants. It can be modified to improve its performance in certain applications. For example, it can be modified to have a superior affinity for certain pollutants, such as HMs or organic substances. This can be accomplished through different techniques, such as surface modification or the addition of chemical groups to the silica gel surface [7]. A study by Ozcelik *et al* [7] reported the creation and characterization of N-(2-aminoethyl) salicylaldimine bonded silica gel. Fourier transform infrared spectroscopy and C, H, and N elemental analyses were used for characterization. The analytical conditions, including the volume of the solution, the pH, the type of eluent, and the flow rates of the sample solution, were optimized to enable simultaneous preconcentration of Cu(II), Ni(II), Cd(II), and Zn(II) utilizing an improved silica-gel-loaded column and a solid-phase extraction technique.

2.2.4 Zeolites

Zeolites are naturally occurring minerals with an exceptionally porous structure that allows them to effectively sorb a variety of pollutants [58]. They have been extensively acknowledged and utilized in industries such as nuclear waste management, petrochemical processing, and water treatment owing to their capability to eliminate HMIs, radioactive isotopes, and organic substances [3, 9, 16, 49, 59]. They possess a remarkable affinity for these waste products, making them valuable tools in the decontamination of liquids and gases. Natural zeolites show great promise as materials for removing HMs from different contaminated water resources. To develop effective water treatment processes, it is critical to understand the chemistry, structure, and performance of natural zeolites in eliminating HMs from water. Various natural zeolites have been shown to sorb HMs such as cadmium (Cd(II)), arsenic (As(III), As(V)), lead (Pb(II)), chromium (Cr(III), Cr(VI)), nickel (Ni(II)) and mercury (Hg(II)). In an assessment [16], clinoptilolite emerged as the most utilized natural zeolite for HM elimination, as it has been shown to be effective in removing Pb, Hg, As, Cd, Cr, and Ni. An interesting observation is that the sorption properties and capacities for HMs may vary among natural zeolites from different geographical origins, implying that natural zeolites from different regions of the world possess distinctive characteristics [16]. In air purification applications, zeolites are particularly effective at sorbing VOCs, which are noxious and contribute to air contamination. VOCs such as benzene, formaldehyde, and toluene are commonly found in indoor air and can be efficiently eliminated by zeolites [60].

The adsorption mechanism of zeolites involves a combination of physical and chemical processes. Physical adsorption occurs when contaminants are trapped within the zeolite pores due to weak van der Waals forces; this removes small, volatile molecules. In contrast, chemical adsorption entails chemical bond formation between the zeolite surface and the adsorbed molecules, allowing for the removal of larger organic compounds and HMs [61, 62]. The factors that influence transfer in adsorption and heat storage include the mass flow rate of the adsorbate, the structural design of the zeolite bed, the cycle time, and the operating temperature; these were summarized from a systems perspective by Feng *et al* [63]. According to Velarde *et al* [16], the main alterations for zeolites include acid, base, and salt treatments, metallic functionalization, and cationic surfactant modification. Acid, base, and salt treatments alter the structure of zeolites, leading to improved microporosity, surface area, and pore size. Cationic surfactant and metallic reagent modifications boost the surface of zeolite, improving its affinity for anionic contaminants.

The characteristics and performance of modified zeolites and zeolite composites has been examined with a focus on their adsorption capabilities and interactions with adsorbates. In addition, the reactive behaviors of various zeolites and adsorbates have been evaluated [63]. The utilization of zeolite adsorption systems in energy storage and heat transfer applications has been explored. Various applications of zeolite-based systems have been examined, including energy storage systems (ESSs), heat exchangers (HXs), dehumidifiers, energy absorption systems (EASs), VOC elimination systems, and hydrocarbon (HC) traps. The research findings demonstrate the great promise of zeolite adsorption systems, whether used individually or in conjunction with other equipment, in the domains of energy efficiency and environmental sustainability.[63]

Zeolites can be regenerated and reused, providing an economical and eco-friendly replacement to outdated air purification techniques that rely on repeated filter replacements. Zeolites also demonstrate excellent potential in water purification applications. They are efficient adsorbents for the removal of HMs such as Cd, Pb, and Hg, which can pose serious health risks when present in drinking water [64]. Zeolites' high cation exchange capacity allows them to adsorb positively charged ions, making them efficient in removing these contaminants. In addition, zeolites can remove organic compounds including pesticides, herbicides, and pharmaceuticals from water due to their high surface area and microporous structure. An important advantage of zeolites in water purification is their ability to be regenerated and reused. Unlike traditional water treatment methods that often rely on chemical treatments, zeolites can be reused multiple times, reducing the need for frequent replacements and minimizing waste generation. Furthermore, zeolites have been explored for their potential to remove greenhouse gases such as CO_2 from the atmosphere. As a key contributor to global warming, the removal of CO_2 is crucial in mitigating climate change. Zeolites can act as catalysts in carbon mineralization processes, in which CO_2 is converted into stable carbonates for safe underground storage.

The use of zeolites to accelerate this reaction can reduce the concentration of CO_2 in the atmosphere, contributing to climate change mitigation efforts. Zeolites are versatile and effective adsorbents with significant applications in environmental sustainability. Their unique porous structure, high adsorption capacity, selectivity for specific contaminants, and ability to be regenerated make them valuable in various industries such as water treatment, air purification, and waste management. Zeolites offer a sustainable and environmentally friendly solution for removing pollutants from liquids and gases, contributing to the preservation and protection of the environment [61].

2.2.5 Polymeric materials

Polymeric adsorbents are generally synthetic materials, with some exceptions such as chitosan, which occurs less frequently in nature in some fungi [65]. These adsorbents find wide application in the removal of liquid and gas pollutants and they possess distinctive properties that enhance their effectiveness and potential as environmentally sustainable materials [49, 66–68]. One of the key attributes of polymeric adsorbents is their large surface area, which facilitates efficient pollutant adsorption, such as in CO_2 capture [69]. This enables them to attract and retain contaminants on their surfaces, effectively eliminating them from the surrounding environment. Polymeric adsorbents exhibit selective adsorption, targeting specific pollutants based on their chemical properties [70]. In addition, the functionalization of these materials can improve their selectivity, which enhances their efficiency in pollutant removal processes.

Amberlite polymeric resins have been found to be suitable for removing as well as immobilizing some organic extractants such as ethylenediaminetetraacetic acid (EDTA) and thenoyltrifluoroacetone (TTA) from aqueous solutions [71]. The transport and separation properties of ultrafiltration membranes are affected by several factors, including the transmembrane pressure, the membrane cut-off, the membrane material, and the concentration of the surface active agent [72]. The properties of polymeric adsorbents can be customized by adjusting their chemical composition, molecular structure, and porosity [10]. These modifications optimize their adsorption capacity, stability, and compatibility with various types of pollutants. The fine-tuning of a polymeric adsorbent material can be achieved using dopant atoms, adatoms, strain and stress, electric potential, etc. which results in the engineering of the material to target specific contaminants, making it highly versatile in diverse environmental applications [10].

To ensure that polymeric adsorbents are managed in an environmentally sustainable way, several considerations should be addressed. Polymeric materials should be benchmarked for optimal operation conditions, including temperature, pressure, and acidity in order to prevent plasticization as well as physical ageing of membranes due to their high free volume; this results in longer operational times [73]. The production approach should embrace sustainable practices, such as minimizing energy consumption and waste generation; as for all adsorbent materials, it should include the use of nontoxic chemicals and waste reutilization [74]. In addition, the adoption of eco-friendly manufacturing processes and reduced operating costs results in reduced environmental impacts [70].

The reusability and recyclability of polymeric adsorbents are important considerations for these systems. The application of proper methods for the regeneration of adsorption capacity after saturation enables effective recycling and reduces the demand for new materials while minimizing waste generation. It is essential to study the permissible immediate reuse of membranes to achieve the desired purity as well as the stacking/addition of extra membrane modules to obtain the desired level of purity [75]. The safe disposal of used polymeric adsorbents is crucial for environmental sustainability. Depending on their specific characteristics, various disposal methods can be employed. For polymeric adsorbents compatible with thermal treatment, combustion and carbonization can be carried out, coupled with appropriate emission control technologies [76]. For non-combustible materials, sending them to landfill can be considered, provided proper containment measures are in place to prevent the leaching of pollutants into the environment. In some cases, alternative approaches such as adsorbent immobilization or encapsulation can be explored. These methods involve incorporating polymeric adsorbents into matrices or encapsulating them within materials that offer structural support.

This immobilization helps prevent the release of adsorbed pollutants during handling, transportation, and disposal.

The management of polymeric adsorbents in support of environmental sustainability requires a comprehensive approach. This entails sustainable production practices that prioritize reusability and recyclability and careful consideration of appropriate disposal methods. By incorporating these strategies, we can harness the potential of polymeric adsorbents while minimizing their environmental footprint, thus contributing to a cleaner and healthier environment [76].

2.3 The properties of commonly used adsorbents

By considering and optimizing these properties, materials scientists can develop adsorbents that meet specific requirements for a wide range of applications. The careful analysis of pore size, adsorption efficiency, surface area, selectivity, and regeneration ability supports the design of highly effective and sustainable adsorbents. Some key aspects for consideration include pore size, adsorption efficiency, surface area, selectivity, and regeneration ability [37] The effectiveness of adsorbents in various applications relies heavily on their specific properties as well as secondary contamination [77]. Among the key properties to consider are selective adsorption, superior surface area, and regeneration ability. A suitable adsorbent must possess high selectivity, allowing it to preferentially adsorb certain substances while disregarding others. Furthermore, it should demonstrate a large absorption capacity to ensure the efficient removal of targeted compounds. Specific consideration should also be given to the longevity and cost-effectiveness of the adsorbent.

It is important to note that different adsorbents exhibit diverse properties due to variations in their pore diameter, active surface, surface functional groups and quality of pore distribution [78]. Comprehending the significance of selectivity, surface area, regeneration ability, and the classification of pore sizes can effectively lead to the development of adsorbents with enhanced performance tailored to specific applications [79]. This understanding plays an important role in optimizing the design and functionality of adsorbents, enabling advancements in the area of environmental protection and sustainability.

2.3.1 Pore size

Pore size plays a vital role in determining an adsorbent's effectiveness. The pore size is classified into three main ranges based on an International Union of Pure and Applied Chemistry (IUPAC) classification. [25, 62]. Macropores are pores with diameters greater than 50 nm; they provide ample space for the adsorption of larger molecules. Mesopores, which can be referred to as transitional pores, have diameters ranging from 2 nm to 50 nm and are important for the efficient uptake of medium-sized molecules. Micropores, with diameters smaller than 2 nm, are particularly effective in capturing small molecules due to their narrow pore size distribution. In addition, it is feasible to maximize adsorption capacity by employing rational design techniques such as adjusting the particle size, pore sizes, and morphologies of the adsorbents, as well as by incorporating functional groups onto their surfaces to target specific adsorbates. These enhancements are necessary for further improvements [80].

2.3.2 Adsorption efficiency

Adsorption efficiency refers to an adsorbent's ability to effectively capture and retain target compounds. It is influenced by factors such as adsorbent dosage (the adsorbent dosage has a critical value until it reaches the optimal value), the concentration of the pollutant (efficiency decreases with reduced amounts of pollutant), the effect of temperature (increasing the temperature beyond the optimal level could lead to damage at the adsorption sites, resulting in decreased efficiency), the reusability of the adsorbent (the more the adsorbent is reduced, the lower the efficiency), pore size, surface area, selectivity, pH, contact time/agitation time, etc [81]. Adsorbents with a larger surface area offer more sites for adsorption, resulting in increased efficiency.

2.3.3 Surface area

The surface area is a crucial parameter in adsorbent design. It refers to the total area available for adsorption and is directly related to the adsorbent's capacity to capture and retain molecules. A higher surface area allows for greater adsorption potential, leading to improved performance. Thus, for effective absorption, an ideal adsorbent should possess porosity and a substantial surface area [82, 83]. As an example, the use of magnetite nanoparticles (Fe_3O_4 NPs) for HMI (such as Hg and Cr) remediation results in improved reactivity because of their extraordinary surface-area-to-volume ratio [28].

2.3.4 Selectivity

Selectivity is another critical property, as it determines an adsorbent's preference for specific substances over others. High selectivity ensures the adsorbent can target and remove the desired compounds with precision [84]. The selectivity for various commonly used adsorbents can be improved via chemical means such as protonation. The surface of nanochitosan/clinoptilolite has been shown to undergo chemical modification following the application of acid or pentaethylenehexamine, resulting in an increased presence of protonated functional groups. This modification led to an increased positive charge density on the adsorbent material. As a result, a robust electrostatic attraction was formed between the protonated active sites and nitrate ions in an acidic environment, thereby improving the nitrate adsorption process [66].

2.3.5 Regeneration ability

Regeneration ability is an essential characteristic of adsorbents. It refers to the adsorbent's capability to be restored and reused after capturing target compounds. According to Fachina *et al* [85], an effective adsorbent should maintain a capacity of at least 75% even after undergoing six cycles of reuse. Gkika *et al* [27] demonstrated

Figure 2.1. The cycle of adsorbent regeneration. Reprinted from [27], Copyright (2022), with permission from Elsevier.

that the disposal of the adsorbent can negatively affect the environment, highlighting the need for mitigatory measures such as regeneration and reuse processes such as desorption. The recycling process is described by the stage of sorbent reusability, the cost of the process, and the potential ecological impacts. The reuse cycle is a nonstop sorption and desorption cycle, which may result in substantial monetary and ecological gains. Figure 2.1 shows the reuse/regeneration phase [27]. This review emphasizes the significance of desorption and regeneration, which are as crucial as the adsorption stage. It also provides an overview of different regeneration methods, along with the factors that influence them, their advantages, and disadvantages. Adsorbents with good regeneration ability can be effectively rejuvenated, extending their lifespan and reducing overall costs.

2.3.6 The modification of adsorbents

The modification of adsorbents is an area that has continuously been explored and is actively being explored to enhance their performance and customize them to target specific contaminants. Techniques such as the introduction of dopant atoms, adatoms, strain and stress, electric potential, and other modifications can be employed to fine-tune polymeric adsorbents, thereby augmenting their versatility and effectiveness in environmental remediation. The physiochemical sorption properties of adsorbents can be improved through modifications and treatments. As a result of enhancements in morphology, pore size, particle size, pore volume, stability, capacity, and possibly magnetic properties, low-cost adsorbents can achieve desirable properties. In addition, developments in adsorbent characteristics (for instance, the surface functionalization/additional of functional groups) can further enhance their effectiveness. These advancements enable the efficient, nontoxic, and economical elimination of pollutants from aqueous solutions [80].

The incorporation of dopant atoms or adatoms into adsorbents can bring about alterations in their chemical properties and surface characteristics. This, in turn, leads to an improved adsorption capacity and selectivity towards specific contaminants. For instance, the introduction of metal ions or nanoparticles through doping can create catalytic sites or enhance surface reactivity, enabling the adsorbent to more efficiently target specific pollutants. Furthermore, subjecting the adsorbent to strain and stress through physical or mechanical means can induce structural modifications, resulting in the creation of additional active sites and an enhancement in adsorption capabilities [22, 86–90]. Applying an electric potential to adsorbents also offers advantages in the realm of contaminant removal. Electrochemical techniques such as electrostatic attraction, electrosorption, and electrochemical regeneration can be harnessed to amplify adsorption capacity and facilitate the regeneration of adsorbents. The application of an electric potential brings about changes in the adsorbent's surface charge and electrostatic interactions, thereby bolstering adsorption efficiency. Moreover, the fine-tuning of adsorbents enables the targeting of specific contaminants in environmental applications. Modifying the surface properties of an adsorbent can enhance its affinity towards HMIs, organic pollutants, dyes, and emerging contaminants, enabling it to effectively address specific challenges associated with pollution [15, 91]. Through the engineering of the adsorbent's characteristics using techniques such as dopant atoms, adatoms, strain and stress, electric potential, and other methods, the adsorption capacity, selectivity, and efficiency can be optimized for specific target pollutants [5].

These advancements in adsorbent engineering hold tremendous promise across a range of environmental applications. They contribute to the development of tailored adsorbents for wastewater treatment, gas separation, HM removal, dye removal, and the elimination of emerging contaminants. The ability to fine-tune adsorbents affords flexibility in addressing various pollution scenarios and enables the implementation of more effective and sustainable environmental remediation strategies.

The design and engineering of adsorbents through techniques such as dopant atoms, adatoms, strain and stress, electric potential, and other modifications presents a pathway that can enhance their versatility and effectively target specific contaminants [62, 92, 93]. These advancements pave the way for improved performance and customized solutions in diverse environmental applications spanning from water treatment to gas separation. Continued research and development in this area will undoubtedly contribute to the advancement of effective and sustainable environmental remediation strategies.

2.4 Management options

Several adsorbent materials as discussed in section 2.1 are commonly used to support environmental sustainability. This is because of their varying unique characters as well as properties. The removal of contaminants from various media using common adsorbents such as carbon-based materials, clay minerals, silica gel, zeolites, and polymeric materials plays an important role In environmental sustainability [12, 37]. The management options entail their reusability, recyclability/regeneration, and final disposal. In this section, reusability through desorption and regeneration and its cost are presented and discussed.

2.4.1 Reusability through desorption

The desorption process involves releasing the adsorbate from the surface or bulk of the adsorbent material. Desorption can be accomplished by employing either heat or a solvent. While heat represents a rapid and efficient method, it is also highly expensive. Furthermore, heat treatments necessitate pretreatment, which may result in the release of hazardous contaminants and may create unsafe process characteristics. In addition, this process can face challenges related to regulation and public perception. Conversely, solvent desorption offers an environmentally viable, time-efficient, and cost-effective alternative that holds great promise for the future [27].

In this scenario, desorption is first carried out to remove the adsorbate and regenerate the adsorbents. It is important to develop techniques for regenerating adsorbents after use to minimize waste and maximize sustainability. The optimal regeneration technique should focus not only on the desorption process but also facilitate the recovery of the adsorbent material, ensuring that the adsorbate can be collected without being destroyed [24]. However, non-destructive techniques for desorption are currently limited, as they may lead to the damage of either the adsorbate or adsorbent, or both. The adsorption and desorption mechanism can be understood by considering that the surface of the adsorbent contains active sites responsible for the adsorption process, which is typically characterized as exothermic. These active sites facilitate the binding forces linking the atoms of the solid structure, leading to the adsorption of the foreign adsorbate molecules. On the other hand, the desorption process (an endothermic process in most cases) requires activation to reclaim the sorbate from the sorbent material. For wastewater treatment, various desorption techniques such as solvents, acids, alkalis, temperature variations, pH modifications, or supercritical CO_2 are employed [79, 94, 95].

2.4.2 Regeneration and cost

Regeneration is an important aspect for adsorbents, and it entails their reusability and efficiency. Regeneration refers to the efficient retrieval and reutilization of depleted adsorbents, accomplished through a technically and financially viable process as well as the effectiveness of the adsorption process. It is a crucial factor to be considered when considering the practical implementation of adsorbents on a large scale. Various factors, such as pH, dose, temperature, and initial composition, affect the removal capacity. Additionally, material properties, including size, porosity, and surface area, are key considerations. For instance, CNTs possess a unique structure and surface that provide many active sites and plentiful space. Kumar *et al* [96] conducted a study that utilized a setup and sensitivity assessment to highlight the significance of regenerating and reusing adsorbents. The findings emphasize the cost of phosphate adsorption required to achieve extremely low phosphate concentrations, which falls between \$100 and \$200 per kilogram of P. It is essential for future research to prioritize investigating the adsorption capacity at low phosphate levels, as well as exploring methods for regenerating and reusing both the adsorbent and the regenerating liquid. Gkika *et al* [27] discussed why spent adsorbents should be reused with a focus on the latest challenges and limitations. To investigate their ability to desorb and regenerate using different desorption media, the adsorbents were divided into four groups: CNTs, AC compounds, graphene, and clays and polymer adsorbents. The study examined various process factors such as concentration, required dose, pH, adsorption ability, and cost of the adsorbents in order to conduct a thorough analysis. The assessment of the desorption process also included the recovery efficiency and the potential for reuse. Among the adsorbents, those based on graphene demonstrated the highest adsorption capacity, ranging from 108 to >480 mg g^{-1}, while AC materials exhibited capacities between 34 and >384 mg g^{-1}. CNTs and polymer-based adsorbents showed lower and more variable sorption capacities, ranging from 1 to >138 mg g^{-1} and 7 to >57 mg g^{-1}, respectively. Nearly all the reviewed cases followed the PSO model. These materials displayed confiscation efficiencies ranging from 71% to 99%. The overall advantage of these adsorbents is their high reusability.

2.4.3 Final disposal

Accurate and precise management of commonly used adsorbents, such as carbon-based materials, clay minerals, silica gel, zeolites, and polymeric materials, is crucial for their successful application. This includes developing strategies as well as policies for the final disposal of adsorbents after reuse and recycling. This final disposal can take place via the chemical dissolution and recycling of the degraded adsorbent followed by chemical separation to separate the adsorbate from the adsorbent. The chemical process would be a more effective management practice but might initially be more costly in comparison to burning or disposal at a landfill site; therefore, further research work should evaluate chemical disposal methods. However, the chemical procedure offers a more environmentally sustainable and balanced approach with a minimal environmental impact.

2.5 Conclusions and recommendations

This chapter evaluated the different adsorbent types, their characteristics, and their management options with an emphasis on the reusability, recyclability, and final disposal of commonly used adsorbents in support of environmental sustainability.

The adsorbent materials considered were carbon-based materials, clay minerals, silica gel, zeolites, and polymeric materials. These materials all have considerable surface areas and a strong affinity for the adsorbate being adsorbed through physical, chemical, or physiochemical means.

An important aspect of various adsorbents that support environmental sustainability is reuse via regeneration; therefore, the efficacy of the regenerated adsorbent materials is crucial. The efficiency of regeneration is often evaluated for different numbers of cycles in order to attain elevated efficiency percentages, thereby minimizing operating and capital costs and preventing the re-release of pollutants into the environment through spent adsorbents. To obtain a more comprehensive understanding and gather information applicable to large-scale applications, the qualities and attributes of restored materials should be the subjects of further study. A prospective study should not only be dedicated to recognizing the equilibrium stage but also to exploring restoration kinetics, desorption effectiveness, and other relevant aspects that enable utilization in larger-scale settings. It is essential to conduct pilot-scale evaluations to gather initial data and then optimize the restoration findings to resemble real conditions and larger-scale settings.

References

[1] Zhao G 2011 Sorption of heavy metal ions from aqueous solutions: a review *Open Colloid Sci. J.* **4** 19–31

[2] Woo S-Y, Lee H-S, Ji H and Kim Y-D 2021 Adsorption isotherm model for analyzing the adsorption characteristics of water vapor to commercially available silica gel adsorbents for adsorption desalination applications *J. Chem. Eng. Data* **66** 1144–56

[3] Pourhakkak P, Taghizadeh M, Taghizadeh A and Ghaedi M 2021 Adsorbent *Interface Sci. Technol.* **33** 71–210

[4] De Gisi S, Lofrano G, Grassi M and Notarnicola M 2016 Characteristics and adsorption capacities of low-cost sorbents for wastewater treatment: a review *Sustain. Mater. Technol.* **9** 10–40

[5] Adeyemo A A, Adeoye I O and Bello O S 2017 Adsorption of dyes using different types of clay: a review *Appl. Water Sci.* **7** 543–68

[6] Das P, Arias V A, Kambala V, Mallavarapu M and Naidu R 2013 Remediation of perfluorooctane sulfonate in contaminated soils by modified clay adsorbent-a risk-based approach *Water Air Soil Pollut.* **224** 1714

[7] Ozcelik G, Imamoglu M, Yildiz S Z and Kara D 2012 Chemically modified silica gel with n-(2-aminoethyl)-salicylaldimine for simultaneous solid phase extraction and preconcentration of Cu(II), Ni(II), Cd(II) and Zn(II) in waters *Water Air Soil Pollut.* **223** 5391–9

[8] Chua H T, Ng K C, Chakraborty A, Oo N M and Othman M A 2002 Adsorption characteristics of silica gel + water systems *J. Chem. Eng. Data* **47** 1177–81

[9] Abdelmoneim A A, Abdul-Moneim M, Geies A A and Farghaly S O 2020 Application of zeolite prepared from natural materials for reducing ground water salinity and sodium ion from Wadi el-Assiuti *Desalin. Water Treat.* **193** 72–82

[10] Zhou J, Wu P, Dang Z, Zhu N, Li P, Wu J and Wang X 2010 Polymeric Fe/Zr pillared montmorillonite for the removal of Cr(VI) from aqueous solutions *Chem. Eng. J.* **162** 1035–44

[11] Lazaratou C V, Vayenas D V and Papoulis D 2020 The role of clays, clay minerals and clay-based materials for nitrate removal from water systems: a review *Appl. Clay Sci.* **185** 105377
[12] Laskar N and Kumar U 2022 Application of low-cost, eco-friendly adsorbents for the removal of dye contaminants from wastewater: current developments and adsorption technology *Environ. Qual. Manag.* **32** 209–21
[13] Sabzehmeidani M M, Mahnaee S, Ghaedi M, Heidari H and Roy V A L 2021 Carbon based materials: a review of adsorbents for inorganic and organic compounds *Mater. Adv.* **2** 598–627
[14] Momina M and Shahadat S I 2018 Regeneration performance of clay-based adsorbents for the removal of industrial dyes: a review *RSC Adv.* **8** 24571–87
[15] Azha S F, Shahadat M, Ismail S, Ali S W and Ahammad S Z 2021 Prospect of clay-based flexible adsorbent coatings as cleaner production technique in wastewater treatment, challenges, and issues: a review *J. Taiwan Inst. Chem. Eng.* **120** 178–206
[16] Velarde L, Nabavi M S, Escalera E, Antti M L and Akhtar F 2023 Adsorption of heavy metals on natural zeolites: a review *Chemosphere* **328** 138508
[17] Mobarak M, Qaysi S, Saad Ahmed M, Salama Y F, Mohamed Abbass A, Abd Elrahman M, Abdel-Gawwad H A and Seliem M K 2023 Insights into the adsorption performance and mechanism of Cr(VI) onto porous nanocomposite prepared from gossans and modified coal interface: steric, energetic, and thermodynamic parameters interpretations *Chinese J. Chem. Eng.* accepted
[18] del A, Borja-Urzola C, García-Gómez R S, Bernal-González M, del M and Durán-Domínguez-de-Bazúa C 2021 Chitosan-calcite from shrimp residues: a low-cost adsorbent for three triazines removal from aqueous media *Mater. Today Commun* **26** 102131
[19] da Costa J S, Bertizzolo E G, Bianchini D and Fajardo A R 2021 Adsorption of benzene and toluene from aqueous solution using a composite hydrogel of alginate-grafted with mesoporous silica *J. Hazard. Mater.* **418** 126405
[20] Liu P and Zhang L 2007 Adsorption of dyes from aqueous solutions or suspensions with clay nano-adsorbents *Sep. Purif. Technol.* **58** 32–9
[21] Li C, Long X, Songfeng E, Zhang Q, Li T, Wu J and Yao Y 2019 Magnesium-induced preparation of boron nitride nanotubes and their application in thermal interface materials *Nanoscale* **11** 11457–63
[22] Obodo K O, Ouma C N M, Obodo J T, Gebreyesus G, Rai D P, Ukpong A M and Bouhafs B 2021 Sn3C2 monolayer with transition metal adatom for gas sensing: a density functional theory studies *Nanotechnology* **32** 355502
[23] Li J, Wang X, Zhao G, Chen C, Chai Z, Alsaedi A, Hayat T and Wang X 2018 Metal-organic framework-based materials: superior adsorbents for the capture of toxic and radioactive metal ions *Chem. Soc. Rev.* **47** 2322–56
[24] Leonzio G, Fennell P S and Shah N 2022 A comparative study of different sorbents in the context of direct air capture (DAC): evaluation of key performance indicators and comparisons *Appl. Sci.* **12** 2618
[25] Kim Y D, Woo S Y, Lee H S and Ji H 2021 Adsorption isotherm model for analyzing the adsorption characteristics of water vapor to commercially available silica gel adsorbents for adsorption desalination applications *J. Chem. Eng. Data* **66** 1144–56
[26] Ravikovitch P I and Neimark A V 2006 Density functional theory model of adsorption on amorphous and microporous silica materials *Langmuir* **22** 11171–9

[27] Gkika D A, Mitropoulos A C and Kyzas G Z 2022 Why reuse spent adsorbents? The latest challenges and limitations *Sci. Total Environ.* **822** 153612

[28] Geneti S T, Mekonnen G A, Murthy H C A, Mohammed E T, Ravikumar C R, Gonfa B A and Sabir F K 2022 Biogenic synthesis of magnetite nanoparticles using leaf extract of thymus schimperi and their application for monocomponent removal of chromium and mercury ions from aqueous solution *J. Nanomater.* **2022**

[29] Pandey S 2017 A comprehensive review on recent developments in bentonite-based materials used as adsorbents for wastewater treatment *J. Mol. Liq.* **241** 1091–113

[30] Ummartyotin S and Pechyen C 2016 Strategies for development and implementation of bio-based materials as effective renewable resources of energy: a comprehensive review on adsorbent technology *Renew. Sustain. Energy Rev.* **62** 654–64

[31] Ansari A and Rodrigues D F 2022 Enhancement of filtration and adsorption processes for water treatment using graphene-based nanomaterials *World Scientific Reference of Water Science* (Singapore: World Scientific) pp 1–30

[32] Das G P and Alibrandi A 2021 Determining distribution coefficients for low cost adsorbents *Remediation* **31** 19–27

[33] Jha S, Gaur R, Shahabuddin S and Tyagi I 2023 Biochar as sustainable alternative and green adsorbent for the remediation of noxious pollutants: a comprehensive review *Toxics* **11** 1–23

[34] Ponnuchamy M, Kapoor A, Senthil Kumar P, Vo D V N, Balakrishnan A, Mariam Jacob M and Sivaraman P 2021 *Sustainable Adsorbents for the Removal of Pesticides from Water: A Review* (Berlin: Springer International Publishing)

[35] Yeow J T W and Wang Y 2009 A review of carbon nanotubes-based gas sensors *J. Sensors* **2009** 493904

[36] Chen Y, Liu Q, Yan Y, Cheng X and Liu Y 2010 Influence of sample cell physisorption on measurements of hydrogen storage of carbon materials using a Sieverts apparatus *Carbon N. Y* **48** 714–20

[37] Kyzas G Z and Kostoglou M 2014 Green adsorbents for wastewaters: a critical review *Materials (Basel)* **7** 333–64

[38] Ko K J, Kim H, Cho Y H, Lee H, Kim K M and Lee C H 2022 Overview of carbon monoxide adsorption performance of pristine and modified adsorbents *J. Chem. Eng. Data* **67** 1599–616

[39] Shukla S K, Al Mushaiqri N R S, Al Subhi H M, Yoo K and Al Sadeq H 2020 Low-cost activated carbon production from organic waste and its utilization for wastewater treatment *Appl. Water Sci.* **10** 1–9

[40] Kozyatnyk I, Yacout D M M, Van Caneghem J and Jansson S 2020 Comparative environmental assessment of end-of-life carbonaceous water treatment adsorbents *Bioresour. Technol.* **302** 122866

[41] Chikri R, Elhadiri N, Benchanaa M and El maguana Y 2020 Efficiency of sawdust as low-cost adsorbent for dyes removal *J. Chem.* **2020** 8813420

[42] Sellaoui L, Knani S, Erto A, Hachicha M A and Ben Lamine A 2016 Equilibrium isotherm simulation of tetrachlorethylene on activated carbon using the double layer model with two energies: steric and energetic interpretations *Fluid Phase Equilib.* **408** 259–64

[43] Bushra R, Mohamad S, Alias Y, Jin Y and Ahmad M 2021 Current approaches and methodologies to explore the perceptive adsorption mechanism of dyes on low-cost agricultural waste: a review *Microporous Mesoporous Mater.* **319** 111040

[44] Ngulube T, Gumbo J R, Masindi V and Maity A 2017 An update on synthetic dyes adsorption onto clay based minerals: a state-of-art review *J. Environ. Manage.* **191** 35–57

[45] Chen L F, Liang H W, Lu Y, Cui C H and Yu S H 2011 Synthesis of an attapulgite clay@carbon nanocomposite adsorbent by a hydrothermal carbonization process and their application in the removal of toxic metal ions from water *Langmuir* **27** 8998–9004

[46] del M, Orta M, Martín J, Santos J L, Aparicio I, Medina-Carrasco S and Alonso E 2020 Biopolymer-clay nanocomposites as novel and ecofriendly adsorbents for environmental remediation *Appl. Clay Sci.* **198** 105838

[47] Diagboya P N, Olu-Owolabi B I, Mtunzi F M and Adebowale K O 2020 Clay-carbonaceous material composites: towards a new class of functional adsorbents for water treatment *Surf. Interfaces* **19** 100506

[48] Pires J and Carvalho S 2022 Adsorbents from natural clays for the separation of hexane isomers *J. Environ. Chem. Eng.* **10** 107689

[49] Oladoye P O 2022 Natural, low-cost adsorbents for toxic Pb(II) ion sequestration from (waste)water: a state-of-the-art review *Chemosphere* **287** 132130

[50] Zhang T, Wang W, Zhao Y, Bai H, Wen T, Kang S, Song G, Song S and Komarneni S 2021 Removal of heavy metals and dyes by clay-based adsorbents: from natural clays to 1D and 2D nano-composites *Chem. Eng. J.* **420** 127574

[51] Tombácz E and Szekeres M 2006 Surface charge heterogeneity of kaolinite in aqueous suspension in comparison with montmorillonite *Appl. Clay Sci.* **34** 105–24

[52] Doi A, Ejtemaei M and Nguyen A V 2019 Effects of ion specificity on the surface electrical properties of kaolinite and montmorillonite *Miner. Eng.* **143** 105929

[53] Deng L, Yuan P, Liu D, Annabi-Bergaya F, Zhou J, Chen F and Liu Z 2017 Effects of microstructure of clay minerals, montmorillonite, kaolinite and halloysite, on their benzene adsorption behaviors *Appl. Clay Sci.* **143** 184–91

[54] Luo L, Guo Q and Cao Y 2019 Uptake of aqueous tungsten and molybdenum by a nitrate intercalated, pyroaurite-like anion exchangeable clay *Appl. Clay Sci.* **180** 105179

[55] Hoel A S and Carusi A 2015 Thinking technology with Merleau-Ponty *Postphenomenological Investigations: Essays on Human–Technology Relations* (Lanham, MD: Lexington Books) 73–84

[56] Grande C A, Morence D G B, Bouzga A M and Anne Andreassen K 2020 Silica gel as a selective adsorbent for biogas drying and upgrading *Ind. Eng. Chem. Res.* **59** 10142–9

[57] Farrukh A, Akram A, Ghaffar A, Tuncel E, Oluz Z, Duran H, Ur Rehman H and Yameen B 2014 Surface-functionalized silica gel adsorbents for efficient remediation of cationic dyes *Pure Appl. Chem.* **86** 177–1188

[58] Roth W J and Čejka J 2011 Two-dimensional zeolites: dream or reality *Catal. Sci. Technol.* **1** 43–53

[59] Esaifan M, Warr L N, Grathoff G, Meyer T, Schafmeister M T, Kruth A and Testrich H 2019 Synthesis of hydroxy-sodalite/cancrinite zeolites from calcite-bearing kaolin for the removal of heavy metal ions in aqueous media *Minerals* **9** 484

[60] Ackley M W, Rege S U and Saxena H 2003 Application of natural zeolites in the purification and separation of gases *Microporous Mesoporous Mater.* **61** 25–42

[61] Aghel B, Mohadesi M, Gouran A and Razmegir M H 2020 Use of modified Iranian clinoptilolite zeolite for cadmium and lead removal from oil refinery wastewater *Int. J. Environ. Sci. Technol.* **17** 1239–50

[62] Dimas Rivera G L, Martínez Hernández A, Pérez Cabello A F, Rivas Barragán E L, Liñán Montes A, Flores Escamilla G A, Sandoval Rangel L, Suarez Vazquez S I and De Haro Del Río D A 2021 Removal of chromate anions and immobilization using surfactant-modified zeolites *J. Water Process Eng* **39** 101717

[63] Feng C, Han J E W, Deng Y, Zhang B, Zhao X and Han D 2021 Key technology and application analysis of zeolite adsorption for energy storage and heat-mass transfer process: a review *Renew. Sustain. Energy Rev.* **144** 110954

[64] Cortés-Martínez R, Martínez-Miranda V, Solache-Ríos M and García-Sosa I 2004 Evaluation of natural and surfactant-modified zeolites in the removal of cadmium from aqueous solutions *Sep. Sci. Technol.* **39** 2711–30

[65] Aranaz I, Alcántara A R, Civera M C, Arias C, Elorza B, Caballero A H and Acosta N 2021 Chitosan: an overview of its properties and applications *Polymers (Basel)* **13** 3256

[66] Eltaweil A S, Omer A M, El-Aqapa H G, Gaber N M, Attia N F, El-Subruiti G M, Mohy-Eldin M S and El-Monaem E M A 2021 Chitosan based adsorbents for the removal of phosphate and nitrate: a critical review *Carbohydr. Polym.* **274** 118671

[67] Tofan L 2022 Polymeric biomass derived adsorbents for Co(II) remediation, recycling and analysis *Polymers (Basel)* **14** 1647

[68] Park C W, Kim B H, Yang H M, Seo B K and Lee K W 2017 Enhanced desorption of Cs from clays by a polymeric cation-exchange agent *J. Hazard. Mater.* **327** 127–34

[69] Abbasian A and Takht Ravanchi M 2023 Polymeric membranes and surfaces for CO_2 capture *CO_2-philic Polymers, Nanocomposites and Solvents: Capture, Conversion and Industrial Products* (Amsterdam: Elsevier) pp 17–55

[70] Haider S, Lindbråthen A and Hägg M B 2016 Techno-economical evaluation of membrane based biogas upgrading system: a comparison between polymeric membrane and carbon membrane technology *Green Energy Environ* **1** 222–34

[71] Sheha R R and Someda H H 2005 Removal of some chelators from aqueous solutions using polymeric ingredients *Chem. Eng. J.* **114** 105–13

[72] Majewska-Nowak K, Kowalska I and Kabsch-Korbutowicz M 2005 Ultrafiltration of SDS solutions using polymeric membranes *Desalination* **184** 415–22

[73] Zhao C, Hussain W, Chlib Alkaaby H H, Al-Khafaji R M, Alghazali T, Izzat S E, Shams M A, Abood E S, Yu A E and Ehab M 2022 Polymeric nanocomposite membranes for gas separation: performance, applications, restrictions and future perspectives *Case Stud. Therm. Eng.* **38** 102323

[74] Goh P S, Othman M H D and Matsuura T 2021 Waste reutilization in polymeric membrane fabrication: a new direction in membranes for separation *Membranes (Basel)* **11** 782

[75] Koutsiantzi C, Mitrakas M, Zouboulis A, Kellartzis I, Stavropoulos G and Kikkinides E S 2022 Evaluation of polymeric membranes' performance during laboratory-scale experiments, regarding the CO_2 separation from CH_4 *Chemosphere* **299** 134224

[76] Lawler W, Bradford-Hartke Z, Cran M J, Duke M, Leslie G, Ladewig B P and Le-Clech P 2012 Towards new opportunities for reuse, recycling and disposal of used reverse osmosis membranes *Desalination* **299** 103–12

[77] Alsawy T, Rashad E, El-Qelish M and Mohammed R H 2022 A comprehensive review on the chemical regeneration of biochar adsorbent for sustainable wastewater treatment *npj Clean Water* **5** 29

[78] Unuabonah E I, Günter C, Weber J, Lubahn S and Taubert A 2013 Hybrid clay: a new highly efficient adsorbent for water treatment *ACS Sustain. Chem. Eng.* **1** 966–73

[79] Jain A, Kumari S, Agarwal S and Khan S 2021 Water purification via novel nano-adsorbents and their regeneration strategies *Process Saf. Environ. Prot.* **152** 441–54

[80] Tony M A 2022 Low-cost adsorbents for environmental pollution control: a concise systematic review from the prospective of principles, mechanism and their applications *J. Dispers. Sci. Technol.* **43** 1612–33

[81] Elkhaleefa A, Ali I H, Brima E I, Shigidi I, Elhag A B and Karama B 2021 Evaluation of the adsorption efficiency on the removal of lead(II) ions from aqueous solutions using *Azadirachta indica* leaves as an adsorbent *Processes* **9** 559

[82] Amrhar O, El Gana L and Mobarak M 2021 Calculation of adsorption isotherms by statistical physics models: a review *Environ. Chem. Lett.* **19** 4519–47

[83] Li Y, Bahamon D, Sinnokrot M, Al-Ali K, Palmisano G and Vega L F 2021 Computational modeling of green hydrogen generation from photocatalytic H_2S splitting: overview and perspectives *J. Photochem. Photobiol. C Photochem. Rev.* **49** 100456

[84] Wang X, Ma C, Xiao J, Xia Q, Wu J and Li Z 2018 Benzene/toluene/water vapor adsorption and selectivity of novel C-PDA adsorbents with high uptakes of benzene and toluene *Chem. Eng. J.* **335** 970–8

[85] Fachina Y J, de Andrade M B, Guerra A C S, dos Santos T R T, Bergamasco R and Vieira A M S 2022 Graphene oxide functionalized with cobalt ferrites applied to the removal of bisphenol A: ionic study, reuse capacity and desorption kinetics *Environ. Technol.* **43** 1388–404

[86] Obodo K O, Ouma C N M, Gebreyesus G, Obodo J T, Ezeonu S O and Bouhafs B 2019 DFT + U studies of the electronic and optical properties of ReS_2 mono-layer doped with lanthanide atoms *Mater. Res. Express* **6** 106307

[87] Ouma C N M, Singh S, Obodo K O, Amolo G O and Romero A H 2017 Controlling the magnetic and optical responses of a MoS_2 monolayer by lanthanide substitutional doping: a first-principles study *Phys. Chem. Chem. Phys.* **19** 25555–63

[88] Uwadiunor E, Kotasthane V, Yesudoss D K, Nguyen H, Pranada E, Obodo K, Radovic M and Djire A 2023 Pt-like catalytic activity from an atomistically engineered carbonitride MXene for sustainable hydrogen production *Chem Catal.* **3** 100634

[89] Obodo K O, Ouma C N M, Obodo J T and Braun M 2017 Influence of transition metal doping on the electronic and optical properties of ReS_2 and $ReSe_2$ monolayers *Phys. Chem. Chem. Phys.* **19** 19050–7

[90] Chettri B *et al* 2021 Induced ferromagnetism in bilayer hexagonal boron nitride (h-BN) on vacancy defects at B and N sites *Phys. E Low-Dimens. Syst. Nanostruct.* **126** 114436

[91] Akpomie K G and Dawodu F A 2016 Acid-modified montmorillonite for sorption of heavy metals from automobile effluent *Beni-Suef Univ. J. Basic Appl. Sci.* **5** 1–12

[92] Wang Q, Gao W, Liu Y, Yuan J, Xu Z, Zeng Q, Li Y and Schröder M 2014 Simultaneous adsorption of Cu(II) and SO_4^{2-} ions by a novel silica gel functionalized with a ditopic zwitterionic Schiff base ligand *Chem. Eng. J.* **250** 55–65

[93] Unuabonah E I, Günter C, Weber J, Lubahn S and Taubert A 2013 Hybrid clay: a new highly efficient adsorbent for water treatment *ACS Sustain. Chem. Eng.* **1** 966–73

[94] Singh K and Gautam M 2017 Development of inexpensive biosorbents from de-oiled mustard cake for effective removal of As(V) and Pb(II) ions from their aqueous solutions *J. Environ. Chem. Eng.* **5** 4728–41

[95] Baskar A V *et al* 2022 Recovery, regeneration and sustainable management of spent adsorbents from wastewater treatment streams: a review *Sci. Total Environ.* **822** 153555

[96] Kumar P S, Korving L, van Loosdrecht M C M and Witkamp G J 2019 Adsorption as a technology to achieve ultra-low concentrations of phosphate: research gaps and economic analysis *Water Res.* X **4** 100029

IOP Publishing

Adsorption Applications for Environmental Sustainability

Kingsley Eghonghon Ukhurebor, Uyiosa Osagie Aigbe and Robert Birundu Onyancha

Chapter 3

Activated biosorbents for the removal of metals from aqueous solutions

Uyiosa Osagie Aigbe, Kingsley Eghonghon Ukhurebor, Robert Birundu Onyancha, Benedict Okundaye, Efosa Aigbe, Kingsley Obodo, Enaroseha O E Omamoke, Lunyanda Lunga Noto, Adelaja Otolorin Osibote and Harrison I Atagana

The release of heavy metals (HMs) into the environment creates a substantial hazard and has a long-lasting effect on ecosystems and communities' well-being. There has been an increase in research into the development of scientific solutions that are eco-responsive and cost-effectively achievable, i.e. those that can both decrease cost and increase effectiveness. In this context, the biosorption process (BP) is a dependable, sustainable, and capable option for the treatment of HMs owing to its inexpensive and nontoxic materials such as agricultural and industrial wastes and dead biomasses of both cellulosic and microbial origins. Hence, this chapter provides a systematic review of the various biosorbents utilized for HM confiscation.

3.1 Introduction

Artificial chemicals' benefits to normal life are incontrovertible, but their methodical and involuntary release into the broader ecosystem is a direct result of industrial advancement. Chemical contaminants have been released since the period of industrialized advancement, but their release and spread have increased significantly in the past half-century. Trillions of masses of chemically active materials are distributed into the ecosystem by mineral-handling, mining, energy production, construction, and farming. Also, as a result of the anthropogenic distribution of geogenic chemicals, humans have fabricated over 1.4×10^5 chemicals and blends of chemicals, most of which did not occur beforehand. It is clear from studies of the current worldwide inventories of chemicals that this number could exceed about 3.5×10^5. Many of these substances are understood to be noxious in minute doses; at times they can blend with other contaminants or occur as breakdown products after

doi:10.1088/978-0-7503-5598-8ch3

their precursors are transported into the geosphere and biosphere. It is estimated that as many as 2×10^2 billion tonnes of chemicals are released yearly, among which greenhouse emissions make up 20% or more. A major pollutant released into the environment, which is of immense concern, is HMs [1, 2].

HM elements existed naturally in the crust of the Earth during the formation of the Earth, but metal mining, chemical fertilizer use and industrial production from anthropogenic actions have resulted in the rise of metallic substances in marine and terrestrial environments. There is an amplified worldwide concern about water pollution and the limited availability of clean water due to HM pollution, as it is a key prerequisite for living beings and social income [3].

In the last decade, swift suburbanization and industrial development have created a massive worldwide water contamination issue. Fast industrialization has led to an increased release of HMs into the ecosystem, thereby contaminating the environment. The different forms of water pollutants are HM ions, bacteria, organic particles, dyes, and viruses. HMs are distinguished from other noxious substances due to their persistence in the ecosystem (non-recyclable) and the prospect of their buildup in existing tissues, which leads to their intensification in the food chain [4, 5]. Metal ions (MIs) are released into the environment by many unpredictable activities and pathways, including runoff from surface water [6].

Well known HMs such as chromium (Cr), cadmium (Cd), nickel (Ni), arsenic (As), lead (Pb), and mercury (Hg) are key contaminants for freshwater resources owing to their significant impact on living organisms. Unlike other natural pollutants, HMs are non-recyclable and tend to persist in the ecosystem. Table 3.1 shows the various HMs, their impacts, and their recommended allowed concentrations (RACs) in water according to global health organizations and environmental protection agencies such as the World Health Organization (WHO) and the United States Environmental Protection Agency (USEPA) [5]. The buildup of HMs in an individual's body can trigger very severe harmful impacts [6–8], as enumerated in table 3.1. HMs are carcinogenic and their residual volume can lead to various health issues in humans [4].

There are snowballing indications of mental, generative, and hormonal syndromes and early fatalities triggered by the chemical pollution present in today's social ecosystem. It was reported in 2012 that the number of people who died as a result of exposure to contaminants in water, soil, and air (both outdoors and indoors) was estimated at 8.4 million fatalities in lower and intermediate per-capita-income nations [1]. Hence, among the various types of water contamination, HMs are key ecosystem pollutants that require special consideration and efficient approaches for their decontamination owing to their noxiousness, persistence, non-recyclable nature, and unlimited distribution [3].

With economic development and water scarcity in mind, different manufacturing developers and manufacturers are currently implementing technologies to guarantee cleaner production and reduce contamination and water consumption [9]. Concern for natural ecosystems is becoming a guiding influence behind the expansion of green industrial technologies via product waste management, forcing scientists to search for and create novel, comparatively accessible materials and procedures to

Table 3.1. HM sources, their impacts, and their RACs, according to the WHO and the USEPA.

HM	RAC according to the WHO (mg l^{-1})	Industrial sources of wastewater containing HM	Noxious impacts
As	0.01	Textiles, paper, glass, insecticides, combustion from mining and coal, and phosphate-based fertilizers	Kidney, lung, and bladder cancer, liver tumor, nausea, muscular weakness, and neurological disorder.
Hg	0.01	The manufacture of pulp, ore, pharmaceuticals, paper, and batteries	Paralysis, feeble immune system, neurological disorder, and blindness
Cu	1.30	Pigments, paints, medical equipment, cleaning products, fertilizer, and tanneries	Cancer of the lungs, liver injury, insomnia, heart disease, migraine headache, seizures, cancer, and osteoporosis.
Cd	0.003	Fuel combustion, metal plating, power plants, fertilizers, batteries, mining, smelting, and solid waste.	Renal disturbance, liver fibrosis, and cancer, malfunction of the kidney, and bone lacerations.
Cr	0.05/0.01	Dye, textile activities, ceramics, and steel production	Mutagenic, teratogenic, cancer, lung tumor, vomiting, hemorrhage, epigastric pain, nausea, and severe diarrhea
Ni	0.015	Coinage, electroplating, mining, paints, glass, refining, and batteries.	Decreased lung function, prolonged cancer, prolonged bronchitis, and asthma, cancer of the lungs, and dermatitis.
Zn	5.00	Galvanization, stabilizers, steel manufacturing, incineration of coal, and mining	Nausea, lethargy, loss of appetite, neurological signs, and gastrointestinal pain.
Pb	0.05	Manufacturing of batteries, bronze, ceramics, roofing materials, glass, pipes, and ammunition.	Anemia, vomiting, and circulatory illness, mental disease, nervous system and brain injury

decrease the unwanted HM accumulation in the ecosystem in the future. Hence, the search for, and realization of the sequestration of HMs using new and inexpensive sorbents is a strategic necessity [4]. In modern times, effluent treatment has become a notable area of scientific study that includes various fields of engineering and science. In the last few years, various physicochemical methods have been suggested for eradicating different contaminants from aqueous solutions using standard techniques for the treatment of HM-containing effluents. Such methods include ion exchange, coagulation, flocculation, sorption, membrane filtration, oxidation/reduction, incineration, precipitation, etc. Most of these standard techniques are limited with respect to their practicality, dependability, effectiveness, conservation

Table 3.2. The associated benefits and drawbacks of the treatment technologies employed for HM confiscation.

Treatment technique	Pros	Shortcomings/cons
Sorption	A broad range of target contaminants, elevated capacity, rapid kinetics, and selectivity, depending on the sorbent	Its performance depends on the sorbent type. Chemical and physical activation are required to improve the sorption capacity.
Chemical precipitation	Simple process, metal insensitive, and reasonable principal cost	The huge amount of metal-containing sludge generated, the huge cost of sludge disposal, and elevated maintenance and initial capital costs
Coagulation/flocculation	Bacterial deactivation capacity, excellent sludge setting and dewatering features	Chemical utilization and heightened production of sludge volume
Ion exchange	Metal selectivity with regulated pH tolerance and high rejuvenation	Elevated preliminary capital and maintenance costs
Flotation	Metal selectivity with minimal retention times and confiscation of tiny particles	Elevated preliminary capital, operating, and maintenance costs
Magnetic separation/purification	Separation and decontamination of magnetically sensitive pollutants	Particle collection varies significantly depending on the huge field gradients created, the particle sizes, and the magnetic properties.
Electrochemical remediation	Moderate metal selectivity, no chemicals are required, and this treatment can be engineered to allow floating solids	Elevated preliminary capital costs, hydrogen creation, and filtration process for flocs
Membrane filtration	Low-solid-content water is produced, low chemical utilization, and minimal space prerequisite	Elevated preliminary capital, operational, and maintenance costs.

impact, noxious by-product management, operational complexity, etc. The associated benefits and drawbacks of the treatment technologies employed for HM confiscation, as reported by [10–12], are summarized in table 3.2.

Sorption is considered to be the best of the standard techniques employed for wastewater treatment, as it is a direct process, inexpensive, extremely capable, and straightforward in the confiscation of excessive MIs, thereby allaying any concerns over elevated working and capital costs, effectiveness, and the requirement for secondary treatment. Since environmental safety and protection are attracting interest worldwide, there is a need to create greener remediation procedures for the confiscation of various contaminants, such as HMs and metalloid ions, via the biosorption method. Biosorption is one of the most studied options for eliminating pollutants from runoff owing to its practical application and economic aspects, since it uses organic materials and renewable biomass as biosorbents. Its name comprises the union of the prefix 'bio' and the suffix 'sorption' [7, 13]. Biosorption denotes the ability of specific biomolecules to capture and remove certain ions from aqueous media. In contrast to the considerably more complex trend of bioaccumulation based on active metabolic transport, biosorption by deceased biomass is inert and is generally based on the attraction between the sorbate and sorbent [13].

The application of biosorbents over the years has attracted much interest, and most biosorbents are generally grouped into marine materials, agricultural wastes, microbial biosorbents, minerals, and industrial by-products. However, the variety of precursor materials used today is distinct and extensive. Essentially, all natural materials with comparatively high carbon contents can serve as unprocessed convectional materials in the synthesis of biosorbents [7, 14].

The advantages of biomaterials include their extraordinary biosorption capacity (BC), the availability of massive quantities of biomaterials, their affordability, the simple desorption of the sorbed pollutants, and biosorbent-specific reusability. The notable inorganic and organic biosorbents that have elevated sorption capacities include fungal biomass, cottonseed shell, chitin and shrimp shell, eggshell waste, bagasse, activated carbon (AC) derived from agricultural plant wastes, vegetables, chitosan, bone ash, walnut shell, used coffee, coal ash, clay, resins, sawdust, cellulose, soybean waste (following oil extraction), tea waste, silica gel, etc. have been utilized as biosorbents for the treatment of effluents containing various kinds of pollutants [7, 15, 16]. This chapter will provide an comprehensive insight into the activation of biosorbents and their effect on the sorption of various HMs and the influence of various factors on the sorption of different HMs.

3.2 Modification procedures

Generally, organic biosorbents are inherently inactive and show weak mechanical properties and physical stability as well as minimal exchange and indiscriminate biosorption capacities (BCs). Hence their chemical modification/functionalization procedure is deemed to be a reliable way of considerably enhancing their biosorbent function by increasing their surface area, creating novel pores, preventing the elution of biosorbent natural compounds into the treated solution via various characteristics, and hosting additional binding sites. Unique binding groups are grafted onto the biosorbent surface via amination, oxidation, carboxylation, thiolation, sulfonation, xanthanation, phosphorylation, and halogenation [17].

The surfaces of various biomasses can be altered to create the desired physicochemical properties using suitable activation technique options. It is even feasible to create biosorbents with specified dimensions of macro-, meso-, and micropores. The modification procedures for carbon materials are grouped into chemical, physical, and biological characteristics; the chemical modification method based on chemical compounds is the method most frequently used to enhance the sorption and the sorption capacity of AC, carbon, or biosorbents, owing to the use of lower temperatures in this procedure and the comprehensive yield of this process. The activation conditions and the nature of the starting material determine the structure, the micropore volume, and the transitional pores created during activation [13].

The usual approach to producing biowaste-based carbon is carbonization; however, in specific circumstances, activation is conducted first [18]. Biomaterials are generally activated through physical and chemical stimulation methods. The chemical activation technique is subdivided into two categories, depending on the operation. This comprises the mixture or immersion of the biomass with the activator at a certain proportion; the carbonization and activation are carried out concurrently at an elevated temperature. Activation via the chemical method is well-known to stimulate certain surface porosity features and functionality which play a critical part in improving the sorptive capability of the created biosorbents [19].

For biosorbent fabrication from various biomaterials, the chemical activation agents utilized are KOH, H_3PO_4, H_2SO_4, NaOH, HNO_3 and $ZnCl_2$. These agents require single-phase reactions, which are produced by infusing appropriate alkali metal-based salts and acids/bases into the raw carbon source to improve the surface area of the AC. In contrast, in the physical activation method (gas activation), a feedstock is carbonized in oxygen (O_2) or low-pressure inert gas at a specific temperature. The gases generally utilized for activation are oxidizing gases such as air, CO_2, and steam. This technique involves the pyrolysis of biomaterials at elevated temperatures, which is generally a two-step process, such as carbonization of the biomaterials in a passive atmosphere followed by material activation via air gasification at a typical activation temperature of 700°C–1100°C. In the activation process, carbon reacts endothermically with H_2O and CO_2, and the degree of reaction is reasonable. The reaction between carbon (C) and O_2 is exothermic (EXO), and the rate of reaction is swift. Hence, the reaction time is reduced when O_2 is combined with the mixture to burn the transitional products H_2 and CO, and heat is released. A highly stable porous structure is created when C_2 reacts with CO_2 and H_2O vapor [19–21]. However, the use of extreme temperatures of 1200°C and above leads to a low carbon yield, the generation of ash, and the destruction of the porous structure, while de-volatilizing the process leads to the formation of pores and enlarged micropores. The biological modification process is a developing technology [22]. The impregnation time and the biological reagent affect the chemical and physical properties of carbon materials. Biomaterials obtained via both activation techniques show a massive specific surface area and their pore size generally corresponds to mesopores and micropores [19].

3.3 The sorption of heavy metals using activated biosorbents

A key benefit of biosorption is that it is based on easily available and easy-to-process biomaterials which have the properties required to confiscate HMs and decrease their concentrations from the parts per million (ppm) range to the parts per billion (ppb) range. To confiscate HMs and improve the BC of the biosorbents, their chemical properties are improved through chemical modifications [23].

It is advisable to utilize biosorbents with better BCs, as this would suggest that an above-average volume of sorbate will be absorbed, while biosorbents with low BCs will, in the long run, require replacement. However, other features such as the sustainability level (with reference to any unwanted impacts of biosorbent use on the ecosystem), the rapidity of the biosorption of target species (which is related to the confiscation efficacy), biosorbent regeneration, and recycling opportunities should be considered while selecting a biosorbent [24].

Due to the structural features of commercial AC, it is an excellent sorbent, even leaving aside its extensive utilization in water and effluent treatment, but its application is generally restricted owing to the substantial capital cost of its preparation and regeneration. This has created an enhanced interest in the creation of reliable and inexpensive sorbents which are highly efficient compared to commercially obtainable AC in the sequestration of HMs from contaminated resources. The advancement of eco-friendly sorbents that are of minimal cost, plentiful in the environment, and available in enormous amounts has been an area of broad study owing to their prospective use in biosorption processes (BPs) for HM confiscation. Organic materials such as agricultural wastes, microbial biomass, industrial wastes, natural remains and other materials such as cellulose, chitosan, etc. have been utilized for this purpose as alternatives to traditional sorbents [6, 25, 26].

Agricultural wastes contain functional groups (FGs) such as carbonyl (–C(=O)–), acetamido, amido, alcohols, esters, sulphydryl, and phenolics. These FGs have a tendency to complex with metals [6]. The potential for biomaterials to be developed into cheaper biosorbents is well recognized, and such materials are considered to be efficient for the treatment of waterborne contaminants based on the increased study of the use of biosorbents for HM confiscation described in the SCOPUS report for the period from 2010 to May 2023 presented in figure 3.1.

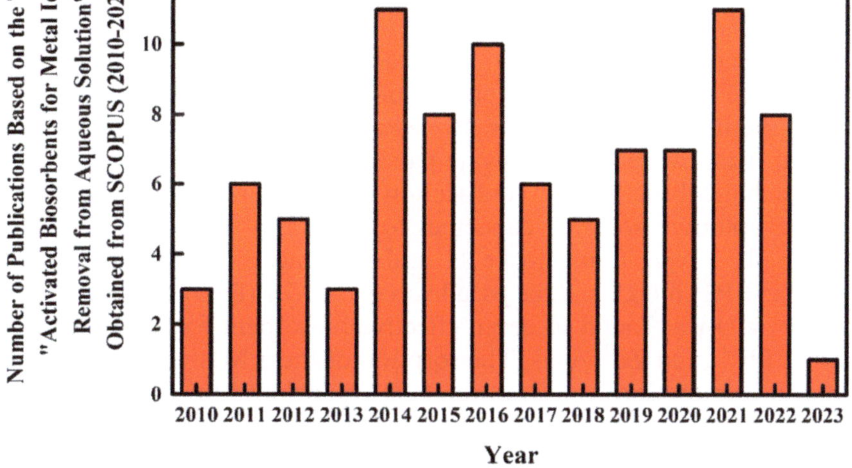

Figure 3.1. A graph indicating the number of reported publications from 2010 to May 2023 using the keywords 'Activated Biosorbents for MI Uptake From Water-Soluble Solutions' obtained from SCOPUS.

In the course of biosorption, the two stages used are the solid stage (biosorbents such as bacteria, fungi, and plants) and the liquid stage (aqueous media containing HM ions which are identified as sorbates). In the confiscation of HMs from aqueous media, algae, bacteria, fungi, and plants are the reported conventional biosorbents; their cell walls (CWs) are the key biosorption sites of the biosorbents and these CWs contains various varieties of FGs which contribute to the BP. These FGs that are capable of HM sorption in biosorbents are hydroxyls ($^{-}$OH alcohols and carbohydrates), amino proteins and nucleic acids, esters (lipids), carboxyls (COOH proteins, organic acids, and fatty acids), carbonyls –C(=O)– (which can be terminal or internal in polysaccharides, aldehydes, ketones, and polysaccharides), phosphates (DNA, RNA, and tissue plasminogen activators), and sulfhydryls (cysteine (amino acid) and proteins) [24].

Recent study has focused on the fabrication of biomass-obtained sorption materials for novel biosorption technologies involving mono- and multi-component sorption procedures, since modern-day agricultural activities create 1×10^6 tons of waste and by-products yearly that have prospects as valuable resources. These agro-developed residues are a promising substitute for conventional sorbents, as they are inexpensive, freely accessible, exceedingly sorptive, readily adaptable, impervious to noxious substances, and straightforward to control during the remediation procedure. To date, a variety of aqueous contaminants have been removed from solutions by various agro-industry-developed sorbents [11]. In addition, the use of agricultural waste products as effective biosorbents is well thought out, as they are enhanced with lignocellulosic materials that confiscate HMs [23].

In a study by Simon *et al*, agro-based industrial pine sawdust, sunflower seed hulls, and corn residue mixed wastes were observed to be applicable in the treatment of effluent containing the HM ions of Ni^{2+}, Cd^{2+}, and Zn^{2+} [27]. In a study by Ahmed *et al*, natural waste residues (NWRs) such as *Lupinus*, *Moringa*, sugarcane straw, and tea filtrate were utilized and assessed for the decontamination of Cu^{2+}, Cr^{6+}, Se^{2-}, Pb^{2+}, and Zn^{2+} ions. MIs were ideally sorbed at pH 6. The Langmuir (LNR) model, the Freundlich (FRH) model and the pseudo first-order (PFO) model were found to perfectly describe the process of MI sorption. The overall confiscation efficiency pattern in this study was *Moringa*>tea>*Lupinus*>sugarcane straw [28]. Microwaved steam was used to activate palm kernel shell, which was subsequently utilized as a biosorbent for the confiscation of Cd^{2+} and Pb^{2+} ions present in a landfill. The optimum sorption capacities (q_m) for 100 mg l^{-1} of Cd^{2+} and Pb^{2+} solutions at pH 3 were found to be 11 and 59.9 mg g^{-1}. The BP was optimally defined by the FRH model and the pseudo second-order (PSO) model [29]. The weed *Rumex acetosella* was found to be an effective biosorbent for the sequestration of As^{3+}, Cd^{2+}, Pb^{2+}, and Zn^{2+} ions. The sorption of these MIs was reported at pH 5. The sorption information best fitted the LNR model and the pseudo second-order model (PSO). From the outcomes obtained, it is evident that the prepared biosorbent was useful and suitable for wastewater remediation [30]. In a study by Jayan *et al* [31], the Periyakulam-2 (PKM-2) variety of *Moringa oleifera* leaves was explored for the treatment of aqueous solutions containing Pb^{2+} and Zn^{2+} ions.

The prepared biosorbent was found to be efficient in the treatment of these HM ions, with optimum removal reported at pH 4.7 and pH 3.2 for Pb^{2+} and Zn^{2+} confiscations, respectively. In addition, the sorption information was observed to be an excellent fit for the LNR model and the PSO model. The determined optimum sorption capacities were 51.71 and 38.50 mg g^{-1} for Pb^{2+} and Zn^{2+}, respectively. The BCs of both MIs were also confirmed to be spontaneous (SPONS), and EXO [31]. Three marine biosorbents, namely the bivalve mollusk *Cerastoderma glaucum* (CG) and the scales of the *Rutilus kutum* (RK) and *Oncorhynchus mykiss* (OM) fish species were used for Pb^{2+} confiscation. These biosorbents were found to be effective in the treatment of Pb^{2+} using the LNR model (RK and OM) and the FRH model (CG). The sorption capacities of RK, OM, and CG were 24.26, 14.39, and 1.29 mg g^{-1} [32]. *Rosa damascene* leaf powder was assessed as a potential biosorbent for Cu^{2+} confiscation, and a quadratic model was observed to capture the drivers of the response based on the Box–Behnken design (BBD). The best conditions for Cu^{2+} confiscation were determined to be pH 5.5, an initial Cu concentration of 55 mg l^{-1}, and a biosorbent dosage of 4.0 g l^{-1}. The PSO model and the Elovich model (ELM) were good fits for the sorption process of Cu^{2+}, while the LNR model and the Dubinin–Radushkevich model (DRM) superlatively matched the experimental information, achieving an optimum BC of 25.13 mg g^{-1}. In. addition, the process of Cu^{2+} sorption was SPONS, feasible, and endothermic (ENDC). The core mechanism of interaction between the *Rosa damascene* biomass and Cu^{2+} was influenced by hydrogen bond formation and ion exchange [33].

From an evolutionary viewpoint, algae are heterogeneous [34]. Algae are non-vascular plants that vary in size from individual-celled or colonial microalgae or phytoplankton-like dinoflagellates and diatoms to huge multicellular macroalgae such as seaweeds or kelp. Most algae lack stems, roots, leaves, and stems, but all algae have chlorophyll. They are categorized based on their pigments, stored food materials, cell walls (CWs), reproductive structures, life history pattern, and morphology into *Pyrrhophyta*, *Rhodophyta*, *Chlorophytata*, *Charophyta*, *Chrysophyta*, *Phaeophyta*, and *Euglenophyta* groups. Bioremediation using algae is a viable technique for the sequestration of HMs from effluents owing to their surface-area-to-volume ratios [35, 36]. Algae are normal biomass and have changing levels of attraction for HMs. Owing to algae's capacity for nutrient confiscation, sorption, and their regenerative and organic nature, they are used to remediate different wastewaters from industries, agriculture, and mining [37].

Algae are commonly used as biosorbents for HMs confiscation owing to their high accessibility, comparatively inexpensive processing cost, and elevated performance. Algae used for the biosorption of HMs are grouped into microalgae (photosynthetic unicellular plants that thrive in fresh or salt water and are grouped according to their pigmentation (diatoms, green algae, blue-green algae (cyanobacteria) and golden algae), photosynthetic membrane arrangement(s), or other morphological characteristics) and macroalgae (seaweeds, which are multicellular plants that are found in fresh or salt water and whose grouping depends on their pigmentation—brown, red, and green algae). The ability of algae to biosorb contaminants is based on their CW arrangement, which is made of chitin, proteins, polysaccharides, and lipids that comprise FGs such as O_2, nitrogen (N_2), sulfur (S), and phosphorus (P) that aid the biosorption of HM ions [24, 38, 39].

The CWs of brown macroalgae are generally made of alginic acid, polymers such as guluronic and mannuronic acids, and cellulose, which are complexed with simple metals such as potassium, calcium, sodium, and polysaccharides such as fucoidan. Fucoidan and alginate (in which COOH groups and secondary sulfate groups are the main biosorption binding sites) have the possibility of binding MIs via ion exchange. The CWs of green algae comprise proteins with FGs of amine groups, COOH, sulfates, and ^{-}OH, which contribute to MI sorption. Red algae CWs contain cellulose, and their BC relies on the sulfated polysaccharides obtained from galactans. In microalgae, the CWs mainly comprise polysaccharides, lipids, and proteins. These compounds consist of ^{-}OH, COOH, sulfate, and phosphate groups, which create a total net negative charge on the cell surface, hence favouring metal cation binding via counter-ion interactions [40–42]. The prospects of three green algae biosorbents (*Cladophora glomerata* (CG), *Enteromorpha intestinalis* (EI) and *Microspora amoena* (MA)) were explored for the elimination of Cr^{6+}. The BC was established to be reliant on the pH, and the optimal BC was observed at pH 2. The FRH model was found to perfectly explain the sorption of Cr^{6+} by the green algae, and dry biomass of *C. glomerata* was found to be the most effective biosorbent for Cr^{6+} removal in this study [43]. Wild algae biomass was tested for the effective removal of Zn^{2+} ions. This study reported that 92.12% of the Zn^{2+} ions were sequestered by the algae biomass activated using NaOH in 2 h at pH 7. Above and below pH 7, a reduction in the elimination of Zn^{2+} ions was reported, which was attributed to the occurrence of metal precipitation from the metal hydroxide form owing to the ease of reaction of MIs with –OH ions and the elevated concentration of H^+, which led to a competitive exchange of H^+ with the cations on the algae surface [44]. Marine and freshwater algae were assessed for the confiscation of wastewater containing Cu^{2+}, Pb^{2+}, Zn^{2+}, and Cd^{2+} ions. The maximum biosorption of the HMs was observed at pH 5 except for Cd^{2+}, for which the highest sorption was observed at pH 11. This study reported that a total metal content of approximately 40 mg l^{-1} was sequestered from the industrial effluents using 50 mg of the biosorbent. This was ascribed to FGs such as ^{-}OH, carboxylic (–COOH), and amine groups present in both biosorbents [45]. Acid-modified *Ulva flexuosa* biomass was explored as a successful biosorbent for the sequestration of Cd^{2+}, Co^{2+}, and Zn^{2+} ions. It was found that the prepared biosorbent was effective in the treatment of wastewater containing these HMs, and optimum biosorption (94.8% ± 3.3%, 87.5% ± 2.3%, and 90.8% ± 1.4%, for Cd^{2+}, Co^{2+}, and Zn^{2+}, respectively [$p < 0.01$)] took place at pH 4 using 0.4 g of biosorbent mass and a contact time of 35–40 min. The LNR model best described the biosorption of these HMs by the prepared biosorbent [46]. Green algae were explored for the confiscation of Cd^{2+}, Cr^{6+}, Cu^{2+}, and Pb^{2+} in a study by Olal [47]. The sorption of these HMs was found to be pH-supported, and optimum confiscations of each MI were observed at pH 5.5, 5.8, 5.9, and 5.0 for Cd^{2+}, Cr^{6+}, Cu^{2+}, and Pb^{2+}, respectively. The sorption processes were optimally defined by the LNR model and the PSO model, and sorption capacities of 22.52, 38.19, and 94.34 mg g^{-1} were observed for Cd^{2+}, Cr^{6+}, Cu^{2+}, and Pb^{2+}, respectively [47]. *Hizikia fusiformis*, green *Gracilaria*, and *Codium fragile* algae were explored for the confiscation of MIs of Cd^{2+}, Cu^{2+}, Pb^{2+}, and Ni^{2+}. The *H. fusiformis* algae were activated using NaOH and HCl; an improvement in the BC for MIs was shown by *H. fusiformis* activated with NaOH, owing to a hydrolysis reaction that created ^{-}OH and –COOH groups. The biosorption process was modeled using the FRH model and the LNR model (Ni^{2+}, Cd^{2+}, Cu^{2+}, and Pb^{2+}), the PFO model, the PSO model and the ELM. The optimum BCs were 167.73, 45.09, 44.38, and 42.08 mg g^{-1} for Pb^{2+}, Cu^{2+}, Ni^{2+}, and Cd^{2+} respectively [48]. Sultana *et al* [49] explored the effectiveness of microalgae (*Chlorella kessleri*) as biosorbents for the confiscation of Pb^{2+}, Co^{2+}, Cu^{2+}, Cd^{2+}, and Cr^{2+}. Utilizing the response-surface methodology (RSM), the most favorable confiscation of Pb^{2+} of 99.54% was reported at pH 6.34. The confiscation efficiency of individual HMs followed the order $Pb^{2+} > Co^{2+} > Cu^{2+} > Cd^{2+} > Cr^{2+}$, and the sorption process for Pb^{2+} was optimally described by the PSOM [49]. *Ulva fasciata* biomass was assessed for Cd^{2+} confiscation, and the Plackett–Burman design was used to establish the most important factor for optimum MI confiscation. The optimum confiscation efficiency for Cd^{2+} was reported to take place at pH 5 [50].

Bacteria are the most copious, prevalent unicellular microorganisms and are also found in symbiosis with other creatures. They have an extremely modest cell structure that lacks a real nucleus and membrane-bound organelles. The outstanding structure of bacteria is their CWs, which provide cell structural integrity [51]. In bacteria, FGs such as COOH, phosphonates, amines, and ^{-}OH groups in the CWs are responsible for binding. Hence, the success of biosorption is dependent on the number of CW structures [52]. The bacteria used as biosorbents are grouped into gram-positive (the bacterial CWs are thicker peptidoglycan layers linked by amino acid bridges) or -negative bacteria; the most significant differences between the two groups are the CW composition and thickness. The gram-positive bacteria have better feasibility for the confiscation of HM cations owing to their considerable electronegative charge density which results from the presence of teichoic and teichuronic acids (polyalcohols which are connected by phosphodiester bonds bonded to the peptidoglycan of the CW). FGs containing N_2, S, O_2, or P are the main contributors to HM biosorption in bacteria [53–56].

Owing to the abundant CW contents (chitin, inorganic salts, cellulose, β-glucan, lipids, chitosans, polyuranides, α-glucan, glucoproteins, and pigments) of fungi, they are receiving increasing consideration as biosorbents for HM confiscation; the above list is suggestive of the presence of a broad array of FGs. Fungal biomasses are easily utilized as biosorbents owing to their generally easy cultivation and their ease of growth via straightforward fermentation methods with largely modest costs [24]. The binding of metal cations is possible due to the negative charge of microorganisms caused by the different anionic structures in their cell surfaces. Hence, bacterial biomasses can be efficiently employed as biosorbents for HM ions sequestration, even at a relatively low concentration [51].

Recently, the practice of using beneficial and capable microorganisms such as plant-growth-promoting bacteria to confiscate HMs has greatly contributed to environmental and agricultural planning due to its considered benefits. Such bacteria utilize chemical pollutants as an energy source via their metabolic processes, but their metabolism threatened by a excessive quantity of inorganic nutrients. Bacterial groups that have contributed to the sequestration of HMs include *Alcaligenes* sp., *Bacillus* sp., *Pseudomonas* sp., *Arthrobacter* sp., *Rhodococcus* sp., *Azotobacter* sp., and methanogens, and *Bacillus* sp. such as gram-positive bacteria are potential agents for the removal of various HMs. Based on evaluations and experiments performed on these numerous groups, different microbial biomasses have be found to possess distinct bioremediation capabilities. Since microbial cell membranes can be disrupted by HMs, bacteria have enzymatic profile features that are vital in order to overcome these noxious effects [52]. Microorganisms' sorption properties are due to the effects of outer cell shielding. Metals are linked to the surface layers of cells by effective groups of compounds. Thus is usually due to an ion transfer reaction between the metal cations and the effective groups, which is takes place because of the negative potential of the exterior cell structures and the microorganisms belonging to different regular units (due to the presence of different chemical groups in the exterior structures); the result is energetic metal linking [57]. The buildup of HMs on microbial biomasses with elevated surface-area-to-volume ratios has an extreme capability for HM sequestration in soluble and particulate states, especially when the HM concentration is <50 mg l^{-1} [58].

Eight resistant indigenous bacteria categorized according to their pattern of resistance (*Enterobacter* sp. (Cu_1), *Enterobacter* sp. (Cu_2), *Stenotrophomonas* sp. (Cd_1), *Providencia* sp. (Cd_2), *Chryseobacterium* sp. (Co_1), *Comamonas* sp. (Co_2), *Ochrobactrum* sp. (Cr) and *Delftia* sp. (M_1)), were isolated from acclimatized activated sludge and employed for the remediation of domestic polluted industrial effluent. All strains were gram-negative, two strains were resistant to Cu, Cd, and Co, and one strain was resistant to a combination of Cr and HMs [59]. Combined bacteria enriched by HM mine soil were employed for the confiscation of Cd^{2+} ions. The main genera in the biomass were *Alcaligenes* (18.49%), *Tissierella* (9.07%), unclassified f *Peptostreptococcaceae* (5.66%), unclassified f *Enterobacteriaceae* (63.95%), and *Terrisporobacter* (1.66%). The living biomass showed a better removal efficiency for Cd^{2+} at pH 6. After the sorption of Cd^{2+}, the main bacterial genera were *Comamonas* (39.94%), unclassified f *Enterobacteriaceae* (34.96%), *Ochrobactrum* (14.07%), *Alcaligenes* (4.84%), *Bordetella* (2.07%), *Serratia* (1.04%), and *Bacillus* (1.01%). The LNR model and PFO model described the sorption of Cd^{2+} by the biomass. The determined sorption capacities for living and autoclaved biomasses were 370 and 278 mg g^{-1}, respectively [60]. Activated sludge consisting of bacteria isolated from a water treatment plant was explored for the confiscation of Cu ions. It was found from the six Cu-resistant bacteria isolated that two Cu-resistant bacteria (C10.4 and C9.4) had the greatest capacities to accumulate Cu of up to 20.45% and 17% using 8.02 and 4.83 mg g^{-1} dry weight of cells, respectively [61]. Selected protozoan and bacterial species were explored for the degradation of Cd, Co, Mn, Pb, Ni, Cu, and Zn ions in industrial effluent samples. Living *Pseudomonas putida* exhibited the greatest elimination efficiency for the various MIs (Co 71%, Ni 51%, Mn 45%, V 83%, Pb 96%, Ti 100%, and Cu 49%), trailed by *Bacillus licheniformis* (Al 23% and Zn 53%) and *Peranema* sp. (Cd 42%) [62]. Sewage sludge and isolated dead cells of *Aeromonas hydrophyla* and *Branhamella* spp. were utilized for the sequestration of Zn^{2+}, Cd^{2+}, Cu^{2+}, and Pb^{2+} ions with an initial MI concentration of 100 mg l^{-1}. From the study outcomes, it was evident that *Aeromonas hydrophyla* and *Branhamella* spp. were effective in the sequestration of Zn^{2+}, Cd^{2+}, Cu^{2+}, and Pb^{2+} ions, achieving 99% removal at pH. The sorption of the MI was best described by the LNR model, which was suggestive of monolayer sorption taking place on the biomass researched. with optimum sorption capacities of 71.91 and 37.52 mg g^{-1} using *Aeromonas hydrophyla* and *Branhamella* spp [63]. Saber *et al* [64] explored immobilized *Pseudomonas alkalinise* for Cr^{6+} confiscation using the Plackett–Burman matrix. To find the model that best described the sorption of Cr^{6+} by biomass, a comparative method was established between a rotatable central composite design (RCCD) and an artificial neural network (ANN). It was found that the ANN model was more specific than the RCCD in predicting the optimum achievable effluent remediation [64]. Table 3.3 summarizes different biosorbent biomasses utilized for the biosorption of different HMs with their optimum sorption capacities.

3.4 Factors impacting the biosorption of heavy metal ions

BPs are influenced by pH, biosorbent dosage, initial HM concentration, contact time, and participating ions or co-ions. The most critical factor among the above is the pH of the solution. For the biosorption of MIs, intensification in pH leads to enhanced sorption but a further increase in the pH of the solution causes precipitation and should be avoided. For various biosorption systems, the ideal pH differs [34].

3.4.1 The effect of pH on the biosorption of heavy metal ions

Adjusting the pH, the contact time between the biosorbent and the MI solution, and the biosorbent dosage can also improve the BC, although the temperature and the pH are the key factors [24]. The pH impacts the ionization of FGs situated on the biosorbent surface and thus the biosorbent surface charge. When the pH is low, cations are practically stable and can bind to the surface of the biosorbent, while metal cations' solubility reduces, increasing the risk of precipitation, at high pH [78].

The pH at which biosorption takes place is exceptionally important; it strongly impacts the BC and in certain circumstances the process by which biosorption takes place. The interaction between the FGs of the biosorbent and the MIs is impacted by the pH value. The form of the MIs present in aqueous solutions and the concentration of these forms are also impacted by the pH, which results in the creation of various species of MI at various pH levels [24]. Biosorption is generally based on an ion-exchange mechanism between the dominant counter-ions on negatively charged ion groups and MIs. The BC continuously improves with intensifying pH until it gets to the ideal pH where the optimum BC is achieved. Upon further intensification of the solution's pH, metal precipitation occurs owing to the creation of metal ^{-}OH or ^{-}OH anionic complexes. The optimum sorption capacity of biosorbents is decreased at low pH values owing to the enhanced

Table 3.3. A summary of the various biosorbent biomasses utilized for HM ion biosorption.

Biosorbent biomass	HMs	Optimum pH	Isotherm model/q_m (mg g^{-1})	Kinetic model	Thermodynamic parameters	References
Oyster shell	F^-	5.5	LNR/27.31	PSO	SPONS and EXO	[65]
Dry haloalkaliphilic archaea *Natronolimnobius innermongolicus*	Cd^{2+}	8	LNR/128.21	PSO	—	[66]
Marine macroalga *Ulva lactuca*	Cu^{2+}, Cr^{3+}, Cd^{2+}, and Pb^{2+}	4.5, 5 and 6	LNR/64.5 and 84.7, 60.9 and 82, 62.5 and 84.6, and 68.9 and 83.3	—	—	[67]
Jatropha biomass	Cu^{2+} and Zn^{2+}	5	LNR/11.54, 20.48 and 22.91 and 14.85, 15.89 and 14.67	PSO	—	[68]
Coconut tree sawdust (CTS), eggshell (ES) and sugarcane bagasse (SB)	Cu^{2+}, Pb^{2+}, and Zn^{2+}	6	FRH/3.89, 25.00 and 23.81, 34.48, 90.90 and 35.71 and 3.65, 21.28 and 40.00	—	—	[69]
Brown alga *Sargassum latifolium* (SAP) and activated carbon (SAC)	Pb^{2+}, Cd^{2+}, and Cu^{2+}	5	LNR/141, 147 and 167	—	—	[70]
Dried microalga *Chlamydomonas* sp.	As^{3+}	4	LNR/52.91	PSO	SPONS and ENDC	[71]
Undaria pinnatifida	Pb^{2+}	—	LNR/250	—	SPONS and ENDC	[72]
Thiobacillus sp. and *Clostridium* sp.	Cu^{2+}	7	90%	—	—	[73]
Bacillus licheniformis sp.	Pb^{2+}	5.5	LNR/42.92	PSO	SPONS and ENDC	[74]
Macrofungus (*Ganoderma lobatum*) biomass	Cu^{2+}	8	FRH/18.38	PSO	—	[75]
Fe_3O_4-loaded AC prepared from plum stone (*Prunus nigra*)	Cu^{2+} and Pb^{2+}	5.5	LNR/48.31 and 80.65	PSO	SPONS and EXO	[76]
Treated ajwa date pits	Cu^{2+}	6.5	LNR/ 1428.57	PSO	SPONS and ENDC	[77]

competition between protons (hydronium ions) and MIs for binding sites. At low pH values, the biosorbent FGs favor hydronium ions as FGs, which occur in protonated structures, thereby leading to electrostatic repulsion between the FGs and the metal cations. With an additional increase in the solution pH, FGs (COOH, ^{-}OH, phosphate groups, and amino and imidazole groups) experience negative charges and turn neutral owing to deprotonation, leading to improved metal cation binding and hence increased BC and swiftness of biosorption, as shown in table 3.3. While a further increase in pH can guarantee that all acidic groups are deprotonated, the free metal cations are then able to precipitate as metal ^{-}OH, thereby decreasing the effectiveness of MI removal [24, 25]. For anionic species such as TcO_4^-, CrO_4^{2-}, SeO_4^{2-}, $PtCl_4^{3-}$, and $Au(CN)_2^-$, biosorption improves at lower pH values. For metals such as Ag^{2+}, Hg^{2+}, and $AuCl_4^-$, biosorption is independent of pH because of the covalent complexes with N and S resulting from ligand formation [79].

3.4.2 The effect of the biosorbent dosage

The biosorbent dosage is a critical factor that impacts confiscation efficiency owing to the presence of supplementary unused biosorption sites [78]. The main factor that determines the biosorption mechanism is the quantity of biosorbent present. Elevated doses increase the creation of cellular aggregates, thereby decreasing the active surface area available for biosorption. In addition, the number of FGs present on the biosorbent surface during the BP is determined by the biosorbent dosage. Owing to the increase in unsaturated biosorbent active sites, an increase in the biosorbent dosage results in a reduction in each MI's capability for biosorption [80]. As reported in the studies of Zafar *et al* [81], Kamilya *et al* [82], and Ahmed *et al* [28], the percentages of MIs confiscated by various biosorbents improved as the dosage of biosorbent was increased, but no significant improvement in the percentages of MIs confiscated was observed after a certain dosage of biosorbent was reached.

3.4.3 The effect of the nitial metal concentration

This is a critical driving force for MIs to overcome the mass-transfer resistance between the bulk and aqueous phases [25]. According to the results of the studies by Azam *et al* [77], Zafar *et al* [81], Kamilya *et al* [82], Jayakumar *et al* [83], and Dulla *et al* [84], it was found that MI BC improved with an increase in the initial MI concentration. Utilizing low metal concentrations did not reveal the actual optimum BC, as various accessible binding sites for MIs were not engaged. A further rise in the initial concentration of the MIs would lead to the full saturation of the biosorbent-accessible binding sites, thereby revealing the maximum metal confiscation potential of the biosorbent. Once the biosorbent reaches saturation, an additional rise in the initial MI concentration is ineffective [24].

3.4.4 The effect of the contact time

The contact time is a significant factor that influences the profile of HM confiscation as well as economic performance. A study of the contact time behavior clearly shows that there is an improvement in HM confiscation at the start of the procedure. This is due to the physical sorption or ion exchange ensuing on the solid sorbent surface. At the beginning, additional binding sites are available; this is the case before the available sites start to reduce and become inactive due to a lack of vacant sites for MI binding on the sorbent surface [37]. The time required for the biosorption of MIs is greatly reliant on the specific type of biosorbent [28]. In the studies by Madala *et al* [85], Ibrahim *et al* [67], Staron and Chwastowski [86], Abd Ali *et al* [87], Nadeem *et al* [88], Abdelfattah *et al* [89], Akbari *et al* [90], Tavana *et al* [91], Cholico-Gonzalez *et al* [92], and Maksoud *et al* [17], the magnitude of the BP increased over time and subsequently reached an equilibrium with the concentration of MI sorbed. In the early stages of the BP, the efficiency of MI confiscation to the biosorbent improved swiftly owing to the profuse accessible active binding sites on the biomaterials. The steady occupation of active biomaterial sites led to reduced confiscation efficiency in the advanced stage of the process; this resulted from the decreased concentration gradient owing to the saturation of the active sites on the biosorbent surface.

3.4.5 Temperature and thermodynamic analysis

The temperature has an effect on the BP. An increase in temperature enhances the tendency for the MIs to be removed from the surface of the biosorbent. As a result, the FG bonds break down, thereby reducing the forces between the MIs and the active accessible sites. In contrast, at low temperatures, elevated confiscation efficiency is ascribed to the physical sorption, which is indicative of an EXO mechanism [37]. In terms of temperature, various thermodynamic properties are assessed in order to comprehend the BP; these include changes in the Gibbs free energy (ΔG°), the entropy (ΔS°), and the enthalpy (ΔH°). The level of spontaneity is indicated by ΔG°, negative values of which imply a SPONS process. A process is also determined to be ENDC or EXO, depending on whether ΔH° is >greater than or less than zero. Its magnitude can also provide information about the characteristics of the biosorption interaction taking place in the system. For physical and chemical biosorption (physisorption), the ΔH° values lie within 2.1–20.9 and 20–800 kJ mol^{-1}. is related to the level of the system after the BP. Positive values of ΔS° indicate extreme instability at the solid–solution interface, which enhances the freedom of the sorbate, whereas negative values of ΔS° indicate that the process is using an associative mechanism and denote reduced disorder at the interface between the solid and the solution, which can lead to sorbate molecules escaping from the surface of the sorbent into the solution [93].

The association between ΔH°, ΔS°, and ΔG° is given by equations (3.1) and (3.2).

$$\Delta G^\circ = \Delta H^\circ - T\Delta S^\circ \tag{3.1}$$

$$\ln K_c = \frac{\Delta G^\circ}{R} - \frac{\Delta H^\circ}{RT} \tag{3.2}$$

From the slope and intercept of the plot of $\ln K_c$ against $1/T$, ΔH° and ΔS° values can be established [33]. Based on the results given in table 3.2, it can be determined from all the reviewed studies that the BPs of HMs by several biosorbents were thermodynamically SPONS, EXOTC, and ENDC.

3.5 Kinetic studies

A key factor in our understanding of the confiscation process is the chemical mechanism, as this allows us to forecast the removal rate of HMs [94]. Sorption kinetics is a key factor to consider when creating a sorption system. Various kinetic models of differing complexity have been employed to assess the mechanism of the sorption process as well as the rates of uptake, mass transfer, and diffusion control. These models are generally grouped into sorption diffusion models and sorption reaction models. Three basic successive steps are employed to formulate diffusion models, and this includes the diffusion of solutes across the liquid film that borders the sorbent material (film diffusion), the diffusion of the solutes in the sorbent pores (intra-particle diffusion), and sorbate sorption on the sorbent's active sites. However, chemical reaction kinetics is based on the entire process of sorption except for the mechanistic stages [17].

Kinetic research gives us an understanding of the outcome data required to model and design an operational biosorption system. During the phase extraction procedure of the solid–liquid interface, kinetic interpretation is required, since it is valuable in describing the equilibrium time, reaction order, and the pathway of the reaction; hence, it defines the mechanism and the rate-limiting step of contaminant capture from the solution by the biosorbent [95].

Hypothetically, a water treatment system based on solid–liquid sorption can be considered in three successive stages, namely the exterior mass sorbate transfer from the bulk solution to the biosorbent surface, the interior diffusion stage, and biosorption to the biosorbent's surface binding sites. Sorbate migration (ions) takes place from the mass solution to the boundary layer where biosorbent particles appear when the biosorbent comes into contact with the sorbate solution. Then, as long as the exterior mass-transfer resistance is overcome (which is due to the peripheral layer surrounding the surface of the biosorbent), the sorbate concentration diminishes in the bulk solution as the sorbate (ions) diffuse to the biosorbent's internal surface. It is clear that a strong interaction takes place between the biosorbent's active sites and the sorbate [17]. The time-dependent biosorption statistics are generally assessed using different kinetic models such as the PFO model, the PSO model, the Weber and Morris model (WMM), the ELM, etc. Generally, the linear formulae of the various abovementioned models are mainly employed to describe the BP (table 3.4) [24, 96].

In kinetic studies, the PFO model and PSO model are mostly used to describe biosorption under non-equilibrium conditions [94]. These models are built on the hypothesis that the sorption rate is related to the number of available sites on the sorbent surface in the appropriate power (first or second) [97]. In contrast, the ELM is employed for comprehensive chemisorption, and its equation is utilized for

Table 3.4. Linearized forms of isotherm and kinetic models.

Isotherm models	Linear form	Plot
LNR model	$\frac{C_e}{q_e} = \frac{1}{q_m K_L} + \frac{C_e}{q_e}$	$\frac{C_e}{q_e}$ vs C_e
FRH model	$\log q_e = \log K_F + \frac{1}{n}\log C_e$	$\log q_e$ vs $\log C_e$
Temkin model (TEMM)	$q_e = \frac{RT}{b}\ln A + \frac{RT}{b}\ln C_e$ or $q_e = B\ln A + B\ln C_e$ $(B = \frac{RT}{b})$	q_e vs $\ln C_e$
DRM	$\ln q_e = \ln q_s - K_D\varepsilon^2$	$\ln q_e$ vs ε^2
Kinetic models		
PFO model	$\log(q_e - q_t) = \log q_e - \frac{k_1 t}{2.303}$	$\log(q_e - q_t)$ vs t
PSO model	$\frac{t}{q_t} = \frac{1}{k_2 q_e^2} + \frac{t}{q_e}$	$\frac{t}{q_t}$ vs t
ELM	$q_t = \frac{1}{\beta}\ln \alpha\beta + \frac{1}{\beta}\ln t$	q_t vs $\ln t$

numerous slow sorption rates and for different chemisorption processes [98]. As can be observed from the papers reviewed in table 3.3, most BPs of HMs by various biosorbents are generally described by the PSO model followed by the PFO model and the ELM. The PSO model is superior to the PFO model and the ELM in terms of fitting, as its q_e values are often much closer to the experimental values than the values given by the PFO model.

3.6 Equilibrium sorption isotherm

Sorption isotherm models are key tools for designing and scaling up sorption systems, as they explain the interaction between the sorbate and the accessible sites on the surface of the biosorbent [99]. Solid–liquid sorption system evaluation is generally based on two kinds of studies, which are based on the equilibrium set of sorption assessments and active uninterrupted flow sorption research. To define experimental sorption data using a set approach, isotherm models such as the LNR model, the FRH model, and the Temkin model (TEMM) are usually employed. The linearized forms of this model are given in table 3.4. It is critical to discover the best tailored isotherm in order to calculate the effectiveness of the fabricated biosorbent and thus create appropriate sorption system designs. Sorption isotherm models are generally categorized into experimental equations and systematic models. The systematic models are founded on MI biosorption mechanisms, which can represent, explain, and forecast the experimental behavior [13].

The LNR model is generally used to represent gas–solid phase sorption and is utilized to measure and compare the sorptive capacities of different biosorbents. This model describes the surface coverage by comparing the relative rates of sorption and desorption. Here, sorption is related to the portion of the biosorbent surface that is accessible, while desorption is related to the portion of the surface of the biosorbent that is covered. This model assumes monolayer coverage of the sorbate such that when a sorption site on the surface of the biosorbent is occupied by the sorbate, no additional sorption can take place at this site. The FRH model is appropriate for sorption processes that happen on diverse surfaces. This model yields a term that describes the surface heterogeneity, the exponential active sites, and the energy distribution. The DRB is an experimental model that is usually utilized in expressing sorption mechanisms with Gaussian energy distributions on various surfaces. This model is only appropriate for an average range of sorbate concentrations, as it shows ideal asymptotic behavior and does not calculate Henry's law at the low point of the pressure. The TEMM describes the impacts of indirect sorbate/sorbate interaction on the sorption process. It is implicit that the molecule's heat of sorption in the layers reduces linearly with enhanced surface coverage [55, 100]. The BPs of most HMs by different biosorbents are mainly defined by the LNR model, followed by the FRH model, according to the studies reviewed, as shown in table 3.3. The linearized forms of the isotherm and kinetic models were reported in [6, 87, 101, 102] and are summarized in table 3.4.

3.7 The mechanism of heavy metal biosorption

The various parameters that can contribute to control mechanisms include the metal of interest's stereochemical, coordination, and chemical features, which are related to the MI's ionic radius, oxidation state, and ion mass. Selected biosorbent properties, such as the number of reactive binding sites, the ease of access and availability of the binding sites, the binding site class, and the attraction between MIs of interest and the biosorbent binding sites also impact the biosorption mechanism [24]. For the potential sequestration of HM ions, various mechanisms have been reported in different studies. Hitherto, biosorption was grouped into metabolism-dependent and -independent biosorption. Today, metabolism- independent and -dependent BPs are called bioaccumulation (active BPs) and biosorption. Bioaccumulation comprises two processes, the first of which is comparable to biosorption (it consists of the attachment of the potentially noxious elements to the biosorbent surface) and the second step is the effective transportation of MIs into cells. Biosorption is a reactive process that happens at a quicker rate than bioaccumulation [34].

Depending on the binding of the MIs to the biomaterial's CWs, the biosorption mechanism is classified as metabolism-dependent or -independent biosorption. Metabolism-dependent biosorption takes place in biosorbents created from living biomaterials, and the mechanisms involved include physical biosorption and chelation [103]. The BP can also be a physicochemical process (based on physical (van der Waals) forces or chemical (covalent) bonding). MIs are immobilized on the surface of the biomaterial to improve the MI uptake through ligands or surface groups (COOH, phosphoryl groups, –C(=O)–, –OH) that are present on the biomaterial. These MI uptakes are impacted by the nature and type of the MIs, the medium's condition and concentration, and the CW composition. On the other hand, metabolism-independent biosorption involves the use of dead biomass for MI uptake. This process may involve ionic interaction or physicochemical biosorption. Normally, living biomaterials are deemed more beneficial than dead biomass, as they can continuously take up MIs and can replenish naturally. MIs entering the biomaterial's CWs experience purification based on the biomaterial's kind and composition. The advantages of using dead cells as biosorbents are that growth conditions, nutritional supplements, and culture media are not mandatory; these biosorbents can be regenerated and reused and are free from noxious substances [104].

During the BP, biosorbents can experience physisorption, chemisorption, microprecipitation, and oxidation or reduction. The term physisorption is associated with electrostatic interaction and van der Waals forces, while chemisorption is linked to chelation, complexation, and ion exchange [37]. Ion exchange is deemed to be the basis of the biosorption mechanism; it occurs via various existing functional group variations on the surface CWs of the biomass. Extracellular polymer substances produced by microorganisms also play a part a critical role in BPs [34]. The surfaces of biosorbent surface comprise FGs which are critical for MI binding in the chemisorption processes. Ion exchange then produces electrostatic interactions between the biosorbent negatively charged CWs and the cations contained in the solution. Chelation is the process by which ligands link with MIs, creating a ring structure that acts as an barrier for the affixed metals and helps to avoid undesirable chemical reactions. The MIs then connect through the complexation process by creating a complex with the active groups in the CW on the biosorbent cell surface through covalent or electrostatic bonding; the resulting compound may be neutral or negatively or positively charged [37].

3.8 Conclusions and recommendations

Today, anthropogenic activities have led to the buildup of different HMs in the environment at a scale that exceeds our capacity to treat or mitigate. To date, biomass-derived activated biosorbents such as microbes and agricultural and industrial wastes have shown potential in the confiscation of HMs owing to their extraordinary specific surface area, FGs, low cost, and rich pore structure. This chapter also discussed the various biosorbent activation processes, the factors impacting the biosorption of HMs by various biosorbents, and the mechanisms of the BPs. The research reviewed showed that the LNR model and the PSO model were identified as best defining the biosorption of most HMs by various biosorbents. In addition, the BPs of these HMs were found to be SPONS and EXOTC or ENDC.

References

[1] Naidu R *et al* 2021 Chemical pollution: a growing peril and potential catastrophic risk to humanity *Environ. Int.* **156** 106616

[2] Briffa J, Sinagra E and Blundell R 2020 Heavy metal pollution in the environment and their toxicological effects on humans *Heliyon* **6** e04691

[3] El-Naggar N, Hamouda R, Abuelmagd M, Alharbi M, Darwish D, Rabei N and Farfour S 2022 A cost-effective and eco-friendly biosorption technology for complete removal of nickel ions from an aqueous solution: optimization of process variables *Green Process. Synth.* **11** 631–47

[4] Miros-Kudra P, Sobczak P and Kopania E 2022 Removal of heavy metals from aqueous solutions with the use of lignins and biomass *Fibres Textiles East. Eur.* **151** 99–111

[5] Muharrem I and Ince O 2017 An overview of adsorption technique for heavy metal removal from water/wastewater: a critical review *Int. J. Pure Appl. Sci.* **3** 10–9

[6] Bayuo J, Rwiza M and Mtei K 2022 A comprehensive review on the decontamination of lead (II) from water and wastewater by low-cost biosorbents *RSC Adv.* **12** 11233–54

[7] Martini S, Afroze S, Roni K, Setiawati M and Kharismadewi D 2021 A review of fruit waste-derived sorbents for dyes and metals removal from contaminated water and wastewater *Desalin. Water Treat.* **235** 300–23

[8] Razzak S, Farooque M, Alsheikh Z, Alsheikhmohamad L, Alkuroud D, Alfayez A, Hossain S and Hossain M 2022 A comprehensive review on conventional and biological-driven heavy metals removal from industrial wastewater *Environ. Adv.* **7** 100168

[9] Ahmed J, Thakur A and Goyal A 2021 Industrial wastewater and its toxic effects *Biological Treatment of Industrial Wastewater* ed M P Shah (London: RSC Chemistry in the Environment)

[10] Bhattacharjee C, Dutta S and Saxena V 2020 A review on biosorptive removal of dyes and heavy metals from wastewater using watermelon rind as biosorbent *Environ. Adv.* **2** 100007

[11] Sarker T, Azam S, El-Gawad A, Gaglione S and Bonanomi G 2017 Sugarcane bagasse: a potential low-cost biosorbent for the removal of hazardous materials *Clean Technol. Environ. Policy* **19** 2343–62

[12] Nnaji N, Onyeaka H, Miri T and Ugwa C 2023 Bioaccumulation for heavy metal removal: a review *SN Appl. Sci.* **5** 125

[13] Gautam R, Mudhoo A, Lofrano G and Chattopadhyaya M 2014 Biomass-derived biosorbents for metal ions sequestration: adsorbent modification and activation methods and adsorbent regeneration *J. Environ. Chem. Eng.* **2** 239–59

[14] Saleem J, Shahid U, Hijab M, Mackey H and McKay G 2019 Production and applications of activated carbons as adsorbents from olive stones *Biomass Convers. Biorefinery* **9** 775–802

[15] Moghadas B, Esmaeili H, Tamjidi S and Geramifard A 2022 Advantages of nano-adsorbents, biosorbents, and nanobiosorbents for contaminant removal *Nano-Biosorbents for Decontamination of Water, Air, and Soil Pollution* (Amsterdam: Elsevier) 105–33 https://shop.elsevier.com/books/nano-biosorbents-for-decontamination-of-water-air-and-soil-pollution/denizli/978-0-323-90912-9

[16] Franco D S P and Dotto G L 2020 A short analysis of biosorbents and its potential removal contaminants from aqueous media *Global J. Eng. Sci.* **5**

[17] Maksoud M, Elgarahy A, Farrell C, Ala'a H, Rooney D and Osman A 2020 Insight on water remediation application using magnetic nanomaterials and biosorbents *Coord. Chem. Rev.* **403** 213096

[18] Aguilar-Rosero J, Urbina-López M, Rodríguez-González B, León-Villegas S, Luna-Cruz I and Cárdenas-Chávez D 2022 Development and characterization of bioadsorbents derived from different agricultural wastes for water reclamation: a review *Appl. Sci.* **12** 2740

[19] Wang B, Lan J, Bo C, Gong B and Ou J 2023 Adsorption of heavy metal onto biomass-derived activated carbon *RSC Adv.* **13** 4275–302

[20] Motitswe M, Badmus K and Khotseng L 2022 Development of adsorptive materials for selective removal of toxic metals in wastewater: a review *Catalysts* **12** 1057
[21] Ghanim B, Leahy J, O'Dwyer T, Kwapinski W, Pembroke J and Murnane J 2022 Removal of hexavalent chromium (Cr (VI)) from aqueous solution using acid-modified poultry litter-derived hydrochar: adsorption, regeneration and reuse *J. Chem. Technol. Biotechnol.* **97** 55–66
[22] Ukanwa K, Patchigolla K, Sakrabani R, Anthony E and Mandavgane S 2019 A review of chemicals to produce activated carbon from agricultural waste biomass *Sustainability* **11** 620411
[23] Afzaal M, Hameed S, Abbasi N, Liaqat I, Rasheed R, Khan A and Manan H 2022 Removal of Cr (III) from wastewater by using raw and chemically modified sawdust and corn husk *Water Pract. Technol.* **17** 1937–58
[24] Ali Redha A 2020 Removal of heavy metals from aqueous media by biosorption *Arab J. Basic Appl. Sci.* **27** 183–93
[25] Charazińska S, Burszta-Adamiak E and Lochyński P 2021 Recent trends in Ni (II) sorption from aqueous solutions using natural materials *Rev. Environ. Sci. Bio/Technol.* **21** 1–34
[26] Okoro H, Pandey S, Ogunkunle C, Ngila C, Zvinowanda C, Jimoh I, Lawal I, Orosun M and Adeniyi A 2022 Nanomaterial-based biosorbents: adsorbent for efficient removal of selected organic pollutants from industrial wastewater *Emerg. Contam.* **8** 46–58
[27] Simón D, Palet C, Costas A and Cristóbal A 2022 Agro-industrial waste as potential heavy metal adsorbents and subsequent safe disposal of spent adsorbents *Water* **14** 3298
[28] Ahmed H, Sobhy N, Hefny M, Abdel-Haleem F and El-Khateeb M 2023 Evaluation of agrowaste species for removal of heavy metals from synthetic wastewater *J. Environ. Public Health* **2023** 7419015
[29] Kong S, Chin C, Yek P, Wong C, Wong C, Cheong K, Liew R and Lam S 2022 Removal of heavy metals using activated carbon from microwave steam activation of palm kernel shell *Environ. Adv.* **9** 100272
[30] Ligarda-Samanez C *et al* 2022 Modified polymeric biosorbents from *Rumex acetosella* for the removal of heavy metals in wastewater *Polymers* **14** 2191
[31] Jayan N, Bhatlu M L and Akbar S 2021 Central composite design for adsorption of Pb (II) and Zn (II) metals on PKM-2 *Moringa oleifera* leaves *ACS Omega* **6** 25277–98
[32] Rezaei M, Pourang N and Moradi A 2022 Removal of lead from aqueous solutions using three biosorbents of aquatic origin with the emphasis on the affective factors *Sci. Rep.* **12** 1–20
[33] Fawzy M, Al-Yasi H, Galal T, Hamza R, Abdelkader T, Ali E and Hassan S 2022 Statistical optimization, kinetic, equilibrium isotherm and thermodynamic studies of copper biosorption onto *Rosa damascena* leaves as a low-cost biosorbent *Sci. Rep.* **12** 8583
[34] Bilal M, Rasheed T, Sosa-Hernández J, Raza A, Nabeel F and Iqbal H 2018 Biosorption: an interplay between marine algae and potentially toxic elements—a review *Mar. Drugs* **16** 65
[35] Bauddh K and Korstad J 2022 Phycoremediation: use of algae to sequester heavy metals *Hydrobiology* **1** 288–303
[36] Salama E, Roh H, Dev S, Khan M, Abou-Shanab R, Chang S and Jeon B 2019 Algae as a green technology for heavy metals removal from various wastewater *World J. Microbiol. Biotechnol.* **35** 1–19
[37] Znad H, Awual M and Martini S 2022 The utilization of algae and seaweed biomass for bioremediation of heavy metal-contaminated wastewater *Molecules* **27** 1275

[38] Menaa F, Wijesinghe U, Thiripuranathar G, Althobaiti N, Albalawi A, Khan B and Menaa B 2021 Marine algae-derived bioactive compounds: a new wave of nanodrugs? *Mar. Drugs* **19** 484
[39] Olabi A, Shehata N, Sayed E, Rodriguez C, Anyanwu R, Russell C and Abdelkareem M 2022 Role of microalgae in achieving sustainable development goals and circular economy *Sci. Total Environ.* 158689
[40] Chugh M, Kumar L, Shah M and Bharadvaja N 2022 Algal bioremediation of heavy metals: an insight into removal mechanisms, recovery of by-products, challenges, and future opportunities *Energy Nexus* **7** 100129
[41] Priya M, Gurung N, Mukherjee K and Bose S 2014 Microalgae in removal of heavy metal and organic pollutants from soil *Microbial Biodegradation and Bioremediation* (Amsterdam: Elsevier) 519–37 https://shop.elsevier.com/books/microbial-biodegradation-and-bioremediation/das/978-0-12-800021-2
[42] Sattayawat P, Yunus I, Noirungsee N, Mukjang N, Pathom-Aree W, Pekkoh J and Pumas C 2021 Synthetic biology-based approaches for microalgal bio-removal of heavy metals from wastewater effluents *Front. Environ. Sci.* **9** 562
[43] Al-Homaidan A, Al-Qahtani H, Al-Ghanayem A, Ameen F and Ibraheem I 2018 Potential use of green algae as a biosorbent for hexavalent chromium removal from aqueous solutions *Saudi J. Biol. Sci.* **25** 1733–8
[44] Maisarah, Adityosulindro S and Wulandari D 2021 Utilization of wild algae biomass as biosorbent for removal of heavy metal Zinc (Zn^{2+}) from aqueous solution *IOP Conf. Ser.: Earth Environ. Sci.* **824** 012017
[45] Utomo H, Tan K, Choong Z, Yu J, Ong J and Lim Z 2016 Biosorption of heavy metal by algae biomass in surface water *J. Environ. Protect.* **7** 1547–60
[46] Lekshmi R, Rejiniemon T, Sathya R, Kuppusamy P, Al-Mekhlafi F, Wadaan M and Rajendran P 2022 Adsorption of heavy metals from the aqueous solution using activated biomass from *Ulva flexuosa Chemosphere* **306** 135479
[47] Olal F 2016 Biosorption of selected heavy metals using green algae, Spirogyra species *J. Nat. Sci. Res.* **6** 22–34
[48] Pham B, Kang J, Lee C and Park S 2021 Removal of heavy metals (Cd^{2+}, Cu^{2+}, Ni^{2+}, Pb^{2+}) from aqueous solution using *Hizikia fusiformis* as an algae-based bioadsorbent *Appl. Sci.* **11** 8604
[49] Sultana N, Hossain S, Mohammed M, Irfan M, Haq B, Faruque M, Razzak S and Hossain M 2020 Experimental study and parameters optimization of microalgae based heavy metals removal process using a hybrid response surface methodology-crow search algorithm *Sci. Rep.* **10** 1–15
[50] El-Naggar N, Hamouda R, Mousa I, Abdel-Hamid M and Rabei N 2018 Statistical optimization for cadmium removal using *Ulva fasciata* biomass: characterization, immobilization and application for almost-complete cadmium removal from aqueous solutions *Sci. Rep.* **8** 12456
[51] Priyadarshanee M and Das S 2021 Biosorption and removal of toxic heavy metals by metal tolerating bacteria for bioremediation of metal contamination: a comprehensive review *J. Environ. Chem. Eng.* **9** 104686
[52] Pham V, Kim J, Chang S and Chung W 2022 Bacterial biosorbents, an efficient heavy metals green clean-up strategy: prospects, challenges, and opportunities *Microorganisms* **10** 610

[53] Kim S, Chang J and Singh M 2015 Peptidoglycan architecture of Gram-positive bacteria by solid-state NMR *Biochim. Biophys. Acta (BBA)-Biomembranes* **1848** 350–62

[54] Vermassen A, Leroy S, Talon R, Provot C, Popowska M and Desvaux M 2019 Cell wall hydrolases in bacteria: insight on the diversity of cell wall amidases, glycosidases and peptidases toward peptidoglycan *Front. Microbiol.* **10** 331

[55] Shamim S 2018 Biosorption of heavy metals *Biosorption* **2** 21–49

[56] Aryal M 2021 A comprehensive study on the bacterial biosorption of heavy metals: materials, performances, mechanisms, and mathematical modellings *Rev. Chem. Eng.* **37** 715–54

[57] Medfu Tarekegn M, Zewdu Salilih F and Ishetu A 2020 Microbes used as a tool for bioremediation of heavy metal from the environment *Cogent. Food Agri.* **6** 1783174

[58] Huang M, Pan J and Zheng L 2001 Removal of heavy metals from aqueous solutions using bacteria *J. Shanghai Univ. (English Edition)* **5** 253–9

[59] Bestawy E, Helmy S, Hussien H, Fahmy M and Amer R 2013 Bioremediation of heavy metal-contaminated effluent using optimized activated sludge bacteria *Appl. Water Sci.* **3** 181–92

[60] Zeng T, Zhang X, Nong H, Qing H, Wang L and Wang A 2022 Efficiency, mechanism and microbial community of Cd (II) removal by mixed bacteria enriched from heavy metals mine soil *Trans. Nonferrous Metals Soc. China* **32** 3404–19

[61] Wahyu I, PINONTOAN R and YUWONO T 2020 Indigenous copper resistant bacteria isolated from activated sludge of water treatment plant in Surabaya, Indonesia *Biodiversitas J. Biol. Divers.* **21**

[62] Kamika I and Momba M 2013 Assessing the resistance and bioremediation ability of selected bacterial and protozoan species to heavy metals in metal-rich industrial wastewater *BMC Microbiol.* **13** 1–14

[63] Kurniawan T, Lo W, Othman M, Goh H and Chong K 2022 Biosorption of heavy metals from aqueous solutions using activated sludge, *Aeromasss hydrophyla*, and *Branhamella* spp. based on modeling with GEOCHEM *Environ. Res.* **214** 114070

[64] Saber W, El-Naggar N, El-Hersh M, El-Khateeb A, Elsayed A, Eldadamony N and Ghoniem A 2021 Rotatable central composite design versus artificial neural network for modeling biosorption of Cr6+ by the immobilized *Pseudomonas alcaliphila* NEWG-2 *Sci. Rep.* **11** 1–15

[65] Hashemkhani M, Rezvani Ghalhari M, Bashardoust P, Hosseini S, Mesdaghinia A and Mahvi A 2022 Fluoride removal from aqueous solution via environmentally friendly adsorbent derived from seashell *Sci. Rep.* **12** 9655

[66] Hegazy G, Soliman N, Ossman M, Abdel-Fattah Y and Moawad M 2022 Isotherm and kinetic studies of cadmium biosorption and its adsorption behaviour in multi-metals solution using dead and immobilized archaeal cells *Sci. Rep.* **13** 2550

[67] Ibrahim W, Hassan A and Azab Y 2016 Biosorption of toxic heavy metals from aqueous solution by *Ulva lactuca* activated carbon *Egypt. J. Basic Appl. Sci.* **3** 241–9

[68] Nacke H, Goncalves Jr A, Campagnolo M, Coelho G, Schwantes D, dos Santos M, Briesch Jr D and Zimmermann J 2016 Adsorption of Cu (II) and Zn (II) from water by *Jatropha curcas* L. as biosorbent *Open Chem.* **14** 103–17

[69] Putra W, Kamari A, Yusoff S, Ishak C, Mohamed A, Hashim N and Isa I 2014 Biosorption of Cu (II), Pb (II) and Zn (II) ions from aqueous solutions using selected waste materials: adsorption and characterisation studies *J. Encapsul. Adsorpt. Sci.* **4** 25–35

[70] Ibrahim W 2016 New trend for removing toxic heavy metals from drinking water by activated carbon based brown algae *Biotechnology* **15** 65–75

[71] Mohamed M, Hozayen W, Alharbi R and Ibraheem I 2022 Adsorptive recovery of arsenic (III) ions from aqueous solutions using dried *Chlamydomonas Heliyon* **8** e12398

[72] Namkoong H, Biehler E, Namkoong G and Abdel-Fattah T 2022 Efficient removal of lead ions from aqueous media using sustainable sources based on marine algae *ACS Omega* **7** 39931–7

[73] Vernans A, Iswanto B and Rinanti A 2019 Removal of heavy metal (Cu^{2+}) by *Thiobacillus* sp. and *Clostridium* sp. at various temperatures and concentration of pollutant in liquid media *J. Phys. Conf. Ser.*

[74] Baran M, Mzahir D, Baran A and Keskin C 2022 Fast and economical biosorption of lead (II) heavy metal in aqueous solutions by *Bacillus licheniformis* sp *Kahramanmaraş Sütçü İmam Üniversitesi Tarım ve Doğa Derg.* **25** 847–58

[75] Yang S, Wang Y and Liu Y 2021 Biosorption of Cu (II) from aqueous solutions by a macrofungus (*Ganoderma lobatum*) biomass and its biochar *Nat. Environ. Pollut. Technol.* **20** 579–85

[76] Parlayıcı Ş and Pehlivan E 2017 Removal of metals by Fe_3O_4 loaded activated carbon prepared from plum stone (*Prunus nigra*): kinetics and modelling study *Powder Technol.* **317** 23–30

[77] Azam M, Wabaidur S, Khan M, Al-Resayes S and Islam M 2022 Heavy metal ions removal from aqueous solutions by treated ajwa date pits: kinetic, isotherm, and thermodynamic approach *Polymers* **14** 914

[78] Qasem N, Mohammed R and Lawal D 2021 Removal of heavy metal ions from wastewater: a comprehensive and critical review *npj Clean Water* **4** 36

[79] Dhankhar R and Hooda A 2011 Fungal biosorption—an alternative to meet the challenges of heavy metal pollution in aqueous solutions *Environ. Technol.* **32** 467–91

[80] Chen Z, Osman A, Rooney D, Oh W and Yap P 2023 Remediation of heavy metals in polluted water by immobilized algae: current applications and future perspectives *Sustainability* **15** 5128

[81] Zafar S, Khan M, Lashari M, Khraisheh M, Almomani F, Mirza M and Khalid N 2020 Removal of copper ions from aqueous solution using NaOH-treated rice husk *Emerg. Mater.* **3** 857–70

[82] Kamilya T, Mondal S and Saha R 2021 Effect of magnetic field on the removal of copper from aqueous solution using activated carbon derived from rice husk *Environ. Sci. Pollut. Res.* **29** 20017–34

[83] Jayakumar V, Govindaradjane S, Rajamohan N and Rajasimman M 2021 Biosorption potential of brown algae, *Sargassum polycystum*, for the removal of toxic metals, cadmium and zinc *Environ. Sci. Pollut. Res.* **29** 41909–22

[84] Dulla J, Tamana M, Boddu S, Pulipati K and Srirama K 2020 Biosorption of copper (II) onto spent biomass of *Gelidiella acerosa* (brown marine algae): optimization and kinetic studies *Appl. Water Sci.* **10** 1–10

[85] Madala S, Nadavala S, Vudagandla S, Boddu V and Abburi K 2017 Equilibrium, kinetics and thermodynamics of cadmium (II) biosorption on to composite chitosan biosorbent *Arab. J. Chem.* **10** S1883–93

[86] Staroń P and Chwastowski J 2021 Raphia-microorganism composite biosorbent for lead ion removal from aqueous solutions *Materials* **14** 7482

[87] Abd Ali G, Salih N, Faroun G and Al-Hamadani R 2023 Adsorption technique for the removal of heavy metals from wastewater using low-cost natural adsorbent *IOP Conf. Ser.: Earth Environ. Sci.*

[88] Nadeem R, Manzoor Q, Iqbal M and Nisar J 2016 Biosorption of Pb (II) onto immobilized and native *Mangifera indica* waste biomass *J. Ind. Eng. Chem.* **35** 185–94

[89] Abdelfattah I, Ismail A, Al Sayed F, Almedolab A and Aboelghait K 2016 Biosorption of heavy metals ions in real industrial wastewater using peanut husk as efficient and cost effective adsorbent *Environ. Nanotechnol., Monitor. Manag.* **6** 176–83

[90] Akbari M, Hallajisani A, Keshtkar A, Shahbeig H and Ghorbanian S 2015 Equilibrium and kinetic study and modeling of Cu (II) and Co (II) synergistic biosorption from Cu (II)–Co (II) single and binary mixtures on brown algae *C. indica J. Environ. Chem. Eng.* **3** 140–9

[91] Tavana M, Pahlavanzadeh H and Zarei M 2020 The novel usage of dead biomass of green algae of *Schizomeris leibleinii* for biosorption of copper (II) from aqueous solutions: equilibrium, kinetics and thermodynamics *J. Environ. Chem. Eng.* **8** 104272

[92] Cholico-Gonzalez D, Ortiz Lara N, Fernandez Macedo A and Chavez J 2020 Adsorption behavior of Pb (II), Cd (II), and Zn (II) onto agave bagasse, characterization, and mechanism *ACS Omega* **5** 3302–14

[93] Ramírez Calderón O, Abdeldayem O, Pugazhendhi A and Rene E 2020 Current updates and perspectives of biosorption technology: an alternative for the removal of heavy metals from wastewater *Curr. Pollut. Rep.* **6** 8–27

[94] Spain O, Plöhn M and Funk C 2021 The cell wall of green microalgae and its role in heavy metal removal *Physiol. Plant.* **1732** 526–35

[95] Kajjumba G, Emik S, Öngen A, Özcan H and Aydın S 2018 Modelling of adsorption kinetic processes—errors, theory and application *Advanced Sorption Process Applications* (London: IntechOpen) 1–19 pp

[96] Hubbe M, Azizian S and Douven S 2019 Implications of apparent pseudo-second-order adsorption kinetics onto cellulosic materials: a review *BioResources* **14** 7582–626

[97] Michalak I, Chojnacka K and Witek-Krowiak A 2013 State of the art for the biosorption process—a review *Appl. Biochem. Biotechnol.* **170** 1389–416

[98] Nayagam J and Prasanna K 2022 Utilization of shell-based agricultural waste adsorbents for removing dyes: a review *Chemosphere* **291** 132737

[99] Yunusa U, Usman B and Ibrahim M 2021 Cationic dyes removal from wastewater by adsorptive method: a systematic in-depth review *Algerian J. Chem. Eng. AJCE* **1** 6–40

[100] Juang R, Yei Y, Liao C, Lin K, Lu H, Wang S and Sun A 2018 Synthesis of magnetic Fe_3O_4/activated carbon nanocomposites with high surface area as recoverable adsorbents *J. Taiwan Inst. Chem. Eng.* **90** 51–60

[101] Hambisa A, Regasa M, Ejigu H and Senbeto C 2023 Adsorption studies of methyl orange dye removal from aqueous solution using Anchote peel-based agricultural waste adsorbent *Appl. Water Sci.* **13** 24

[102] Ouettar L, Guechi E, Hamdaoui O, Fertikh N, Saoudi F and Alghyamah A 2023 Biosorption of triphenyl methane dyes (malachite green and crystal violet) from aqueous media by Alfa (*Stipa tenacissima* L.) leaf powder *Molecules* **28** 3313

[103] Javanbakht V, Alavi S and Zilouei H 2014 Mechanisms of heavy metal removal using microorganisms as biosorbent *Water Sci. Technol.* **69** 1775–87

[104] Yaashikaa P, Kumar P, Saravanan A and Vo D 2021 Advances in biosorbents for removal of environmental pollutants: a review on pretreatment, removal mechanism and future outlook *J. Hazard. Mater.* **420** 126596

Chapter 4

Functionalized biosorbents for the sequestration of dye from aqueous solutions

Uyiosa Osagie Aigbe, Kingsley Eghonghon Ukhurebor, Robert Birundu Onyancha, Benedict Okundaye, Efosa Aigbe, Kingsley Obodo, Lunyanda Lunga Noto, Adelaja Otolorin Osibote and Harrison I Atagana

Dyes are visible contaminants, and an excessive amount of colored material in water makes it undesirable owing to its appearance. Hence, the sequestration of colored materials from dye-bearing wastewater is a critical issue owing to the difficulty of remediating wastewater containing dye molecules using traditional treatment techniques. Concerns such as ecological safety, biodegradability, and climate change are becoming exceedingly important, and concern for the ecosystem is reigniting the need for organic and biodegradable materials for the decontamination of dyes using biological sequestration techniques. This chapter studies the effectiveness of functionalized biomass for the decontamination of dyes.

4.1 Introduction

Ecosystem contaminants continue to be a global concern and a huge challenge faced by society worldwide. Contaminants can be compounds of organic origin or imported matter that cause adverse alterations when they come into contact with the ecosystem [1]. Water, soil, and air contamination are critical global issues today, and considerable concerns are being raised about water supply pollution in particular. Notwithstanding water scarcity, the release of a substantial amount of effluent has imposed a great burden on human civilization. The recent evolution of the human population, which has led to overpopulation, innovations in modernization, as well as economic and industrial expansion, has unquestionably adversely impacted society in terms of water, soil, and air contamination and economic development [2].

Animate and inert contaminants distributed into the ecosystem can cause serious alterations in the atmosphere, such as water pollution, forest and grassland loss,

doi:10.1088/978-0-7503-5598-8ch4

wildlife complications, food chain pollution, undesirable health impacts on fauna and flora, serious health issues in humans, and the creation of different kinds of contaminants. This is because these contaminants are noxious, carcinogenic, and mutagenic, and their occurrence in the environment degrades environmental conditions. Various household activities and industrial discharges are the main sources of contaminants [3]. Mining, biochemical and chemical industries, battery manufacture, smelting expansion, and the extensive use of biochemicals or chemical materials such as herbicides, pesticides, and fertilizers for agricultural purposes have resulted in an enormous release of dyes into both surface and subsurface water sources [2, 4, 5].

Dyes are colored compounds which are employed as textile, wood, and fiber colorants in different industries such as food processing and the production of textiles, inks, printer's inks, leather, plastics, cosmetics, and pharmaceuticals [6]. They improve the appearance of textiles and are employed as finishes in the textile industry, which requires the use of a large water volume and a range of pigments and dyes. They are chemically categorized, depending on their structures, into cationic (basic dyes, which are characterized by a protonated amine group), non-ionic (known as dispersed dyes because of their dissociation in aqueous solutions), and anionic (reactive, direct, and acidic dyes with a negative charge primarily caused by an SO_3^- group) types. They have two different forms, namely organic and synthetic dyes. Up to the late nineteenth century, colors were more or less organic, as their main sources were animals, mollusks, and plants, and they were primarily produced in minuscule amounts. However, organic dyes have been employed for over 150 years. The discovery of mauveine (aniline purple and Perkin's mauve) in 1856 by Henry Perkin while studying a means to fabricate quinine (a pharmaceutical used to treat malaria) resulted in a new generation of dyes [6–8].

Based on their composition, they are grouped based primarily on the functional groups (FGs) to which they owe their color, such as azo, phthalein, nitro, acridine, nitroso, indigoid, triphenylmethane dyes, etc. In addition, depending on its use, the color of a dye on a certain fiber depends on the nature of the dye and the fiber. The dyeing process can use mechanisms such as covalent hydrogen, ionic bonds, and van der Waals forces. Dyes can also be categorized based on their applications, such as acid dyes (these dyes' sodium (Na) salts comprise carboxylic acid (–COOH) and sulfonic acid (–SO_3H) groups), basic dyes (colored basic salts comprising amino or dialkylamino groups such as autochrome aniline yellow dye), substantive or direct dyes (water-soluble dyes such as Congo red (CR) and Martius yellow (MY) dyes), disperse dyes (insoluble dyes which are utilized in dispersed form in stabilizing agents such as phenol, benzoic acid presence to dye fiber, e.g. Celliton fast pink B and Celliton fast blue B dyes), fiber-reactive dyes (which include a reactive group which permanently binds directly with a hydroxyl (OH^-) or amino group of the fiber), ingrain or insoluble azo dyes (which comprise approximately 60% of all dyes utilized; they are obtained by combining arylamines, aminophenols, naphthols, and phenols (para red dye)), vat dyes (insoluble in water, e.g. indigo dye; these are employed in cosmetics, products, drugs, and indicators used for chemical studies) and mordant dyes (which do not bind directly to the fabric without the aid of a mordant) [9–11].

The manufacture of dyes gives an explanatory illustration of their prospective effect on the management of overwhelming pollutant flows. Even at minimal concentrations, they are optically functional and noticeable, and their effect on water resources is very acute [8]. Since dyes inhibit photosynthesis, they block sunlight penetration, contaminate aquatic ecosystems, and pose health risks for humans. Dyes and their by-products are carcinogenic, mutagenic, and teratogenic. Raw dye effluents are extremely challenging to handle owing to their varying, poorly biodegradable, and unsafe contents. It has also been reported that chemical dyes (particular contaminants) cause 2.2×10^5 to 4.3×10^5 health-related incidents year on year. Disability-adjusted life years (DALYs) are used to express comparable losses and are utilized to compute the disease burden ascribed to the effects of different kinds of contamination. Hence there is a requirement for dyes to be confiscated after use [6, 7, 12].

Different forms of dye confiscation technologies are available with differing levels of performance. The methods employed to confiscate and recover noxious pollutants such as dyes from industrial wastewater include oxidation, chemical precipitation, coagulation, reverse osmosis, flocculation, filtration, electrolysis, solvent extraction, sorption, etc. All the abovementioned techniques have their advantages and drawbacks, which require extensive consideration and exploration for their utilization in the dye confiscation process [6]. Sorption has been shown to be the most efficient and productive method owing to its suitability, cost-usefulness, operational ease, elevated sorption capacities, and productivity, which explain why this technique is more cost-effective for dye confiscation [13].

Concerns such as ecological safety, biodegradability, and climate change are becoming exceedingly important, and concerns for the ecosystem are reigniting the need for organic and biodegradable materials for the decontamination of dyes from industrial wastewater using biological sequestration techniques. Biological dye sequestration is an economical procedure that produces less waste, employs fewer chemical reagents, conserves energy, and is reasonably sustainable in developing countries [7]. Recently, there has been great scientific interest in the application of organic-based materials for the confiscation of contaminants from various types of wastewater, owing to the existence of various amino, carboxyl (COOH), and OH^- groups on the surfaces of organic-based materials, which facilitate the biosorption process (BP). There has been an increased probable utilization of agricultural waste and biowaste as low-cost biosorbents owing to their profusion and accessibility, minimal cost, excellent sorption capacity, ease of chemical alteration, effortless restoration, and reduced disposal issues after sorption [13]. Biosorption denotes chemical retention or sorption by a microbial mass, which involves the accretion of waste matter on cell surfaces such as those of live or dead bacteria, fungi, algae, lichen, or organic materials in aqueous solutions [14].

The expanded use of biomaterials as less pricey biosorbents is well documented as a potential resource for pollutant treatment based on the intensified research into the use of biosorbents for dye confiscation highlighted by a SCOPUS report for the

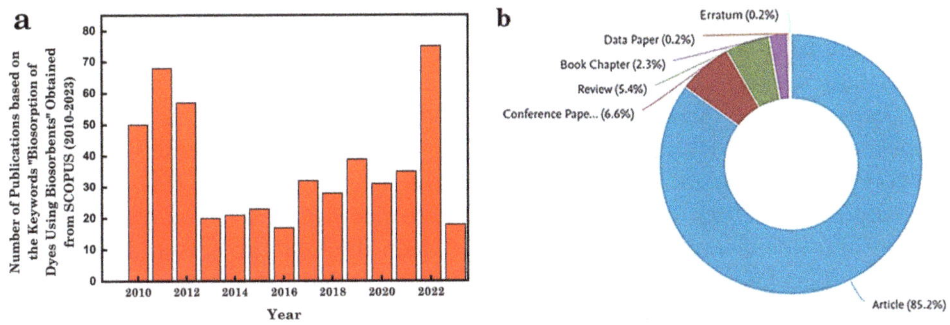

Figure 4.1. (a) The numbers of published titles recoded in SCOPUS matching the keywords 'Biosorption of Dye Using Biosorbent' for the period from 2010–May 2023. (b) A breakdown of the total number of published works recorded in SCOPUS (2010–May 2023).

period from 2010 to May 2023, as shown in figure 4.1(a). Also shown in figure 4.1(b) is an analysis of published works in the SCOPUS database. As can be seen from this figure, 85.2% of the total number of titles published matching the keywords 'biosorption of dyes using biosorbents' from 2010 to May 2023 were articles, while 2.3% of the titles were books (2010–May 2023). This is low compared to the number of articles published from 2010 to May 2023. Hence, the objective of this chapter is to review existing literature on the use of functionalized biosorbents for the confiscation of dye molecules, factors impacting the biosorption of dye, and the mechanisms of dye BPs.

4.2 The functionalization of biosorbents

The utilization and selection of a biomaterial for use as a biosorbent depends on its sorption capacity and reusability over time. A key challenge is the identification of high-performance biosorbents from the large variety of favorable and inexpensive biomaterials. While nearly all lignocellulosic biomaterials have a small surface area, there have been intensive attempts to enhance their use as biosorbents via various methods. These include enhancements of the sorption capacity and regeneration. Hence, the modification of biomaterials via pretreatment is required to prevent the release of organic compounds into treated water by untreated biosorbents. Untreated biosorbents may exhibit elevated chemical and biological oxygen demands and total biological carbon, which are accompanied by a reduction in the oxygen content of the water. Thus, biosorbent modification is anticipated to be extensive in order to avoid the release of unnecessary organic compounds into remediated water and to improve the efficiency of biosorbents in contaminant elimination [15].

Most plant-based agricultural waste is a complicated mixture of hemicellulose, lignin, and cellulose, hence such wastes are referred to as lignocellulosic biomasses, as their components are complexly fixed in the lignocellulosic configuration. Cellulose is linearly created from D-glucose subunits and is known for its elevated crystallinity. Conversely, hemicellulose (hexose and pentose) has unevenly aligned cellulose chains that negatively influence crystallinity. Their chain distance is much smaller than regular cellulose chain distance. Lignin is the exterior coating that keeps hemicellulose and cellulose together and has multifaceted cross-linked phenolic subunit structures. Lignocellulosic materials are extremely hydrophilic and best for the biosorption of dye molecules due to the abundant OH^- groups in the lignocellulosic structure, hence their surfaces are electronegative and attract cationic dyes owing to the electrostatic attraction between the OH^- groups and the cationic ions of the dye. Thus, their properties must be modified before they can effectively biosorb and confiscate anionic dyes [16].

To obtain extraordinary biosorption capacity (BC), biomasses are functionalized with different organic or inorganic chemical reagents to enhance the number of accessible sites either via a superficial hydrolysis procedure or by binding the biomass surface to chemical molecules with a huge number of FGs. The creation of new FGs on the biosorbent surface through the modification of biomass with organic reagents is more effective than using activation with inert substances [17, 18].

The surfaces of carbon materials that lack FGs are known to be nonpolar/hydrophobic, thus the necessity for the surface functionalization (FUNZTN) of carbon materials. The hydrophobicity of a typical biomaterial's surface can be improved to target certain pollutants via chemical or electrical interactions. The different FUNZTN procedures used to enhance contaminant confiscation comprise magnetization, oxidation, and FG grafting and formation with inorganic substances. Modification via surface FUNZTN is a critical phase in promoting a biomaterial's biosorption of a specific group of pollutants and hence improving its BC. This is done utilizing various strategies for biomaterial surface FUNZTN, which generally involve OH^- functional group chemistry [19, 20]. The fabrication technique utilized to transform the exterior FGs on the surfaces of carbon materials comprises post-treatment with chemical agents to manipulate the biosorbent's attributes. The chosen FUNZTN technique depends on the target contaminant's chemical characteristics and anticipated usage. The common modification methods used to establish FGs such as nitrogen, sulfur, and oxygen heteroatoms on the surfaces of carbon materials are nitrogenation (the production of nitrogen in the form of FGs such as –C=N, –C–N, –NH, and $-NH_2$ on the surface of biomaterials, which is accomplished via carbon reactions with ammonia, amines, or urea, which significantly improve surface polarity and particularly the interaction with polar sorbates), sulfuration (the introduction of sulfur in the form of FGs such as C–S, S=O, or C=S to the surface of the carbon sorbent, which is achieved via the reaction of carbon atoms with sulfurizing agents such as elemental H_2S, Na_2S, S, K_2S, SO_2, and dimethyl disulfide), and oxidation (the production of oxygen-containing FGs such as –COOH, OH^-, –C=O, and –C–O on the surface of the biosorbent under refluxing conditions in the presence of one or more inorganic acids such as H_2SO_4 and HNO_3 and oxidizing agents such as H_2O_2, $KMnO_4$, and NaOCl) as shown in figure 4.2, which was adopted from [19, 21, 22].

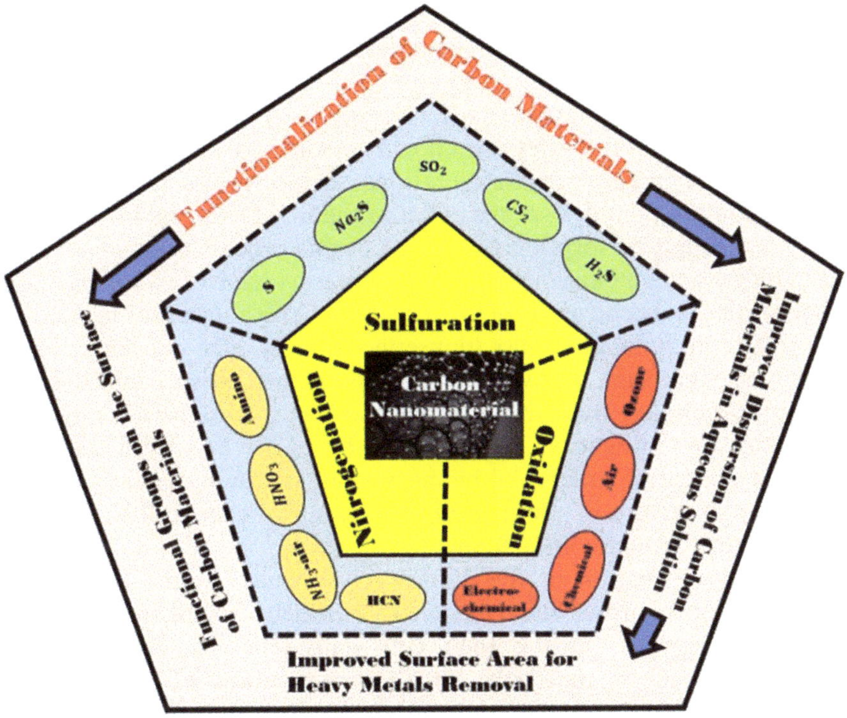

Figure 4.2. The procedures used to functionalize carbon materials. Reprinted from [19], Copyright (2021), with permission from Elsevier.

The fabrication technique utilized to modify the exterior FGs comprises the post-treatment of the biomaterial with chemical agents to regulate the biosorbent features. The preferred technique of modification depends on the envisioned utilization and the selected contaminant's chemical characteristics. Appropriate activation methods yield biomaterials with an elevated surface area and thus outstanding usable confiscation efficiency, enhanced BC, and dye selectivity [19].

Carbon surface chemistry is assessed to a large degree by the amount and nature of surface FGs on the carbon material's surface. The driving mechanism of FUNZTN lies in the introduction of different FGs into biomaterials via manipulating heteroatoms on the biomaterial's surface. The additional benefits of attaching FGs to the surfaces of biomaterials include an enhancement of the electrostatic strength, which improves carbon material exfoliation owing to the charges created [21].

4.3 The biosorption of dyes using functionalized biosorbents

The surface charge of a sorbent is attractive to dyes, and generally, negatively charged biosorbents efficiently sorb positively charged dye molecules [20]. In the search for effective dye confiscation alternatives, inexpensive and environmentally friendly technologies have been in the vanguard of current research into biomaterials/biosorbents. Today, in the search for green safety and sustainability, scholars are evaluating various biomaterials for polluted industrial discharge remediation [8].

Different varieties of biomaterials developed from bacteria, fungi, algae, and industrial and agricultural wastes are known to sorb dye molecules. Their surface properties such as their cell wall configurations, surface-to-volume proportions, and attraction for contaminant dye species enhance the BC. Dead and living cells take up contaminants through biosorption (the scope of this process is not restricted by the use of dead biomass) and bioaccumulation (the preservation of living organisms on a growth-supporting concentration of the poisonous substance and green variations hindering their contaminant confiscation effectiveness). The concentration of contaminants, their significant recovery, effectiveness, and biosorbent cost determine the dead biomass choice. Industrial waste is the best material with which to create biosorbents for dye mitigation because it is free [23, 24]. With regards to the biosorption of pollutants, microbial biomasses (algae, fungi, bacteria, etc.) outperform macroscopic materials (plant products) owing to their cell wall components and their FGs that attract contaminants [25].

The efficient confiscation of dye molecules depends on the exclusive surface chemistry of microorganisms which include FGs (aldehydes, ether, ketones, alcohol, –COOH, and phenolics) in the their cell walls, thereby making biosorbents extraordinarily attractive materials for the confiscation of dye molecules. In achieving high levels of dye biosorption, the critical factors are the biosorbent pore volume and the FGs. A large pore volume allows the optimum quantity of dye molecules to be bound to the biosorbent's active surface sites. In addition, the FGs on a biosorbent's surface determine the hydrophilicity and hydrophobicity of the biosorbent [6]. The different processes involved in biosorption comprise sorption, complexation, ion exchange, precipitation, and absorption. This is a self-regulating passive metabolic procedure that proceeds because of the chemical and physical interactions between the biomass's surface and the contaminant [18, 26].

The capability of rice husk (RH) to confiscate brilliant green (BG), crystal violet (CV), CR, and methylene blue (MB) was explored in a study. The results of this study showed effective confiscation of these dye molecules using RH [27]. Brown algal biomass functionalized using citric acid was also assessed for the sequestration of CV dye molecules. The Langmuir (LNR) model and the pseudo second-order (PSO) model provided excellent fits to the experimental data, with an optimum BC of 279.1 mg g^{-1}. The thermodynamic factors determined confirmed that the process of CV biosorption was spontaneous (SPONS) and exothermic (EXO) [28]. A study by Essekri *et al* [29] reported the biosorption of CV dye by red seaweed modified with citric acid. The BP was superlatively defined by the LNR model and the PSO model, and the optimum monolayer sorption capacity was found to be 217.4 mg g^{-1}. The BP was also found to be thermodynamically endothermic (ENDC) and SPONS. The biosorption mechanism of CV dye by the functionalized biosorbent was via an electrostatic driving force pathway. A response surface methodology (RSM) assessment illustrated that the optimum CV biosorption efficiency of 93.40% was obtained at pH 7, a biosorbent dosage of 1.5 g l^{-1}, an initial CV concentration of 20 mg l^{-1}, and a biosorption contact time of 90 min at 20°C. Sawdust functionalized with monosodium glutamate and crude sawdust (CS) has been evaluated for malachite green (MG) sorption. The optimum confiscation efficiencies were reported at pH 6 and pH 8 for FS and CS biosorbents, respectively. The BPs for the experimental data were described well by the LNR model and the PSO model, and the optimum BC range was 85.47–196.08 mg g^{-1}. In addition, the sorption procedure was thermodynamically SPONS and EXO [30]. The microalgae *Chlorella vulgaris* has been assessed for the confiscation of erionyl turquoise azoic (ETA) textile dye. The biosorption of ETA dye was found to be pH-dependent, and the ideal biosorption of ETA dye was observed at pH 9 [31]. *Cucumis sativus* peel waste has also been explored as a biosorbent for the sequestration of CV dye molecules. The sorption process was perfectly described by the LNR model and the PSO model, and the optimum BC was 149.25 mg g^{-1}. The thermodynamic factors showed that the BP was ENDC and SPONS for CV removal [32]. Functionalized stink bean pod (*Parkia speciosa*) powder has been explored for the sequestration of Remazol Brilliant Blue R (RBBR) and Remazol Brilliant Violet 5R (RBV5R) dyes. The functionalized biosorbent was found to be effective in the sequestration of the various dye molecules and achieved confiscation efficiencies of 83.35% and 80.07% for RBBR and RBV5R, respectively. The BPs were superlatively defined by the Freudlich (FRH) model and the PSO model, achieving BCs of 51.04 and 40.65 mg g^{-1} for RBBR and RBV5R, respectively [33]. The effectiveness of cauliflower leaf powder for the confiscation of MB dye molecules has been examined in a study. It was realized that the sorption process was best defined by the FRH model and the PSO model, and a BC of 149.22 mg g^{-1} was observed. In addition, the determined thermodynamic parameters showed that the BP was SPONS and EXO [34]. Cassava and peach palm agro-wastes have been assessed as possible biosorbents for tartrazine yellow dye (TAR). The optimum biosorption efficiency for the confiscation of this dye was observed at pH 2. The LNR and FRH models were a good fit for the experimental data obtained using cassava and peach palm agro-wastes [35]. Locust bean husk stimulated with orthophosphoric acid has been assessed for the efficient confiscation of rhodamine B (RhB) dye. It was found that the activated biosorbent was effective for the confiscation of RhB dye molecules at pH 3 and the BP was appropriately defined by the PSO model and the LNR model; the established optimum BC was 1111.1 mg g^{-1}. Thermodynamically, the BP was SPONS and ENDC [36]. Watermelon peel has been explored for BG dye confiscation and was found to be effective in the treatment of aqueous solutions containing BG dye. According to the isotherm data, the LNR reasonably depicts the BP of BG dye by the biosorbent, and an optimum BC of 25 mg g^{-1} was observed [37]. Sawdust and periwinkle shells have also been found to be efficient in the treatment of wastewater containing indigo dye (ID). The optimum confiscation of ID to the biosorbents was reported at pH 6 using a biosorbent dosage of 0.3 g and an interaction time of 60 min. The BP was ideally defined using the LNR and FRH models in this study [38]. A microalgae (*Chlamydomonas reinhardtii*), which was immobilized by a polysulfone nanofibrous web, was assessed as an efficient biosorbent for RB5 and Reactive Blue 221 (RB221) dyes. The immobilized microalgae were found to be an efficient and promising biosorbent for the treatment of both dye wastes [39]. Green algae (*Nannochloropsis* sp.) functionalized with silica and covered with magnetite particles (Fe_3O_4) were assessed as an efficient biosorbent for MB dye confiscation. The prepared biomaterial was confirmed to be useful in the confiscation of MB dye at pH 7. The BC was determined to be 1.39×10^{-1} mmol g^{-1} [40]. Activated green macroalgae (*Cladophora* sp.) have also been found to be effective in the treatment of safranin O dye waste, according to a study by Salmam *et al* [41]. A microbial consortium which was extracted from wastewater and functionalized with silver-doped hydroxyapatite has been used for the confiscation of reactive orange 84 (RO84). The effective removal of RO84 dye was reported for this functionalized biosorbent, which achieved a confiscation efficiency of 95% [42]. Isolated bacteria which were linked to the Firmicute (genus *Enterococcus*) and Proteobacteria phyla (γ-group) were assessed as an efficient biosorbent for reactive violet 5R (RV5R). The strain *Enterococcus* sp. L2 was found to be effective for the sequestration of the dye molecules at pH 8 and up to 2% w/v salinity [43]. Coffee-dyed bacterial cellulose has been established to be efficient in MB confiscation at pH 6 [44]. Using ionic cross-linking, *Lentinus concinnus* biomass was immobilized on carboxylmethyl cellulose in the presence of $FeCl_3$ and used for the effective decontamination of Disperse Red 60 (DR60). The prepared biosorbent was found to be effective in the confiscation of DR60 dye at pH 6. The sorption process of DR60 dye by the prepared biosorbent was exceptionally explained by the PSO model, the LNR model, and the Temkin model (TMM). The optimum BC was determined to be 92.6 mg g^{-1} [45]. The impact of lignin on azo dye (RBV5R, Direct Black 38 (DB38), Direct Black 22 (DB22), and Direct Red 5B (DR5B)) remediation was explored *in vivo* using *Echinodontium taxodii*. The optimum percentages of these dyes confiscated by the biosorbent in submerged cultures were 91.8% for RBV5R (control 66.0%), 76.9% for DB38 (control 43.8%), 43.4% for DB38 (control 17.0%) and 44.75% for DB22 (control 12.2%) [46]. The white rot fungus *Trametes pubescens* was immobilized with sodium alginate and assessed for the sorption of CR dye; it was thought to be applicable in the treatment of CR dye at pH 2. The experimental data was in perfect agreement with the PSO and FRH models. The optimum BC was determined to be 357.1 mg g^{-1} [47]. The fungi *Aspergillus niger*, *Aspergillus terreus*, and *Rhizopus oligosporus* were found to be effective in the decontamination of Acid Blue 161 (AB161) and Procion Red MX-5B (PRMX-5B) at pH 4. Based on the sorption of AB161 and PRMX-5B dyes by the pelletized fungal biomass, it was found that the FRH model best depicted the sorption process [48]. Table 4.1 summarizes the different functionalized biosorbents employed for the confiscation of dye molecules.

4.4 Factors impacting the biosorption of dyes

Because the chemical structures of dyes are relatively diverse, the sorption of dyes by a functionalized biosorbent is strongly impacted by the biosorbent features and different experimental parameters such as the biosorbent's dosage, contact time,

Table 4.1. A summary of various prepared functionalized biosorbents used for the removal of dye molecules.

Biosorbent biomass	Dyes	Optimum pH	Isotherm model/q_m (mg g^{-1})	Kinetic model	Thermodynamic characteristics	References
Clove leaf agrowaste biosorbent activated with potassium hydroxide	MG	6	Dubinin–Radushkevich (DR)/131.6	PSO and Elovich (ELH)	SPONS and EXO	[49]
Alfa (*Stipa tenacissima* L.) leaf powder	MG and CV	8	LNR and FRH/90.1 and 110.9	PSO	SPONS and ENDC	[50]
Mandarin-CO-TETA derived from mandarin peels	Acid Brown 14 (AB14)	1.5	LNR/416.7	PSO	—	[51]
Ozone, purified and sonicated sawdust biochars	MB	12	FRH (ozone sawdust and sonicated sawdust biochars), and LNR (purified sawdust)/200, 526, and 769	PSO	—	[52]
Clove leaf waste modified with sodium hydroxide	MB	6	LNR and FRH/9.8	Pseudo-first-order (PFO) models	Non-SPONS and ENDC	[5]
Green algae (*Bracteacoccus* sp.) isolated from North Jordan	MB	7	LNR/96%	PSO	—	[53]
Fallen leaves of *Prunus dulcis* (almond) activated by sodium hydroxide (NaOH)	Acid Blue 113 (AB113)	6.5	LNR and TM/97.09	PSO	SPONS and EXO	[54]
Pineapple plant stem (*Ananas comosus*) biosorbent	Basic Blue 3, (BB3) and CR	10 and 4	FRH and LNR and TM/58.98 and 11.97	PSO	Non-SPONS and EXO	[55]
Date palm leaves	CV	10.10	FRH/37.74	—	—	[56]
	CR	4–5	LNR/665	PFO		[57]

Wheat straw functionalized with cetyltrimethyl ammonium bromide					SPONS and EXO	
Functionalized *Glossogyne tenuifolia* leaves	CR and MG	2 and 10	FRH/37.9 and 434.8	PSO	SPONS and ENDC	[58]
Banana and durian peel biosorbents	Acid Blue 25 (AB 25)	3	LNR/70.0 and 89.7	PSO	SPONS and EXO	[59]
Biomass-based activated carbon	MB and CR	4	FRH and LNR/769.2 and 453.4	PSO	SPONS and EXO	[60]
Prickly bark (peel) of cactus fruit	MB	12	FRH/222.2	PSO	SPONS and EXO	[61]
Anchote peel biosorbent	Methyl orange (MO)	8	FRH/103.0	PSO	SPONS and EXO	[62]
MgO-impregnated clay, and *Mucor* sp. composites	MG	7	LNR/13.8 and 17.2	PSO	SPONS and EXO	[63]
Banana peel biosorbent	MB	—	—	PSO	—	[64]
Pelagic *Sargassum* and *Sargassum muticum*	MB	—	279	—	—	[65]
Marine algae Ulva *lactuca* biosorbent	MB	8	—	—	—	[66]
Date palm fibers, sawdust, and rice husk biosorbent	MO	2	LNR/48.8, 50.5 and 48.8	—	—	[67]
Activated *Gmelina arborea* leaves	RhB	3	LNR/1000.0	PSO	SPONS and ENDC	[68]
Yeast-MnO_2	MG	10	LNR/243.9	PSO	SPONS and ENDC	[69]

initial dye concentration, pH, and temperature [3, 14, 70]. The impacts of these environmental factors on the BPs are presented in the subsequent sections.

4.4.1 The influence of pH on dye biosorption

A critical parameter of the BP is the solution's pH, which impacts the chemistry of dye solutions and the action of the FGs present in biosorbents [71]. The pH has an direct impact on the sorption capacity as well as the biosorbent's surface chemistry and the dye molecules' ease of access to binding sites. The optimal percentage of dye molecules confiscated under the best possible conditions tends to reduce swiftly in alkaline or acid mediums [3]. The pH denotes the alkalinity or acidity of an aqueous solution. It is important to study the impact of pH on dye biosorption. The pH sensitivity also influences a sorbent material's surface charge and the degree of ionization of acidic and basic compounds. To find the optimal pH for sorption process optimization, the point of zero charge is utilized (pH_{PZC}). The pH_{PZC} is a critical factor that impacts the surface sorption capability and the kinds of surface active centers. The pH_{PZC} is generally utilized to determine or describe the electrokinetic attributes by assessing the pH at which the surface charge of the biosorbent is zero. In systems in which pH is utilized to represent pH_{PZC}, the most important ions are H^+ and OH^-. Hence, the rate of pollutant uptake is increased or reduced depending on the initial pH value. Consequently, OH^- and hydrogen (H^+) ions are sorbed with huge effectiveness, which improves the performance of the sorption process as the FGs at the sorption sites break, thereby leading to a variation in the reaction kinetics and equilibrium features. The biosorption of anionic and cationic species is favorable when the biosorbent surface charge is positive or negative because the pH of the solution is less or more than the pH_{PZC} [51, 72]. As reported in various studies in which different biosorbents were employed for the sequestration of dye molecules (see table 4.1), at reduced pH values, dye uptake rates are normally impacted; they are reduced for cationic dye sorption, but for anionic dyes, the uptake percentage is enhanced.

4.4.2 The influence of the initial dye concentration on the biosorption of dyes

This is an essential factor in the biosorption process, as a given biosorbent mass can sorb a specific quantity of dye. The impact of the initial concentration depends on the instantaneous association between the biosorbent surface's accessible binding sites and the dye concentration [73]. The confiscation of dye molecules reduces with increasing initial dye concentrations. Accordingly, the quantity of dye sorbed is fixed for a given mass of biosorbent. Hence, at elevated dye concentrations, a smaller volume can be confiscated. At reduced dye concentrations, there are enough vacant functioning sites for dye molecules to occupy but with a subsequent increase in the dye concentration, the number of functioning sites for the sorption of dye molecules is reduced. An increase in the initial concentration also leads to an intensification in the sorption capacity. This phenomenon is ascribed to increased dye molecule mobility, which creates a strong driving force that supports the dye molecules' transmission from the solution to the biosorbent surface until equilibrium is attained due to the complete occupancy of the functional sites on the biosorbent surface by dye molecules [45, 50, 60, 74–76].

4.4.3 The influence of the biosorbent dosage on the biosorption of dyes

Biosorbent dosage optimization is critical in defining the smallest possible amount needed to attain optimum biosorption [77]. In general, the percentage of dye molecules sorbed improves with increasing biosorbent dosage. First, the degree of increase in the percentage of dye molecules confiscated has been found to be swift, but this slows down as the biosorbent dosage is increased. At low biosorbent dosages, dye molecules can easily reach the accessible active sites on the surface of the biosorbent and hence the percentage confiscation per unit weight of biosorbent is elevated. With a further increase in the dosage of the biosorbent, there is a smaller corresponding improvement in the sorption of dye molecules due to the fact that numerous sites are left unsaturated during the sorption process. The percentage of dye molecules confiscated saturates after a certain increase in dosage. This is ascribed to the fact that at elevated sorbent dosage, there is fast but insignificant sorption by the biosorbent surface, resulting in a smaller solute concentration in the solution than at lower biosorbent dosages. Hence, improving the biosorbent dosage leads to a decrease in the amount of dye sorbed per unit mass of the biosorbent (q_e, in mg g^{-1}) [60, 66, 67, 69, 73, 78–80].

4.4.4 The influence of the contact time on the biosorption of dyes

The contact time plays a critical part in the attainment of an equilibrium state between an aqueous solution and a biosorbent; it directly impacts the sorption percentage as well as the redox reaction. Different kinetic models are utilized to describe the impact of contact time on the biosorption process [4, 56]. Up to a particular point, the rate of dye confiscation improves with increasing contact interaction time owing to dye deposition on the accessible sorption sites of the biosorbent. Beyond this point, increasing the interaction contact time does not enhance the uptake of dye molecules [81]. The sorption rates of different dyes by biosorbents are swift; they reach their maximum point at the experimental phase and progressively slow down until equilibrium is attained owing to the deposition of dye on the available sorption sites of the biomaterial [78, 82–85]. At this stage, the quantity of dye desorbed from the biosorbent is in a state of active equilibrium with the quantity of dye being sorbed to the biomaterial. The time needed to achieve a state of equilibrium and the quantity of dye sorbed at the equilibrium time reflect the optimum sorption capacity of the biosorbent [4, 73].

4.5 Biosorption isotherm models and kinetic models

4.5.1 Isotherm models

Kinetic and isotherm models are utilized to better describe the biosorption activities of biomaterials. The sorption isotherm models are useful in analyzing the sorption capacities of biosorbents. The sorption isotherm refers to the establishment of a sorption equilibrium relationship that is determined by the quantity of sorbate on the sorbent and the equilibrium concentration of the sorbate at a constant temperature. The LNRM (Langmuir Model), FRHM (Freundlich Model) and DRM (Dubinin-Radushkevich Model) respectively. While TM is Temkin Model and Sips models are different models for assessing sorption isotherms, among which the LM and FH models are the most often employed [86]. The sorption isotherm offers required data such as the sorption mechanism, sorbent affinity, surface properties, and the inner bonds once various parameters are supplied. The relationship between the quantity of sorbed contaminants and the residual contaminants in the solution is defined by the sorption isotherm curves. In most cases, the equilibrium-state behavior of a biosorbent is researched using the LM and FH models [87].

The LM assumes monolayer sorption by a surface comprising a limited quantity of sorption sites with identical sorption behaviors and no migration of the sorbate in the surface plane (i.e. sorption is reversible). While the FH adopts diverse surface energies, the energy term in the LM equation diverges as a function of surface coverage. The linear forms of both models are depicted by equations (4.1) and (4.2):

$$\frac{C_e}{q_e} = \frac{C_e}{Q_m} + \frac{1}{Q_m K_L} \tag{4.1}$$

$$\ln q_e = \ln K_f + \frac{1}{n} \ln C_e, \tag{4.2}$$

where C_e, q_e, q_m, K_L, n, and K_f represent the equilibrium dye concentration (mg l^{-1}), the quantities of dye sorbed at equilibrium and saturation (mg g^{-1}), and the LM and FH constants which measure both intensity and BC [62, 73]. As indicated by the studies reviewed in table 4.1, the LNR and FRH models best explained the BPs of most dyes by various biosorbents.

4.5.2 Biosorption kinetic models

The sorbate–sorbent interaction and the sorption conditions depend on the sorption kinetics. The suitability of sorption kinetics for wastewater control applications has been studied, and two key evaluation elements for the BP are the rate of reaction and its mechanism. The rate of solute uptake defines the residence time required to realize the sorption reaction and can be obtained by examining the kinetics. Most sorption research employs the PFO and PSO models to study the sorption kinetic. The first-order (FO) rate equation was presented by Lagergren in 1898; it defines the kinetics of the liquid–solid phase sorption of malonic and oxalic acids by charcoal. Lagergren's FO rate equation was called the PFO model to differentiate it from kinetic equations that determine the sorption capacity from the solution concentration and can be linearly expressed as (equation (4.3)):

$$\log (q_e - q_t) = \log q_e - \frac{k_1 t}{2.303}, \tag{4.3}$$

where q_t and k_1 represent the sorption capacity at time t (in mg g^{-1}) and the PFO model rate constant (in 1 min^{-1}). A plot of $\log(q_e - q_t)$ against t gives a linear association, from which k_1 and q_e can be established from the slope and intercept of the plot, respectively [88]. The PSO model equation was first proposed by Blanchard *et al* [50]. This model was proposed based on the postulation that a sorption system could be a chemical process controlled by a chemical sorption mechanism involving the sharing or transfer of electrons between dye molecules and the biosorbent. According to this model, the number of available sorbent sites is linked to the solute sorption rate, and the quantity of solute on the sorbent's surface impacts the rate of reaction [55, 82]. The linear form of this model is given by equation (4.4):

$$\frac{t}{q_t} = \frac{1}{k_2 q_e^2} + \frac{t}{q_e}, \tag{4.4}$$

where k_2 is the rate constant of the PSO model (in $g.mg^{-1}.min^{-1}$). This model is applicable when t/q_t is plotted against t and a linear association is obtained; this allows k_2 and q_e to be established from the intercept and slope of the plot, respectively [50]. As can be observed from the studies reviewed in table 4.1, the PSO model is the most successful model at describing the BPs of most dyes by various biosorbents, followed by the PFO model.

4.6 Thermodynamic studies

Various thermodynamic parameters are linked to the temperature dependence of the BP. The thermodynamic characteristics of the BP determine whether or not the process is spontaneously established. The spontaneity of a chemical reaction is indicated by the Gibbs free energy change (ΔG°), and this is determined from both its enthalpy (ΔH°) and entropy (ΔS°). Negative and positive values of ΔG° signify spontaneity and non-spontaneity of the BP, respectively. The ΔG° value can be determined using equation (4.5):

$$\Delta G^\circ = -RT \ln K_c \tag{4.5}$$

where K_c, R, and T signify the sorption equilibrium constant, the universal gas constant, and the absolute temperature. The relationships between ΔH°, ΔS°, and ΔG° are given by equations (4.6) and (4.7).

$$\Delta G^\circ = \Delta H^\circ - T\Delta S^\circ \tag{4.6}$$

$$\ln K_c = \frac{\Delta G^\circ}{R} - \frac{\Delta H^\circ}{RT} \tag{4.7}$$

The values of ΔH° and ΔS° are determined from the slope and intercept of the plot of $\ln K_c$ versus $1/T$. Negative and positive values of ΔH° are indicative of exothermic and endothermic BPs, respectively. A positive value of ΔS° indicates improved randomness of the solid–solution interface during the BP [89–91]. Based on the results in table 4.1, it can be deduced from almost all of the reviewed studies that the BPs of dyes by various biosorbents were thermodynamically SPONS and EXO or ENDC.

4.7 The mechanisms of the biosorption process

Biosorption mechanisms are either metabolism-independent (physical and chemical adsorption, electrostatic interaction, ion exchange, complexation, chelation, and microprecipitation—which generally take place in microbial cells and plant cell walls) or metabolism-dependent (physical sorption and chelation), depending on the manner in which dye molecules bind to the cell walls of the biomaterials [11, 18]. Mechanistic studies of the biosorption of various water pollutants are important in order to assess the influences on their effectiveness and are favorable for improving the confiscation process conditions. Normally, the process of biosorption is based on the physicochemical properties of the biosorbent, such as its chemical composition, hydrophobicity, solubility, molecular, reactivity, and surface charge. In addition, various FGs are responsible for the sorption processes of dyes. The common mechanisms for the biosorption of dyes comprise complexation/coordination, electrostatic interaction, ion exchange, aggregation, microprecipitation, oxidation, and reduction. The interactions between water contaminants and biosorbents are based on surface and interstitial sorption. Surface sorption is achieved via van der Waals forces, hydrogen bonding, or dipole interactions, while interstitial sorption involves the diffusion of contaminants (ions or molecules) toward the meso-, macro-, and micropores of biosorbents and hence the biosorption of contaminants by the interior surfaces of the biosorbent [8, 18, 81].

4.8 Conclusions

This chapter provided an overview of the recent development in the FUNZTN of biosorbents and their utilization in the treatment of effluent-containing dyes. There has been increased research over the years into the use of biological treatment methods for dye confiscation owing to their eco-friendly treatment processes and cost-effectiveness. The use of this technique in the treatment of dye waste depends on the exceptional surface chemistry that results from the presence of FGs. The findings of the studies reviewed here show that the BPs of dye molecules by various biosorbents depend on the pH of the solution. This chapter examined the BPs of dye sequestration by the various functionalized biosorbents and found that they are mostly described or defined by the LNR, FRH, PSO, and PFO models. In addition, the BPs were found to be thermodynamically SPONS and EXO or ENDC. The future focus of research should be the use of functionalized biosorbents in the real-time remediation of dye-containing wastewater as well as the optimization of the FUNZTN process.

References

[1] Masindi V and Muedi K 2018 Environmental contamination by heavy metals *Heavy Metals* **10** 115–32

[2] Aigbe U, Ukhurebor K, Onyancha R, Osibote O, Darmokoesoemo H and Kusuma H 2021 Fly ash-based adsorbent for adsorption of heavy metals and dyes from aqueous solution: a review *J. Mater. Res. Technol.* **14** 2751–74

[3] Praveen S, Jegan J, Bhagavathi Pushpa T, Gokulan R and Bulgariu L 2022 Biochar for removal of dyes in contaminated water: an overview *Biochar* **4** 1–16
[4] Aigbe U, Ukhurebor K, Onyancha R, Okundaye B, Pal K, Osibote O, Esiekpe E, Kusuma H and Darmokoesoemo H 2022 A facile review on the sorption of heavy metals and dyes using bionanocomposites *Adsorpt. Sci. Technol.* **2022** 8030175
[5] Kusuma H *et al* 2023 Biosorption of methylene blue using clove leaves waste modified with sodium hydroxide *Res. Chem.* **5** 100778
[6] Aragaw T and Bogale F 2021 Biomass-based adsorbents for removal of dyes from wastewater: a review *Front. Environ. Sci.* **9** 558
[7] Islam T, Repon M, Islam T, Sarwar Z and Rahman M 2022 Impact of textile dyes on health and ecosystem: a review of structure, causes, and potential solutions *Environ. Sci. Pollut. Res.* **30** 9207–42
[8] Elgarahy A, Elwakeel K, Mohammad S and Elshoubaky G 2021 A critical review of biosorption of dyes, heavy metals and metalloids from wastewater as an efficient and green process *Clean. Eng. Technol.* **4** 100209
[9] Kumar A, Dixit U, Singh K, Gupta S and Beg M 2021 Structure and properties of dyes and pigments *Dyes and Pigments: Novel Applications and Waste Treatment* (London: IntechOpen)
[10] Gregory P 1990 Classification of dyes by chemical structure *The Chemistry and Application of Dyes* (Boston, MA: Springer) 17–47 pp
[11] Asgher M 2012 Biosorption of reactive dyes: a review *Water Air Soil Pollut.* **223** 2417–35
[12] Naidu R *et al* 2021 Chemical pollution: a growing peril and potential catastrophic risk to humanity *Environ. Int.* **156** 106616
[13] Khatoon H and Rai J 2016 Agricultural waste materials as biosorbents for the removal of heavy metals and synthetic dyes: a review *Octa J. Environ. Res.* **4** 208–29
[14] Argun Y, Karacalı A, Çalışır U, Kılınç N and Irak H 2016 Biosorption method and biosorbents for dye removal from industrial wastewater: a review *2nd Int. Conf. on Science, Ecology and Technology (ICONSETE'2016)*
[15] Adewuyi A 2020 Chemically modified biosorbents and their role in the removal of emerging pharmaceutical waste in the water system *Water* **12** 1551
[16] Haque A, Sultana N, Sayem A and Smriti S 2022 Sustainable adsorbents from plant-derived agricultural wastes for anionic dye removal: a review *Sustainability* **14** 11098
[17] Bulgariu L and Bulgariu D 2018 Functionalized soy waste biomass: a novel environmental-friendly biosorbent for the removal of heavy metals from aqueous solution *J. Clean. Prod.* **197** 875–85
[18] Yaashikaa P, Kumar P, Saravanan A and Vo D 2021 Advances in biosorbents for removal of environmental pollutants: a review on pretreatment, removal mechanism and future outlook *J. Hazard. Mater.* **420** 126596
[19] Aigbe U and Osibote O 2021 Carbon derived nanomaterials for the sorption of heavy metals from aqueous solution: a review *Environ. Nanotechnol., Monitor. Manag.* **16** 100578
[20] Norrrahim M *et al* 2021 Nanocellulose: a bioadsorbent for chemical contaminant remediation *RSC Adv.* **11** 7347–68
[21] Yang X, Wan Y, Zheng Y, He F, Yu Z, Huang J, Wang H, Ok Y, Jiang Y and Gao B 2019 Surface functional groups of carbon-based adsorbents and their roles in the removal of heavy metals from aqueous solutions: a critical review *Chem. Eng. J.* **366** 608–21

[22] Tee G, Gok X and Yong W 2022 Adsorption of pollutants in wastewater via biosorbents, nanoparticles and magnetic biosorbents: a review *Environ. Res.* **212** 113248
[23] Singh P and Srivastava S 2022 A review on biosorbents *Int. J. Res. Anal. Rev.* **9** 370–382 https://ijrar.org/viewfull.php?&p_id=IJRAR22A2002
[24] Lellis B, Fávaro-Polonio C, Pamphile J and Polonio J 2019 Effects of textile dyes on health and the environment and bioremediation potential of living organisms *Biotechnol. Res. Innov.* **3** 275–90
[25] Dhankhar R and Hooda A 2011 Fungal biosorption–an alternative to meet the challenges of heavy metal pollution in aqueous solutions *Environ. Technol.* **32** 467–91
[26] Derco J and Vrana B 2018 *Biosorption* (London: IntechOpen)
[27] Abbas F 2013 Dyes removal from wastewater using agricultural waste *Adv. Environ. Biol.* **7** 1019–26 https://www.aensiweb.com/old/aeb/2013/1019-1026.pdf
[28] Essekri A *et al* 2021 Novel citric acid-functionalized brown algae with a high removal efficiency of crystal violet dye from colored wastewaters: insights into equilibrium, adsorption mechanism, and reusability *Int. J. Phytorem.* **23** 336–46
[29] Essekri A, Aarab N, Hsini A, Ajmal Z, Laabd M, El Ouardi M, Ait Addi A, Lakhmiri R and Albourine A 2022 Enhanced adsorptive removal of crystal violet dye from aqueous media using citric acid modified red-seaweed: experimental study combined with RSM process optimization *J. Dispersion Sci. Technol.* **43** 1359–72
[30] Gong R, Feng M, Zhao J, Cai W and Liu L 2009 Functionalization of sawdust with monosodium glutamate for enhancing its malachite green removal capacity *Bioresour. Technol.* **100** 975–8
[31] Moreno Z and Herrera J 2022 Biodegradation of Erionyl Turquoise azoic textile dye from *Chlorella vulgaris* microalgae in simulated wastewater *J. Textile Eng. Fashion Technol.* **8** 121–4
[32] Shakoor S and Nasar A 2019 Utilization of *Cucumis sativus* peel as an eco-friendly biosorbent for the confiscation of crystal violet dye from artificially contaminated wastewater *Anal. Chem. Lett.* **9** 1–19
[33] Chang T, Hadibarata T and Syafrudin M 2020 Functionalized stink bean pod (*Parkia speciosa*) powder for adsorption of reactive dye *Biointerface Res. Appl. Chem.* **11** 11616
[34] Ansari S, Khan F and Ahmad A 2016 Cauliflower leave, an agricultural waste biomass adsorbent, and its application for the removal of MB dye from aqueous solution: equilibrium, kinetics, and thermodynamic studies *Int. J. Anal. Chem.* **2016** 8252354
[35] Dos Santos L, Porto C, Bulla M, Batistela V and Barros B 2021 Peach palm and cassava wastes as biosorbents of tartrazine yellow dye and their application in industrial effluent *Sci. Plena* **17**
[36] Bello O, Adegoke K, Sarumi O and Lameed O 2019 Functionalized locust bean pod (*Parkia biglobosa*) activated carbon for Rhodamine B dye removal *Heliyon* **5** e02323
[37] Badr S and Isra'a S 2021 Using agricultural waste as biosorbent for hazardous brilliant green dye removal from aqueous solutions *J. Eng. Sci. Technol.* **16** 3435–54 https://jestec.taylors.edu.my/Vol%2016%20Issue%204%20August%202021/16_4_45.pdf
[38] Okoya A and Diisu D 2021 Adsorption of indigo-dye from textile wastewater onto activated carbon prepared from sawdust and periwinkle shell *Trends Appl. Sci. Res.* **16** 1–9
[39] San Keskin N, Celebioglu A, Uyar T and Tekinay T 2015 Microalgae immobilized by nanofibrous web for removal of reactive dyes from wastewater *Ind. Eng. Chem. Res.* **54** 5802–9

[40] Wijayanti T and Ansori M 2021 Application of modified green algae *Nannochloropsis* sp. as adsorbent in the simultaneous adsorption of Methylene Blue and Cu (II) cations in solution *Sustain. Environ. Res.* **31** 1–12
[41] Salman J, Grmasha R and Al-sareji O 2022 Potential efficiency of macroalgae *Cladophora* sp. to remove toxic industrial dye (safranin O) from aqueous solution *AIP Conf. Proc.* **2386** 020048
[42] Rajendran S *et al* 2022 Functionalized microbial consortia with silver-doped hydroxyapatite (Ag@ HAp) nanostructures for removal of RO84 from industrial effluent *Crystals* **12** 970
[43] Rathod J and Archana G 2022 Functional characterization of bacterial isolates from dye decolorizing consortia and a step-up metabolic engineering based on NADH-regeneration *bioRxiv* 2022–04
[44] Kim H and Kim H 2022 Production of coffee-dyed bacterial cellulose as a bio-leather and using it as a dye adsorbent *PLoS One* **17** e0265743
[45] Bayramoğlu G, Ozalp V and Arıca M 2017 Removal of Disperse Red 60 dye from aqueous solution using free and composite fungal biomass of *Lentinus concinnus Water Sci. Technol.* **75** 366–77
[46] Han Y, Shi L, Meng J, Yu H and Zhang X 2014 Azo dye biodecolorization enhanced by *Echinodontium taxodii* cultured with lignin *PLoS One* **9** e109786
[47] Si J, Yuan T and Cui B 2015 Exploring strategies for adsorption of azo dye Congo Red using free and immobilized biomasses of *Trametes pubescens Ann. Microbiol.* **65** 411–21
[48] Almeida E and Corso C 2019 Decolorization and removal of toxicity of textile azo dyes using fungal biomass pelletized *Int. J. Environ. Sci. Technol.* **16** 1319–28
[49] Sudarni D, Aigbe U, Ukhurebor K, Onyancha R, Kusuma H, Darmokoesoemo H, Osibote O, Balogun V and Widyaningrum B 2021 Malachite green removal by activated potassium hydroxide clove leaves agro-waste biosorbent: characterization, kinetics, isotherms and thermodynamics studies *Adsorp. Sci. Technol.* **2021** 1–15
[50] Ouettar L, Guechi E, Hamdaoui O, Fertikh N, Saoudi F and Alghyamah A 2023 Biosorption of triphenyl methane dyes (malachite green and crystal violet) from aqueous media by alfa (*Stipa tenacissima* L.) leaf powder *Molecules* **28** 3313
[51] Eldeeb T *et al* 2022 Biosorption of acid brown 14 dye to mandarin-CO-TETA derived from mandarin peels *Biomass Convers. Biorefinery* **1–21**
[52] Eldeeb T, Aigbe U, Ukhurebor K, Onyancha R, El-Nemr M, Hassaan M, Ragab S, Osibote O and El Nemr A 2022 Adsorption of methylene blue (MB) dye on ozone, purified and sonicated sawdust biochars *Biomass Convers. Biorefinery* **2022** 1–23
[53] Al-Fawwaz A, Shra'ah A and Elhaddad E 2023 Bioremoval of methylene blue from aqueous solutions by green algae (*Bracteacoccus* sp.) isolated from North Jordan: optimization, kinetic, and isotherm studies *Sustainability* **15** 842
[54] Jain S and Gogate P 2017 Acid Blue 113 removal from aqueous solution using novel biosorbent based on NaOH treated and surfactant modified fallen leaves of *Prunus dulcis J. Environ. Chem. Eng.* **5** 3384–94
[55] Chan S, Tan Y, Abdullah A and Ong S 2016 Equilibrium, kinetic and thermodynamic studies of a new potential biosorbent for the removal of basic blue 3 and congo red dyes: pineapple (*Ananas comosus*) plant stem *J. Taiwan Inst. Chem. Eng.* **61** 306–15
[56] Ghazali A, Shirani M, Semnani A, Zare-Shahabadi V and Nekoeinia M 2018 Optimization of crystal violet adsorption onto Date palm leaves as a potent biosorbent from aqueous solutions using response surface methodology and ant colony *J. Environ. Chem. Eng.* **6** 3942–50

[57] Zhang R, Zhang J, Zhang X, Dou C and Han R 2014 Adsorption of Congo red from aqueous solutions using cationic surfactant modified wheat straw in batch mode: kinetic and equilibrium study *J. Taiwan Inst. Chem. Eng.* **45** 2578–83

[58] Yang J and Hong G 2018 Adsorption behavior of modified *Glossogyne tenuifolia* leaves as a potential biosorbent for the removal of dyes *J. Mol. Liq.* **252** 289–95

[59] Samat J, Shahri N, Abdullah M, Suhaimi N, Padmosoedarso K, Kusrini E, Mahadi A, Hobley J and Usman A 2021 Adsorption of Acid Blue 25 on agricultural wastes: efficiency, kinetics, mechanism, and regeneration *Air, Soil Water Res.* **14** 11786221211057496

[60] Nizam N, Hanafiah M, Mahmoudi E, Halim A and Mohammad A 2021 The removal of anionic and cationic dyes from an aqueous solution using biomass-based activated carbon *Sci. Rep.* **11** 1–17

[61] Abdelkarim S, Mohammed H and Nouredine B 2017 Sorption of methylene blue dye from aqueous solution using an agricultural waste *Trends Green Chem.* **3** 1–7

[62] Hambisa A, Regasa M, Ejigu H and Senbeto C 2023 Adsorption studies of methyl orange dye removal from aqueous solution using Anchote peel-based agricultural waste adsorbent *Appl. Water Sci.* **13** 24

[63] Moustafa M 2023 Preparation and characterization of low-cost adsorbents for the efficient removal of malachite green using response surface modeling and reusability studies *Sci. Rep.* **13** 4493

[64] Bakar A, Mazlan W, Akbar N, Badrealam S and Ali K 2019 Agriculture waste from banana peel as low cost adsorbent in treating methylene blue from batik textile waste water effluents *J. Phys.: Conf. Ser.* **1349** 012078

[65] Nielsen B, Maneein S, Anghan J, Anghan R, Al Farid M and Milledge J 2022 Biosorption potential of Sargassum for removal of aqueous dye solutions *Appl. Sci.* **12** 4173

[66] Pratiwi D, Prasetyo D and Poeloengasih C 2019 Adsorption of methylene blue dye using marine algae *Ulva lactuca IOP Conf. Ser.: Earth Environ. Sci.* **251** 012012

[67] Dakhil I 2020 Recycling of agriculture wastes for efficient removal of methyl orange dye using batch adsorption unit *IOP Conf. Ser.: Mater. Sci. Eng.* **881** 012186

[68] Bello O, Alabi E, Adegoke K, Adegboyega S, Inyinbor A and Dada A 2020 Rhodamine B dye sequestration using *Gmelina aborea* leaf powder *Heliyon* **6** e02872

[69] Santos B, Cossolin A, Reis H, Castro K, Silva E, Pereira G, Sousa Junior P, Dall'Oglio E, Vasconcelos L and Morais E 2019 Baker's yeast–MnO_2 composites as biosorbent for Malachite green: an ecofriendly approach for dye removal from aqueous solution *Rev. Ambient. Água* **14** e2254

[70] Razzak S, Farooque M, Alsheikh Z, Alsheikhmohamad L, Alkuroud D, Alfayez A, Hossain S and Hossain M 2022 A comprehensive review on conventional and biological-driven heavy metals removal from industrial wastewater *Environ. Adv.* **7** 100168

[71] de Freitas G, da Silva M and Vieira M 2019 Biosorption technology for removal of toxic metals: a review of commercial biosorbents and patents *Environ. Sci. Pollut. Res.* **26** 19097–118

[72] Badawi A, Abd Elkodous M and Ali G 2021 Recent advances in dye and metal ion removal using efficient adsorbents and novel nano-based materials: an overview *RSC Adv.* **11** 36528–53

[73] Bharathi K and Ramesh S 2013 Removal of dyes using agricultural waste as low-cost adsorbents: a review *Appl. Water Sci.* **3** 773–90

[74] Zein R, Purnomo J, Ramadhani P, Alif M and Putri C 2023 Enhancing sorption capacity of methylene blue dye using solid waste of lemongrass biosorbent by modification method *Arabian J. Chem.* **16** 104480

[75] Liang C, Shi Q, Feng J, Yao J, Huang H and Xie X 2022 Adsorption behaviors of cationic methylene blue and anionic reactive blue 19 dyes onto nano-carbon adsorbent carbonized from small precursors *Nanomaterials* **12** 1814

[76] Abou Oualid H, Abdellaoui Y, Laabd M, El Ouardi M, Brahmi Y, Iazza M and Abou Oualid J 2020 Eco-efficient green seaweed codium decorticatum biosorbent for textile dyes: characterization, mechanism, recyclability, and RSM optimization *ACS Omega* **5** 22192–207

[77] Aigbe U, Das R, Ho W, Srinivasu V and Maity A 2018 A novel method for removal of Cr (VI) using polypyrrole magnetic nanocomposite in the presence of unsteady magnetic fields *Sep. Purif. Technol.* **194** 377–87

[78] Grabi H, Lemlikchi W, Derridj F, Lemlikchi S and Trari M 2021 Efficient native biosorbent derived from agricultural waste precursor for anionic dye adsorption in synthetic wastewater *Biomass Convers. Biorefinery* **13** 171–88

[79] Amar I, Zayid E, Dhikeel S and Najem M 2021 Biosorption removal of methylene blue dye from aqueous solutions using phosphoric acid-treated balanites aegyptiaca seed husks powder *Biointerface Res. Appl. Chem.* **12** 7845–62

[80] Adeogun A, Akande J, Idowu M and Kareem S 2019 Magnetic tuned sorghum husk biosorbent for effective removal of cationic dyes from aqueous solution: isotherm, kinetics, thermodynamics and optimization studies *Appl. Water Sci.* **9** 1–17

[81] Nayagam J and Prasanna K 2022 Utilization of shell-based agricultural waste adsorbents for removing dyes: a review *Chemosphere* **291** 132737

[82] El-Nemr M, Hassaan M and Ashour I 2023 Fabrication of N-doping activated carbons from fish waste and sawdust for Acid Yellow 36 dye removal from an aquatic environment *Sci. Rep.* **13** 5892

[83] Silva F, Nascimento L, Brito M, da Silva K, Paschoal W and Fujiyama R 2019 Biosorption of methylene blue dye using natural biosorbents made from weeds *Materials* **12** 2486

[84] Arjona González A, Canal Arias J and García Raurich J 2018 A new biosorbent with controlled grain (I). Efficient elimination of cationic dyes from textile dyeing wastewater *Int. J. Environ. Agri. Res. (IJOEAR)* **4** 14–27 https://ijoear.com/assets/articles_menuscripts/file/IJOEAR-MAR-2018-6.pdf

[85] Akram M, Salman M, Farooq U, Saleem U, Tahir S, Nazir H and Arsalan H 2020 Phthalate-functionalized *Sorghum bicolor* L.: an effective biosorbent for the removal of Alizarin Red S and Bromophenol blue dyes from simulated wastewater *Desalin. Water Treat.* **190** 383–92

[86] Yang J, Hou B, Wang J, Tian B, Bi J, Wang N, Li X and Huang X 2019 Nanomaterials for the removal of heavy metals from wastewater *Nanomaterials* **9** 424

[87] Farhadi A, Ameri A and Tamjidi S 2021 Application of agricultural wastes as a low-cost adsorbent for removal of heavy metals and dyes from wastewater: a review study *Phys. Chem. Res.* **9** 211–26

[88] Khan A, Gul J, Naqvi S, Ali I, Farooq W, Liaqat R, AlMohamadi H, Štěpanec L and Juchelková D 2022 Recent progress in microalgae-derived biochar for the treatment of textile industry wastewater *Chemosphere* **306** 135565

[89] Aksakal O and Ucun H 2010 Equilibrium, kinetic and thermodynamic studies of the biosorption of textile dye (Reactive Red 195) onto *Pinus sylvestris* L *J. Hazard. Mater.* **181** 666–72

[90] Fawzy M, Al-Yasi H, Galal T, Hamza R, Abdelkader T, Ali E and Hassan S 2022 Statistical optimization, kinetic, equilibrium isotherm and thermodynamic studies of copper biosorption onto *Rosa damascena* leaves as a low-cost biosorbent *Sci. Rep.* **12** 8583

[91] Lim L, Priyantha N, Tennakoon D, Chieng H, Dahri M and Suklueng M 2017 Breadnut peel as a highly effective low-cost biosorbent for methylene blue: equilibrium, thermodynamic and kinetic studies *Arab. J. Chem.* **10** S3216–28

IOP Publishing

Adsorption Applications for Environmental Sustainability

Kingsley Eghonghon Ukhurebor, Uyiosa Osagie Aigbe and Robert Birundu Onyancha

Chapter 5

The sequestration of heavy metals from soil using functionalized biosorbents

Uche Eunice Ekpunobi, Daniel Elaigwu Enenche, Charles Ifeanyi Aghanwa, Abimbola Bankole Akinyele, Patrick Ugochukwu Umennadi, Uzochukwu Abraham Onuigbo and Austin Ozukwe

The quantities of heavy metals present in soils have a significant impact on how readily available the ions of these metals are to plants. At higher concentrations, heavy metal ions become very damaging to physiological functions. Recently, environmental contamination from these metals has been strongly linked to rising ecological challenges, leading to global public health concerns. Soil remediation remains a major solution to the problem of soil contamination. The most acceptable soil remediation method should be transparent, dynamic, and sustainable. This chapter presents different types of soil remediation that are innovative, sustainable, and environmentally friendly.

5.1 Introduction

Heavy metals are naturally present at trace levels in the soil (less than 1000 mg kg^{-1}), but they are rarely present at toxic levels because of the weathering of the parent materials during pedogenetic processes [1]. In general, 'heavy metals' refers to elements with an atomic density greater than 5.0 g cm^{-3} that have metallic properties [2]. It also includes transition metals, some metalloids, lanthanides, and actinides, all of which can be linked to contamination and have the potential to be toxic or environmentally harmful [2]. Two categories of heavy metals can be distinguished: nonessential, such as cadmium (Cd), lead (Pb), chromium (Cr), and mercury (Hg), which are not known to have any biological functions, and essential, such as iron (Fe), zinc (Zn), copper (Cu), manganese (Mn), and nickel (Ni), which are needed in small amounts for plant nutrition. These essential heavy metals directly participate in redox reactions and contribute to the biochemistry and physiology of plants and animals by being crucial components of a number of

doi:10.1088/978-0-7503-5598-8ch5

enzymes [3]. Although the nonessential heavy metals have a greater effect on living things, even heavy metals that are micronutrients can become toxic if a certain concentration is exceeded [4]. At higher concentrations, all heavy metal ions become harmful to physiological functions because they form complex compounds that have toxic effects inside the body's cells. Nonessential metals react by displacing essential metals from their natural binding sites or by interacting with ligands [5, 6]. For instance, the sulfhydryl (SH) groups of enzymes vital to microbial metabolism are bound by Hg^{2+}, Cd^{2+}, and Ag^{2+} metal ions, thereby inhibiting sensitive enzyme activity [5]. The biological significance of heavy metals such as Hg, Cd, and Pb is negligible. Potential carcinogens such as Pb, Hg, and Cd are linked to ailments such as brain damage and kidney, blood, nervous system, and bone disorders [7]. However, a deficiency of essential metals can also affect the human body. A diarrheal state can contribute to deficiencies caused by malabsorption. On the other hand, the critical amounts of these metals that cause their toxic effects depend on a series of variables, including the rate at which such metals enter the body through ingestion or inhalation, their distribution in the tissues, the concentration attained, and finally, the rate at which the metal is excreted [8].

5.2 Natural sources of heavy metals

The principal origins of heavy metals are natural geological processes such as weathering, erosion, and changing geological materials in the subsurface soil [9]. The parent material from which soils develop is the surface deposits left over from the underlying geologic processes. The physical, chemical, and mineralogical properties of a soil are the result of the weathering of its parent material. When soil is formed, weathering causes the physical and chemical dissolution of minerals, releasing metals from the parent material into the soil as well as releasing their solutions into the soil water and groundwater. Numerous studies have found a close connection between the parent material's metal content and the soils that form during weathering [10].

The minerals and rocks that make up the soil's parent material come from various types of bedrock that weather over time, releasing trace elements into the soil. Unlike some metals that are predominant in the Earth's crust as well as in the mineral structure of the soil, such as silicon, iron, aluminum, calcium, and magnesium, most metals in rocks and minerals only occur in trace amounts [11]. According to Kabata-Pendias [12], mineral concentrations vary within the same soil matrix, such that Cu and Cr are more abundant in topsoils than in the source materials minerals, whereas Ni is more abundant in the source material minerals than in the surface horizon minerals. This may be because Cr and Cu are more mobile during chemical weathering. While some elements, such as Mn, tend to accumulate in surface horizons, subsurface horizons frequently have higher concentrations of other elements, such as Cr.

Lithogenic metals are mainly passed down from the parent substance. pedogenetic metals have lithogenic and anthropogenic origins, but mineral transformation and other pedogenetic processes change how they are distributed in soil profiles.

Anthropogenic heavy metals are deposited in soils because of human activity [13]. Their origins are easy to track because they are typically mobile compared to lithogenic and pedogenetic metals, which have indistinguishable origins [12]. The types and proportions of soils found in a given landscape are greatly influenced by the length and intensity of the pedogenetic processes as well as the characteristics of the parent material. By weathering and leaching minerals and organic matter and by reorienting (clay illuviation) and remodeling material, pedogenetic processes unmistakably alter soil (*in situ* clay formation). Surface enrichment, leaching, translocation, and mobilization are some of the pedological factors that affect how trace elements are distributed in a soil profile [14].

5.2.1 Anthropogenic sources of heavy metal contamination

The exponential rise in the use of heavy metals in the home and in various applications such as industry, agriculture, and technology has significantly increased human exposure [15]. In addition to these sources, environmental contamination can also be caused by air deposition, metal corrosion, sediment re-suspension, soil erosion of metal ions, leaching of heavy metals, and metal evaporation from water resources into soil water and groundwater [16].

Numerous studies have shown that anthropogenic sources account for the majority of the heavy metal pollution in the environment. The anthropogenic sources of heavy metals in soils include sewage sludge, inorganic contaminants in fertilizers and manure, and heavy-metal-containing pesticides [17]. Other major sources include vehicle exhaust particles, the deterioration of car parts, emissions from industrial worksites and power plants, domestic emissions, and atmospheric deposits [18].

Unpredictable pollution levels also rise because of unchecked urbanization and the rapid expansion of industry around many cities and coastal areas (concentration). The persistent and bioaccumulative nature of heavy metals such as As, Cr, Cu, Pb, Hg, and Zn makes them a major concern compared to other pollutants. Since they bioaccumulate throughout the food chain, they may reach a level of contamination that may have a dangerous impact on aquatic life as well as human health, especially for people who consume seafood [19].

According to a report by the Interstate Technology and Regulatory Council's Small Arms Firing Range Team [20], some shooting ranges in the United States had Pb concentrations ranging from 27 000 to 233 142 mg kg^{-1}. Stauffer *et al* [22] raised concerns about the use of ammunition as sources of outdoor contamination by reporting intermittent concentrations of Pb, Cu, Zn, and Sb up to 25 m from the shooting position. According to the work of Staszewski *et al* [22], heavy metal concentrations have also been reported along transport lines, especially railway tracks.

5.2.2 The bioavailability (speciation) of heavy metals

The availability of these metal ions to plants depends on the concentration of heavy metals in the soil solution. Bioavailability, according to Newman and Jagoe [23], is the extent to which a contaminant in a potential source is available for absorption by an organism. Because of their instability and high bioavailability, heavy metals that originate from anthropogenic sources are more hazardous [24, 25].

In addition, Hund-Rinke and Koerdel [26] noted that metal speciation, the type of exposure, and the complexity of the process strongly influence bioavailability. 'Chemical species' refers to the distinct isotopic composition, molecular structure, oxidation, or electronic state(s) of the element(s) present. The bioactivity and bioavailability of trace metals are strongly influenced by their coexisting chemical forms and thus their speciation [27]. A good illustration of the significance of the oxidation state is chromium. Cr^{4+} is genotoxic and carcinogenic, whereas Cr^{3+} is regarded as an essential element. The oxidation state has an impact on bioavailability as well. Following dermal and oral exposure, Cr^{4+} is more effectively absorbed than Cr^{3+} [28].

The concentration of metals in the soil influences their bioavailability, movement, and possible toxicity. These factors also depend on how metals associate with soluble species and how successfully the soil can release them from the solid phase into the soil solution [29]. There are several different forms in which metals can be found in soil, including chelated metal complexes in soil solutions, precipitates, complexes, adsorbates on organic or inorganic cation exchange surfaces, and inorganic crystals [30].

'Total metal concentration' refers to the portion of metal in the soil that is removed by highly effective extractants such as nitric and perchloric acids, rather than the actual total value, since not all lattice-bound metals may be extracted. Along with the readily exchangeable ions, this also refers to the metals that are more firmly bound to the soil's solid phases and hence not typically available to soil organisms for uptake [31].

The link between the concentrations of heavy metals in the soil and the amounts of heavy metals absorbed by plant tissues is now the basis for expressing soil heavy metal bioavailability, and several techniques are employed to measure soil heavy metal concentrations following extraction. Given their relative ease and reduced time requirements, single extraction methods—such as those using hydrochloric acid (HCl), diethylenetriaminepentaacetic acid (DTPA), and ethylenediaminetetraacetic acid (EDTA), among others, are the most widely used [32].

Three bioavailability terms have been reported. The first, which is environmental availability, focuses on the amount of the overall concentration in the soil that is actually dissolved in the pore water as well as the total content that is potentially available and is adsorbed by the soil matrix, such as the amounts that are specifically and non-specifically adsorbed, organically bound, and surface precipitated. There is a maximum amount of metal that can be desorbed from the soil matrix while sustaining equilibrium with the dissolved proportion in the groundwater under pressurized conditions or throughout the full exposure period. This is referred to as the potentially available fraction [33, 34].

The fraction of dissolved metal species that can be absorbed by plant roots and other soil organisms is known as environmental bioavailability (the second term). Hence, this kind of bioavailability is controlled by physiological processes and depends on the movement of metals from pore water to plant roots and absorption mechanisms that may only be found in specific plants and heavy metals. Metal species may enter the root cell actively through metabolic processes, mass flow, or diffusion [35].

The biological impact of heavy metals on plants and other organisms is referred to as toxicological bioavailability, and it is dependent on a number of complex interactions, including translocation, metabolism, and detoxification. Toxicological bioavailability, also known as accumulation in roots, shoots, and grains, growth inhibition, or toxic effects on specific or metabolic levels in plants, can be determined using bioassays [36]. Such accumulations have the ability to transform into more toxic species and can enter the environment in a bioavailable form [37].

5.3 Factors affecting the speciation of heavy metals

Through their impact on metal extractability, soil characteristics may affect metal bioavailability. Additionally, different studies have found different ways in which soil characteristics affect the concentrations of extractable metals in the soil. Understanding the mechanisms determining metal bioavailability in soils is also necessary to resolve the mismatch between the information required to create relevant environmental quality targets (such as assessments of soil solutions or bioavailable metals) and the information available (total metal concentrations). Theoretically, abiotic factors such as pH, Eh, organic matter, clay fractions, and hydrous oxides can affect the solubilities of metals in soils [31].

In soils with pH values ranging from fairly acidic to slightly alkaline, heavy metals are typically adsorbed at the hydroxyl surfaces of oxides or clay minerals [33]. This type of adsorption is primarily associated with the covalent-bonding-mediated development of inner-sphere complexes of metal hydroxides. Metals with a heightened level of specific adsorption are typically thought to be less readily available to the environment [33, 38, 40]. Tiller *et al* [41] and Herms and Bruemmer [42] have reported that increasing the hydroxide constant increased the specific adsorption of metals. A reduction in metal mobility triggered an increase in the hydroxide constant, according to research by Baes and Mesmer [43].

In contrast, heavy metals are mostly non-specifically adsorbed at the binding sites of cation exchange of clay minerals and oxides at low pH through electrostatic linkages, which are connected to the exchangeable fraction of metals and are believed to be easily (environmentally) obtainable [33, 40]. Because of this, heavy metals are more available and mobile at reduced pH levels than they are at soil pH values ranging from slightly alkaline to moderately acidic.

Metals hardly form insoluble precipitates with phosphates at low pH because most of the phosphate ions have been protonated [44]. This is because, at an acidic pH, extra protons (H^+) are present that fill up metal-binding sites. Normal conditions allow metal ions to replace protons to form various species, such as hydroxo-metal complexes. Hydroxo-metal complexes formed with cadmium, nickel, and zinc can occasionally be soluble, whereas those made with chromium and iron are insoluble. Metal solubility and bioavailability can be drastically reduced by a minute pH change. Owing to Mn's solubility, for instance, it can still form organic ligands or anionic complexes at alkaline pH levels [45]. However, the sensitivity of metal bioavailability to pH varies depending on the metal [46].

The division of soils into clay, silt, and sand fractions provides the clearest explanation of how soil morphology affects metal solubility in the soil. These terms are further defined by the soil's particle size fractions, which are divided into three categories: 2 μm, 2–50 μm, and >50 μm [47]. The clay fraction, which is primarily made up of clay minerals, is notable for having a high capacity to bind heavy metals. Heavy metals are, however, weakly absorbed by sandy soils due to their low sorption capacity and acidity, which causes them to move into groundwater [48]. According to Sharma *et al* [49], copper has high soil mobility and is, therefore, less absorbed by soil particles.

Organic matter plays a significant role in soils' ability to store heavy metals in a form that can be exchanged. In addition, organic matter adds organic chemicals to the soil solution that can act as chelates and improve plants' access to metals [50]. The functional groups that make up most of a soil's organic matter are primarily COOH or OH groups, which are high in oxygen and can act as exchange sites. Moreover, metal complexes are bound to the extremely reactive oxygen groups that hold the metals in place, which can support complicated reactions in organic materials [15].

When combined with heavy metals, organic substances produce simple or complex chelate compounds. They stop toxic substances from moving, rendering them immobile in the soil. It should be emphasized, though, that the quantity of active fixation points in soil humus is correlated with the ability of this medium to fix heavy metals. Since the precise nature of the fixation points for humus substances is unknown, determining changes in metal solubility in a particular soil environment under the influence of organic substances is a good indicator of the likelihood that plants will be able to absorb those substances [51].

The ability of metal ions to bind to abiotic surfaces and ligands such as soil organic matter is influenced by the pH value as well as other elements such as the soil's organic matter, metal ion content, and possibly the concentration of metal ions present. These elements may compete for the same binding site [15]. Fine particles in surface soils typically have higher concentrations of heavy metals due to their greater organic matter and clay contents, increased surface area, and presence of Mn oxide phases [52].

5.4 The toxicity of heavy metals

Heavy metals are considered to be pollutants because they are harmful to humans, animals, and plants. The ability of heavy metals to bioaccumulate in our bodies is their main issue. Bioaccumulation describes the process by which metals build up in specific tissues and do not naturally leave the body. Due to bioaccumulation, heavy metals are transferred from smaller species (such as fish) to humans as they move up the food chain. The liver, kidneys, bowel, brain, nervous system, spleen, and eyes are among the primary tissues that heavy metals attack [53].

The type of contamination caused by a metal may be neurotoxic, carcinogenic, mutagenic, teratogenic, toxic (acute, chronic, or subchronic), or any combination of these [54]. Heavy metals have been demonstrated to interact with nuclear proteins and DNA, causing site-specific harm. They are capable of causing both direct and indirect damage. In the case of 'direct' damage, the metal alters the structures of biomolecules. However, heavy metals can generate endogenous oxidants that cause 'indirect' harm, such as hydroxyl and superoxide radicals, hydrogen peroxide, nitric oxide, and other reactive oxygen and nitrogen species. Similarly, it has been observed that heavy metals stimulate signaling pathways [55].

As a result of metal toxicity, free radicals form; these damage DNA, alter sulfhydryl homeostasis, and cause lipid peroxidation. The redistribution of intracellular calcium stores was observed to be a nonlethal effect of high levels of inorganic mercury in kidney cells [55]. Many metals are produced using manganese as a purifying agent. Overexposure to manganese can cause pneumonia and harm to the central nervous system, among other symptoms. Rigidity of the trunk, stiffness, awkwardness of the limbs, tremors in the hands, weakness, weariness, bewilderment, hallucinations, an odd or awkward gait, muscular spasms (dystonia), and psychiatric disorders are additional signs and symptoms of manganese toxicity [56].

Lead is extremely toxic to living things. It has an impact on red blood cells, harms various organs such as the liver, kidneys, heart, and male gonads, and impairs the immune system [57]. When the kidneys and liver are unable to remove enough silver from the body due to a capacity overload, silver toxicity becomes apparent. Argyria is a condition in which there is an accumulation of inert silver sulfide and silver selenide precipitates in the connective tissue surrounding the glands and vascular tissue in the papillary layer of the dermis but not in the epidermis; it is caused by silver toxicity [58].

Excessive exposure to gold can cause skin rashes and yellowing, headaches, stomach and intestinal bleeding, vomiting, bone marrow depression, myokymia, mucous membrane irritation, and jaundice [56]. In laboratory animals, oral exposure to, or inhalation of cadmium causes kidney damage [8]. Reduced development, cancer, organ damage, nervous system impairment, and, in extreme situations, death are further impacts to animals [59].

5.5 Soil remediation

Soil remediation remains a major solution to the problem of soil contamination. It is the process of eliminating contaminants from the soil so as to purify and revitalize it [60]. In this way, the health of the soil and its inhabitants are protected. The ultimate outcome of any attempt to remediate soil is that the contaminants contained within are reduced in concentration to a safe level if not completely removed.

Whatever method is chosen to remove contaminants, it must be open and dynamic enough to qualify as 'sustainable' remediation [61]. This serves as a fundamental guide in decision-making. The goal is that there should be no associated risks for the ecosystem and human health due to the remediation process, but instead, it should provide short- and long-term solutions [61, 62].

5.5.1 Types of remediation

There are different types of soil remediation techniques and methods, but the choice of any method in any given situation is dependent on a number of factors. These include:

- **The soil matrix:** a major consideration in the choice of remediation technique is the composition of the soil, which includes its physical form (clay, sand, or silt) as well as its chemistry (pH level, acidity, salinity, mineral content).
- **The type of pollutant:** The interactions between some pollutants and the soil or materials used for remediation vary. Consequently, some remediation techniques are more effective at tackling certain contaminants than others, and so they may require a specific approach if effectiveness is to be maximized and safety ensured.
- **The concentration of pollutants.** The type of technique is usually reliant on the overall mass and concentration of the pollutant in the soil. This is so because some techniques favor low concentrations, while others can work with higher ones.
- **The goal of remediation:** The objective of remediating a particular soil or site will help in making a final decision. For example, soil that is to be used for agricultural purposes may require a more effective technique than soil that is intended to be used for construction purposes.

Consequently, two broad types of soil remediation are specified, depending on where treatment mechanisms are applied: *ex situ* and *in situ* treatments.

Ex situ: Here the soil of concern is excavated and treated outside the contaminated site [63]. This technique is preferred for immobilization or containment processes.

In situ: This is type of treatment in which the soil remains within the subsurface and pollutants are treated at the place of origin. This remediation technique generally provides a higher cost advantage than *ex situ* process remediation, as well as minimizing the contact between people and the contaminated medium. The choice of method is largely dependent on economic considerations, especially the size of the area in relation to the cost-benefit ratio. For instance, *in situ* processes of remediation are the only available option for the treatment of huge contaminated sites. This is because the operation is relatively simple with minimal disturbance to the site, and the cost is less than that of *ex situ* treatment. The disadvantage, however, is that *in situ* treatments may be slower and more complex to manage, and they are only successful at sites with permeable soil.

Soil remediation techniques can be further classified based on the procedure employed for remediation: chemical, biological, physical, thermal, or mixed methods [64]; they can also be divided according to their mode of action: containment, transport, and transformation [65, 66].

Chemical techniques are the preferred methods for soil treatment and contaminant removal. These techniques employ adsorption, retaliation, load exchange, oxidation, and reduction as their modes of action. Sometimes, the mechanism used can be a combination of two or more processes [63, 67]. Physical processes immobilize or withdraw contaminants from the soil using physical means. Surface capping, vaporization, and electrokinetics are among the technologies used. Biological processes use living things, such as plants, animals, or microorganisms, for decontamination. Their modes of action may include adsorption, transformation, or degradation [67]. Thermal processes can mobilize, volatilize, or destroy contaminants in the soil when the subsurface is heated either by conductive heating, electric resistance heating, radio-frequency heating, or steam heating [68]. The efficiency of these technologies may be enhanced if necessary by combining them. Mixed methods that combine physical, chemical, or biological technologies can solve problems that may arise from the use of a single technology [67].

It is important to note that there are certain conditions in which the soil contamination is overwhelming and an attempt to fully remediate it would either be impossible, impractical, or expensive [63]. In such cases, the pollutants can be neutralized and immobilized to prevent them from spreading beyond their boundaries. This is known as stabilization or containment. In principle, three main soil remediation technologies are in general use; these include soil washing, thermal desorption, and bioremediation. Water and surfactants are used in the process of washing soil to eliminate pollutants. In this process, pollutants are either dissolved or suspended in the wash solution to separate the soil by particle size [69]. Sometimes the pH of the wash water may be adjusted to aid in the removal of heavy metals and organics. In addition, chelating agents or basic leaching agents can be used to augment the process. This method has been successfully used for the remediation of hydrocarbon contamination. The process consists of two units that use heat to increase the volatility of contaminants so that they can be separated from the solid material, after which they are collected or destroyed. Contaminated solids are heated in the first unit until they reach the pollutants' boiling points. The heated particles are then pumped to the second stage of the process, in which the vapor is either eliminated by a thermal oxidizer or condensed in a vapor recovery unit. In practice, direct-fired plants are mostly used for low-level contamination; however, indirect-fired units are more commonly used because of their versatility and their ability to recapture the hydrocarbons.

5.6 Bioremediation

Bioremediation occurs naturally and involves the use of living microorganisms, such as bacteria and fungi, or other classes of naturally occurring biomaterials, due to their bioconcentrating and metabolic properties, to break down, sequester, or remove pollutants from the soil [70]. The contaminants are then either collected or destroyed without the use of any chemicals or catalysts. This process occurs only under ideal environmental conditions, that is, at the right temperature, pressure, pH, and moisture. Most studies of bioremediation technology have emphasized how biosorption and bioaccumulation are the main functional mechanisms of action. Both mechanisms have attracted enormous interest due to their high efficiency and affordability and the accessibility of biomass [71–73].

In general, the biosorption process is fueled primarily by microorganisms or dead biological materials, especially agricultural wastes, which constitute the majority of the sources studied. These nonliving materials interact with soluble chemicals in the environment. When this happens, the chemicals bind themselves to the cellular surfaces through various mechanisms [74]. This mechanism, which is an undirected physiochemical process, may include but is not limited to, ion exchange, surface complexation, microprecipitation, and electrostatic attraction [75, 76].

The majority of the studies of biosorption have focused on the elimination of the common toxic metals utilized in industrial processes. These include Cu, Cd, Pb, Cr, Zn, and Ni, which are found in most industrial effluents. The efficiency of certain plant and microbial biomasses in removing pollutants through absorptive mechanisms yields a cost-effective technology for effluent detoxification. In the case of organic pollutants, the nature of the organic compounds involved, such as their hydrophobicity or hydrophily, or their polar or nonpolar character, contributes largely to the success of the sorption process [60].

All in all, before biomass can be used for an industrial-scale biosorption process, it must possess sufficient binding capacity and selectivity [77, 78]. Although microorganism-derived biomasses are the best match for these conditions, agricultural wastes are the materials most commonly used for biosorption purposes. Agricultural waste materials consist of cellulose, hemicelluloses, and lignin, which contain functional groups needed for the binding and removal of pollutants. These functional groups may include carbonyls, phenolics, alcohols and acetamido, amido, amino, and sulfhydryl groups, which form chelates and metal complexes with heavy metal ions. Biomasses made of lignocellulosic materials possess natural sorption capabilities and have the benefits of being inexpensive, abundant, and renewable. Reports have described the use of recycled lignocelluloses to make fiber board [79]. A high percentage of cellulose is present in many agricultural wastes, increasing their potential for metal uptake. Consequently, many agro-wastes have been successfully used to treat contaminated soils [80]. Several studies have highlighted the need to enhance functional groups as a means to increase the number of adsorptive active sites. This is usually done using various pretreatment methods [81, 82].

5.6.1 Factors affecting the biosorption process

When considering the potential of biosorbents, it is important to analyze the factors that influence the biosorbent process. Some of these factors include the class and nature of the biomass, the concentration or dosage of biomass or biosorbent, the initial solute concentration, the equilibrium time between the two phases and the contact time; other considerations include physicochemical factors such as the pH, temperature, and ionic strength [83].

At certain intervals, the temperature can have a negative or positive impact on biosorption, as it is directly connected to the kinetic energy of the metal ions [84]. The process of diffusion can be explained in terms of temperature, which also affects the thermodynamics of biosorption. Thus, a change in temperature greatly affects the quantity of metal absorbed by a biosorbent [84]. The solution's pH influences metal ion chemical speciation and solubility, the biosorbent functionality and functional groups, and the performance of charges on the active sites of the biosorbent surface. Metal ions, being positively charged, attract negatively charged biosorbents in acidic conditions indicated by low pH values, whereas an increase in pH causes metal ions to react with OH^- ions and precipitate at higher pH values.

Biosorption capacity is most important in dilute solutions, given that many pollutants may be present at low concentrations. In general, there is an increase in biosorption capacity at lower concentrations. For solutions in which several metallic species are present, there is competition among individual metals to occupy active sites.

5.6.2 The preparation of biosorbents

In the preparation of biosorbents from plants, it is important to note that some plants or their biomass may contain many toxic compounds that can be harmful or can trigger poisonous effects in humans [85]. As a result, before it is processed, the material to be used as a biosorbent must be detoxified. Typically, four main steps are used to prepare plant samples as raw biomass [73]. To begin with, plant biomasses such as leaves, stems, flowers, bark, roots, or entire plants are collected, sorted, and washed to remove dirt. Second, the cleaned biomass is commonly air-dried, sun-dried, or dried in an oven at temperatures ranging between 50 °C and 105 °C to remove moisture. The third step is size reduction, which may be done by chopping or grinding using manual methods or a hammer mill before sieving the material to obtain small particles using selected mesh sizes. The final step, which is chemical pretreatment, is optional, as many reported studies use the sieved biomass as is. Studies have reported that the use of the additional step enhances the material's adsorption capacity.

5.6.2.1 Cleaning the biomass

The biomass is generally washed with distilled water at room temperature before it is dried [86]. The cleaning of biomass by washing not only removes impurities and pathogens or their carriers but also removes inorganic contaminants that may interfere with the adsorptive capacity of the biomass [73]. These inorganics may be extraneous, such as nonmetals, alkali metals, alkaline earth metals, and transition metals [87], or they may be indigenous, such as insects' vectors [88] and fungal pathogens. Sometimes, such elements are present in negligible proportions; however, they can lower the biosorption capacity of raw biomass even in their small concentrations [89]. Regardless of how cleaning is done, it has an important role in the fabrication of biosorbents [91].

5.6.2.2 The size reduction of biomass

The efficiency of biomass in removing contaminants is a function of the surface area available for adsorption [90]. Size reduction further shortens the diffusion path by increasing the interaction between the pollutants and the biosorbent [91]. In most processes, the raw biomass is chopped or broken, followed by the use of a hammer mill [92], a mixer [93], or a blender [94]. This is closely followed by sieving through the use of a mesh of an appropriate size. Size reduction is a step that cannot be avoided in the preparation of biosorbents, even though it requires a significant quantity of resources and costs in large-scale investigations.

5.6.2.3 Chemical pretreatment

Many plant biomasses are composed of cellulose, hemicellulose, and lignin, which are made up of several complex bonds and functional groups [95]. These intrinsic functionalities may lead to complex behaviors, so there is a need for chemical treatment in most cases. Chemical pretreatments aim to increase the material's surface area and porosity by disrupting the lignocellulosic biomass. These adjustments help to optimize the adsorption capacity of the functionalized biosorbents for contaminant removal. The chemicals most explored and recorded for the purpose of pretreatment include acids [71], bases [96], and oxidizing agents [97]. The treatment of raw biomass with diluted acid usually leads to its hydrolysis, as acids may break down the interchain bonds of hemicellulose [85], while basic solutions facilitate a reduction in the lignin content [98]. While the costs of chemicals and facilities, as well as further drying processes, are disadvantages of chemical pretreatments [99], they enhance the pollution elimination efficiency of biomass by reducing the soluble organic material. According to Saxena *et al* [100], there are several steps in the general preparation method for biosorbents, as shown in figure 5.1.

5.6.3 The mechanism of biosorbent action in the sequestration of heavy metals

Heavy metals are bound chemically or physically to ligands on the biosorbent during the biosorption process. The organism's metabolism may or may not be affected in the case of living biomass. Ion exchange, along with surface complexation, microprecipitation, diffusion, etc, are the main mechanisms of action of the biosorbents used for metal biosorption. Through these various methods, heavy metals are absorbed with the aid of various functional groups found on the biosorbents' cell walls or surface. The sequestration of metals by biosorbents involves a number of functional groups, including carboxyl, hydroxyl, sulfate, phosphoryl, sulfhydryl, amino amine, amide, imidazole, thioether, sulfonate, phosphodiester, phenolic, imine, and phosphate groups [101–103].

Regarding the various biological materials (biomasses) employed as sorbents, which can be living or dead, the biosorbents' modes of action during biosorption can either be metabolism-dependent or non-metabolism-dependent [104].

5.6.4 Metabolism-dependent biosorption

This activity, which only takes place in living cells, is essential to their defense mechanisms and exhibits sensitivities to hazardous metals [104]. The addition of the biomass's metabolic activity to the passive (non-metabolism-dependent) mechanisms makes the process more complicated (figure 5.1). This metabolic activity enables the active transport of heavy metals across the cell membrane and into the interior of the cell. Thus, heavy metals may assemble within the cell (bioaccumulation). In addition,

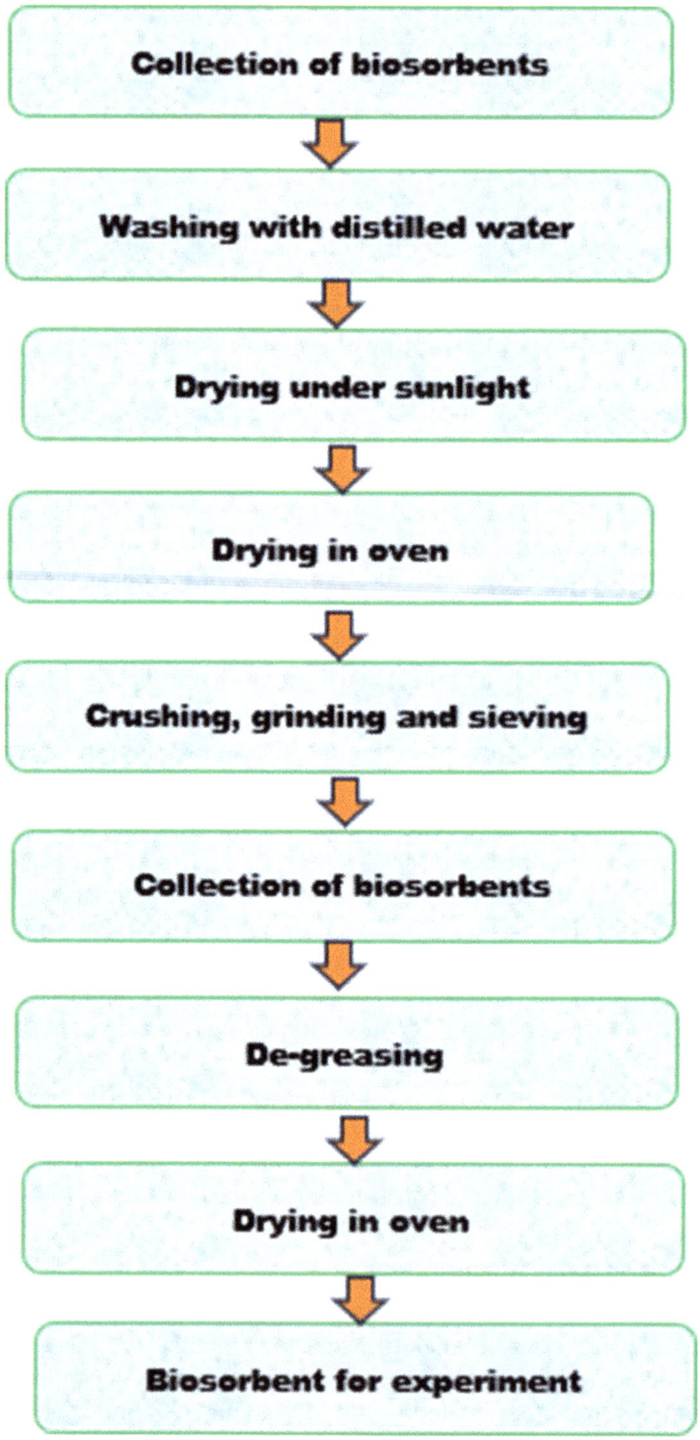

Figure 5.1. The general preparation method for biosorbents.

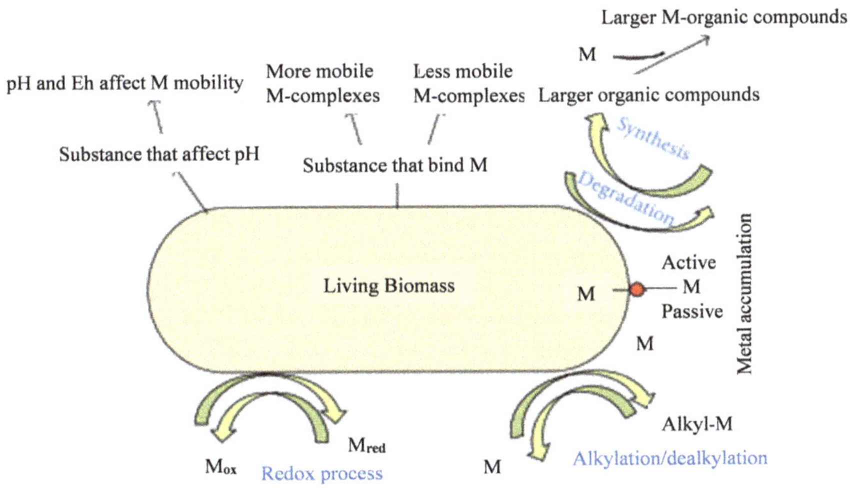

Figure 5.2. Metal accumulation process by microorganisms and its influence on metal mobility. Reproduced from [109]. CC BY 4.0.

because living biomass exhibits enzymatic activity, it is possible for multiple enzymatic processes to change the condition of the pollutant (biodegradation and biotransformation), which could lead to precipitation (metabolism-dependent precipitation) [105, 106]. Metabolism-dependent precipitation is usually linked to the protective mechanism of living cells, in which their reaction to the existence of dangerous metals produces chemicals that facilitate precipitation. As mentioned earlier, the cell wall surfaces of living biomasses contain several functional groups (carboxyls, phosphates, sulfates, etc.) because the cell wall's is composed of polysaccharides, proteins, and lipids. The types of heavy metal ions, the medium, the makeup of the cell walls, and other factors can also affect how much metal live biomass can absorb [107]. When living biomass absorbs substances, those substances adhere to the cell walls and enter the cytoplasm. If the attachment of the metal to the cell wall depends on metabolism, adenosine triphosphate (ATP) is required in order to provide energy [107, 108]. One advantage of active systems is their ability to perform biotransformation or biodegradation, which can improve their capacity to destroy a larger quantity of pollutants in some situations. The degradation of heavy metals by active cells is important because it offers a practical and affordable way to get rid of hazardous substances, especially environmental pollutants [109–114]. Figure 5.2 shows the process of metal accumulation by microorganisms and its influence on metal mobility [109].

5.6.5 Non-metabolism-dependent biosorption

This process usually takes place in biosorbents made from dead cell biomass or processed from such biomass. It involves an interaction between metals and the functional groups present on the biosorbents' surface. The contaminants bind to the biosorbent passively (without metabolic activity) via ionic, chemical, or physical means [109]. The following are the main elements governing and defining these mechanisms: the biosorbent's structure and nature (in the case of microorganisms, living or nonliving); the chemical and coordination properties of the metal ions; the kind of binding site (biological ligand); the process variables, such as the pH, the temperature, and the concentrations of the sorbate, the sorbent, and other competing metal ions; and the availability and accessibility of the binding site [70, 101, 115]. Reversible adsorption–desorption reaches a dynamic equilibrium in this situation, and it is not reliant on energy [116]. The non-metabolic-dependent mode of action of biosorbents is achieved through physical adsorption, ion exchange, complexation, and precipitation (figure 5.2).

Physical adsorption: In this process, the molecules or metals from a gas, liquid, or dissolved solid are adsorbed and create a film over the solid surface of the absorbent or biosorbent, which is the mode of action of biosorbents. This is a physical occurrence that is influenced by electrostatic and van der Waals forces. The physical adsorption mechanism is influenced by the biosorbent's surface area and, in certain cases, the pH of the solution [117, 118]. The two subtypes of this process are physisorption and chemisorption. In contrast to chemisorption, which is characterized by covalent bonding, adhesion in physisorption is caused by weak van der Waals forces or electrostatic attraction. Through electrostatic interactions, dead biomasses of algae, fungi, and yeasts have been shown to adsorb heavy metals such as copper, uranium, cobalt, zinc, and cadmium [104].

Ion exchange: Ion exchange is the systematic replacement of an ion that is readily exchangeable yet bonded to the solid phase by another ion in solution. As a result of an exchange of metal ions for the counterions of the polysaccharides that make up the cell walls of the biosorbent, metal is absorbed [119]. Since cationic species are among the most prevalent forms of most metals in aqueous solutions, the more negative the charge of the biosorbent, the more metal it absorbs. Microbial cell walls' polysaccharides serve as counterions and promote the exchange of bivalent metal ions. The alginates of marine algae, which are found as salts of Mg^{2+}, K^{+}, Na^{+}, and Ca^{2+}, can bind heavy metals by a process of exchange with counterions such as Co^{2+}, Cu^{2+}, Cd^{2+}, and Zn^{2+} [104, 120]. It has been discovered that *Ganoderma lucidum* biosorbs copper ions via an ion exchange mechanism, which is similar to the biosorption that occurs in microorganisms such as bacteria and fungi. The composition of microorganisms' cell walls makes this possible, as they contain polysaccharides that can switch ions with their counterions [121]. According to a study, rice straws can act as a non-metabolism-dependent biosorbent through the ion exchange of Na, K, Mg, and Ca to biosorb Cd^{2+} ions [101]. A cyanobacterium called spirulina has shown the capability to biosorb metals including chromium (III) (Cr^{3+}), Cd^{2+}, and Cu^{2+} via ion exchange by means of functional groups such as carboxyls, phosphates, and hydroxyls [101].

Complexation: Here, metal ions and organic molecules acting as ligands with the ability to donate electrons create complexes by electrostatic attraction or covalent bonding. The more complex process of complexation known as chelation involves the simultaneous coordination of numerous bonds between the organic ligand and the metal ion to produce a more stable complex [115, 118]. The interaction between the metals and active groups increases their removal from the solution by forming complexes on the cell surface. By solubilizing and leaching hazardous metals, organic acids made by active systems also play a crucial role in chelating metals, leading to the production of metallo-organic compounds. Carboxyl groups in living cell polymers and polysaccharides cause metal complexation and adsorption. According to Cimá-Mukul *et al* [123], polyphenols with phenolic hydroxyl ions can chelate metal ions. Plants such as *Leucaena leucocephala*, *Sagittaria trifolia*, *Prosopis juliflora*, etc. may contain polyphenols that can be extracted. Cd(II) and Pb(II) have been successfully extracted from an aqueous medium using such extracts. The hard and soft acids and bases (HSAB) theory postulates that elements are divided into metals and nonmetals, which result in hard or soft acids and hard or soft bases, respectively. It also notes that hard acids have a stronger affinity for hard bases than weak acids and bases. The affinity between the metal ion and the organic ligand, which is the most important factor in complexation, is based on this theory [124]. Borderline ions such as manganese, cadmium, copper, and zinc ions also have a high affinity to form complexes with nitrogen- or sulfur-containing organic compounds. However, lead, which forms a soft acid, would preferentially bind covalently to the organic ligands of biosorbents containing nitrogen or sulfur, which are soft bases [115]. Nevertheless, complexation is influenced by the accessibility and availability of the binding site that contains the donor atom (base).

Precipitation: One of the few mechanisms that contribute to metabolism-dependent biosorption is precipitation; however, precipitation can also lead to metabolism-independent biosorption. Precipitation refers to the development of insoluble metal in the form of a precipitate. When microorganisms are exposed to a dangerous metal ion environment, precipitation occurs in metabolism-dependent biosorption as a consequence of their active defence mechanism [121]. In metabolism-independent biosorption, precipitation results from chemical reactions between the functional groups of the biosorbent's cell wall and the metal ion; these reactions may involve oxidation–reduction processes [101, 104]. According to a study, the two major mechanisms by which iron (III) ions are removed from aqueous solutions with the help of green tomato husks are precipitation and ion exchange. At a pH of 6.0, iron (III) ions precipitate as iron (III) hydroxide [125]. Figure 5.3 shows the classification of the modes of interaction of biosorbents with heavy-metal-polluted media [118].

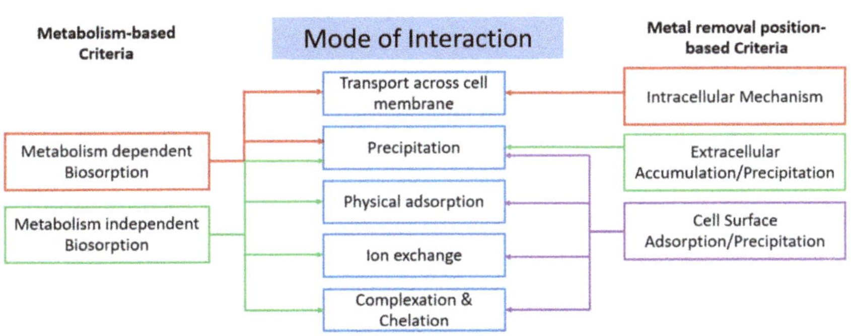

Figure 5.3. Classification of the modes of interaction of biosorbents with heavy-metal-polluted media. Reproduced from [118]. CC BY 4.0.

Location-dependent biosorption can be classified according to the location of the removed metals in the solution, as shown in figure 5.3. The resulting categories are extracellular accumulation or precipitation, intracellular accumulation, and cell surface sorption. These can be further divided in a similar way to the division of metabolism-dependent and independent biosorption [104].

5.6.6 Factors affecting the different types of biosorbents in the sequestration of heavy metals

In the sequestration of heavy metals, many variables may have an impact on various types of biosorbents. These include the metal, the biomass, and the environmental conditions [126, 127]. The following are the main factors that influence biosorbents in the process of sequestering heavy metals (biosorption):

(1) **pH**

The most crucial factor to take into account during biosorption processes is the pH. The biosorption process is comparable to but distinct from ion exchange, and biomass can be thought of as a naturally occurring ion exchange medium with a predominance of weakly acidic and basic groups. The pH of the solution can have an effect on some biosorption systems, including ion exchange [117]. Therefore, the nature of the biomass binding sites and the solubility of the metals can both be determined by the pH of the solution. The potential activity of the functional groups in the biomass and the competition between metallic ions can both be impacted by it, as can the solution chemistry of the metals. As a result, metal biosorption has typically shown strong pH dependence in the systems that have been studied. Possible competition between cations and protons for binding sites, which is frequently reduced at low pH values, is eliminated by the biosorption of metals such as Cu, Cd, Ni, Co, and Zn [128, 129]. In general, when the pH of the metal solution is reduced from pH 6.0 to pH 2.5, the uptake of heavy metals by different biomass types drops quickly. At pH 2 or lower, metal ion removal from such solutions is extremely poor or negligible. As the pH rises from roughly 3.0–5.0, the metal intake is enhanced. Further increases in the pH value result in a reduction in capacity during the process, so this pH range must be optimal. At low pH, ion exchange has been used to achieve the biosorption of lead ions by Hami melon peels; however, at higher pH, electrostatic interactions between carboxylate and hydroxylate groups on the biomass and lead ions were used to achieve the same results [117]. Using specific functional groups, such as carboxyl, phosphate, and hydroxyl groups, spirulina (a biomass of cyanobacteria) was able to biosorb Cr^{3+}, Cd^{2+}, and Cu^{2+} by ion exchange [101].

(2) **Contact time**

Contact time is the term used to describe the period of time allowed for the biosorption process to take place. The amount of time that the biosorbent is in contact with the sorbate does not directly affect the biosorption capacity, but it can be a limiting factor. Increasing the contact period enables the biosorbent material to reveal its maximum biosorption capacity under test conditions. After the biosorbent reaches its optimum biosorption capacity under the specified conditions, extending the contact time would have no extra effect because its binding sites would already be fully saturated [131].

(3) **Initial metal ion concentration**

The effects of increasing the initial metal concentration and contact time are identical. The full maximum biosorption capacity of the biosorbent would not be revealed by using a low metal concentration, since many metal-binding sites might not be filled. The full capability for metal removal would be revealed by increasing the metal concentration until the binding sites of the biosorbent were fully saturated. Increasing the initial metal concentration after the biosorbent is saturated does not help [118]. A significant amount of force is required during initial concentration to overcome all of the metal's mass transfer resistance between the solid and aqueous phases [132]. Because the amount of metal absorbed by the biomass will almost certainly rise with the original concentration of metals, an optimal percentage of metal removal can be approximated for a low starting metal concentration [133]. Therefore, for a given biomass concentration, the likelihood of metal uptake rises as the starting concentration increases [108]. A study examining the potential of banana peel as a biosorbent for heavy metals discovered that as the initial metal concentration rose, so did the metal biosorption [134]. The initial metal concentration employed in simulated conditions ranged from 5–200 mg l^{-1} in the majority of published studies [118, 127].

(4) **Environment**

Biosorption efficiency is unaffected between 20 °C and 35 °C, in contrast to the bioaccumulation process, but at high temperatures (such as 50 °C), biosorption often increases; however, these high temperatures can also permanently destroy microbial living cells, which results in a reduction in metal uptake [127]. This is not the case for dead biomass [135–137]. The extent of adsorption undoubtedly increases with decreasing temperature, since adsorption reactions are exothermic in nature. When the temperature was raised to 40 °C, *Saccharomyces cerevisiae*'s biosorption ability for Ni and Pb reached its maximum and then began to gradually decline [138]. This temperature effect is less pronounced with dead biomass than with living biomass. The amount of biosorbed metal increases more noticeably with living biomass than with dead biomass as the temperature rises [139, 140]. The metal is more thoroughly integrated into the interior of the cell as a result of the increased metabolic activity of live cells that occurs when the temperature rises to an ideal level.

A decrease in ionic strength results in reduced metal biosorption because other cations compete with the metal for binding sites on functional groups [141]. This is a significant problem when using biosorbents on actual effluents, which frequently include complicated mixtures of several cations [118].

(5) **The nature of the biosorbent**
One of the most important factors, in addition to the nature of the biosorbent or biomass, is the type of application used, such as immobilized preparations, freely suspended cells, living biofilms, and so on. The number of vacant sites on the surface of the adsorbent can be increased by breaking up solid crystals into tiny pieces, heating charcoal to an elevated temperature, powdering solid lumps, or using other methods suitable for a specific adsorbent [109]. The binding qualities are impacted by physical treatments such as drying, boiling, autoclaving, and mechanical disruption, but the biosorption capacity is improved by chemical treatments such as alkali or acid treatments. According to infrared spectral data and potentiometric titrations, the surface of a treated biosorbent changed from positively charged to negatively charged as a result of the dissociation of weakly acidic oxygen-containing functional groups that contribute to biosorption, leading to a greater biosorption capacity [144].

(6) **Biomass dosage**
The selective absorption of metal ions has been seen to increase at low biomass doses [121, 143]. However, the use of a low biomass dosage in a medium with complex contamination might increase competition for the biosorbent's binding sites and reduce its biosorption capacity. The competition between metal ions to bind to the functional groups would be reduced if there were more than one metal ion present.

(7) **The affinity of the metal for biosorbents**
Both the surface charges and the permeability of the biomass are affected by pretreatment, either physical or chemical, which opens up the metal-binding groups in preparation for binding. This can be changed by pretreating the biomass with alkalis, acids, detergents, or heat, which can increase the amount of potential metal uptake [127, 135]. The first significant challenge in the biosorption process may be the selection of the best biomass or biosorbent from the wide range of readily available, less costly biomaterials. Although many substances of biological origin have the ability to bind heavy metals, only biomaterials with a high enough metal-binding capacity and selectivity for heavy metals are relevant for large-scale biosorption processes [77].

5.6.7 Advantages (the strengths of various biosorbents)

The following are some general benefits of using biosorbents in the biosorption process over traditional heavy metal removal methods [145]:

- They are readily available renewable biomaterials with low production and acquisition costs.
- Because of their rapid kinetics, they are effective at treating large volumes of wastewater.
- They have high selectivity for the recovery and elimination of specific heavy metals.

- They are capable of handling mixed wastes that contain multiple heavy metals and can often reduce the residual metal concentration to less than one part per billion (ppb).
- They reduce the necessity for additional expensive reagents, which usually cause disposal and space issues.
- They significantly decrease the production of hazardous waste;
- They operate in a wide variety of physiochemical environments, such as temperature, pH, and the presence of other ions.
- They greatly increase the recovery of bound heavy metals from the biomass.

Plant biosorbents minimize chemical and/or biological sludge [146], and their use is generally compatible with sustainable development since biosorbents can be regenerated by desorption. In addition, some metals can also be recovered for further industrial use. Utilizing dead biomass provides many benefits when compared with the use of living microorganisms [147, 148], which can be summarized as follows:

- There is no need for culture media and nutrients to be present in the feed solution.
- They have better storage capacities and can be used for longer periods.
- They do not impose toxicity constraints.
- They offer the possibility of the reuse and recovery of the sorbed contaminants and the saturated biosorbent, respectively.
- The mathematical and statistical modeling of pollutant uptake is simplified.

As already mentioned, dead biomass has advantages over living microbes. Many of the characteristics of living microbes, however, are still underutilized in industrial settings; they are all deserving of additional study, since they might be useful in certain applications. Excellent sources of biosorbents obtained from the natural environment include bacteria, fungi, yeast, and algae [77]. Due to their metal-sequestering properties, these biosorbents can be utilized to lower the concentration of heavy metal ions in solution from the parts per million (ppm) level to the ppb level. They can efficiently remove dissolved metal ions from dilute, complicated solutions with high concentrations.

The most significant biosorbents of microbial origin can be generally divided into the following groups:

5.6.7.1 Bacteria

The most numerous and adaptable microbes, bacteria, make up a sizeable fraction of the total live terrestrial biomass, which weighs around 1018 g [149]. Bacteria have a clear morphology. Although they have is a vast diversity of shapes due to genetic and ecological variations, bacteria are most frequently found in three fundamental shapes: ovoid or spherical (cocci), rod (bacilli, which have a cylindrical shape), and spiral (spirilla). Bacteria come in a variety of shapes and sizes [148]. Among different bacterial species, there is a great degree of heterogeneity in the quantity of surface binding sites, the strength of the binding for different ions, and the binding strategies. Bacteria readily produce amino, carboxyl, phosphate, and sulfate groups for metal biosorption due to their polysaccharide slime layers [149]. Bacterial biomass is often produced as a waste by-product of industrial operations, but it can also be intentionally generated on a large scale [150, 151]. Bacteria are used as biosorbents due to their small size, ability to thrive in controlled environments, and tolerance to a wide range of variable environmental conditions [152–154]. Various bacterial genera, including *Bacillus*, *Pseudomonas*, *Streptomyces*, *Escherichia*, and *Micrococcus*, have been examined for the uptake of metals or organics. The system by which heavy metals bind to the surfaces of bacterial cell walls typically involves two steps. The interaction of metal ions with reactive groups on the cell surface takes place in the first stage, and the deposition of further metal species at higher concentrations occurs in the second stage [155].

5.6.7.2 Algae

Algae are considered to be one of the most promising types of biosorbents. They are of interest in the search for and development of new biosorbent materials since they have a high sorption capacity and are readily available in huge quantities in the seas and oceans [156–161]. Algae create a significant amount of biomass and, in contrast to other biomass and microorganisms such as bacteria and fungi, have low nutritional requirements. They do not typically produce any harmful compounds. Three different kinds of algae are recognized: microalgae (also known as freshwater or green algae), macroalgae (also known as brown or marine algae), and red algae. According to Abbas *et al* [116], the ionic charge of the metal ion, the species of algae, and the chemical makeup of the metal ion solution are all factors that affect how well metal ions bind to algal surfaces. Amine, carboxyl, sulfate, imidazole, sulfhydryl, phosphate, hydroxyl, and chemical functional groups found in cell sugar and cell proteins are examples of potential metal-cation-binding sites of algal cell components [162]. The same biosorption processes used by other microbial biomasses can be used by microalgae to sequester heavy metal ions, and they can also do so by producing phytochelatins in response to hazardous heavy metal stress. In microalgae biosorption, there are two main mechanisms at work: ion exchange, in which heavy metal ions displace lighter ions such as Na, Mg, and Ca; and (ii) complexation between metal ions and different functional groups such as carboxyls, aminos, thiols, hydroxyls, and phosphates, which can interact with heavy metal ions in a coordinated manner. The process of metal biosorption involves both covalent bonding and ionic charge. The main substances known to be involved in biosorption are proteins and polysaccharides. Given the carboxyl and sulfate groups connected to these components, covalent bonding and ionic charge bonding may be anticipated [103].

5.6.7.3 Fungi

One of the biomasses used in industrial fermentation waste that produces particularly good metal sorbents is fungi. Thus, interest in fungi, especially yeasts, has increased. According to Kiran and Abbas *et al* [103], fungi are effective and affordable in the biosorption of heavy contaminated metals from dilute aqueous solutions due to (i) their high percentage of cell wall materials; (ii) their excellent metal-binding properties; (iii) the fact that they are widely accessible as by-products of the food and antibiotic industries; and (iv) the fact that they offer an environmentally friendly solution.

Molds, yeasts, and mushrooms are three kinds of fungi that are very important in practical terms. Metallic elements have been bound by filamentous fungi and yeasts in numerous wastewater treatment processes. Fungi play a significant role in industrial processes and can be found only in natural habitats. Fungi have a wide variety of morphologies, from filamentous and polymorphic fungi to unicellular yeasts, and many of them have intricate macroscopic fruiting bodies.

Although the active mode, particularly for filamentous fungi, may not considerably add to metal uptake at high metal concentrations, it can greatly contribute to metal elimination for yeast. Intracellular uptake has a complicated and poorly understood mechanism. Chitins, glucans, mannans, and proteins make up the majority of the complex macromolecular structures found in fungi, but fungi also contain additional polysaccharides, lipids, and colors such as melanin [163–165].

The cell wall has been identified as the main location of heavy metal ion sequestration by fungal biomass. The majority of the fungal cell wall is made up of several kinds of polysaccharides. Carboxyl groups on uranic acids, proteins, and phosphate groups, nitrogen-containing ligands on proteins, as well as on chitin or chitosan, all have an impact on the metal uptake capacity in the fungal cell wall [166]. A greater quantity of heavy metals amasses in yeasts through the bioaccumulation process than through the biosorption process. However, the main mode of heavy metal uptake by many filamentous fungi is generic biosorption.

5.6.8 Disadvantages (limitations of biosorbents for heavy metal sequestration)

One of the drawbacks of using biosorbents is early experimental saturation and lack of control over the biosorbent's biological characteristics. Powdered nonliving biomass has some limitations that make it challenging to utilize in continuous and batch systems, including the mass loss during regeneration, difficulty in removing the biomass from the reaction system, poor mechanical strength, and reduced particle size [167]. Moreover, it can result in redox interactions between cells and the medium, raising the pH of the system. Also, because this method necessitates nutritional absorption, the biological and chemical demands for oxygen increase [104, 163].

The elimination of one or two metals via biosorption has been the subject of numerous investigations; however, the conclusions drawn from these studies may not be valid in practical settings, in which contaminated media may contain a variety of heavy metals. Organic pollutants may really have an impact on the nature of metal ions and their biosorption process [118].

5.7 Conclusions, recommendations, and the way forward

It is challenging to impose restrictions on studies that attempt to identify the qualities of any substance that is intended to be used for biosorption. The obvious explanation for this is the abundance of available biosorbent materials, both living and dead. These investigations must go further, because biosorption will not be competitive in the absence of a substance with the right properties. But in order to accept that another novel material is a better option, it is also necessary to compare its properties to those of other potential materials that have already been established as sorbents (i.e. commercial sorbents). For biosorbents to be fully accepted, further action is required [169]. It is advisable for biosorption investigations to include a phase that investigates the applicability of laboratory results to actual polluted conditions. Because there are so many other environmental elements that can alter the pH, it would be difficult to optimize the pH in real-world settings. As a result, the biosorption capabilities of biosorbents that were established for certain pH values under laboratory conditions might not be achievable in real-world settings.

References and further reading

[1] Kabata-Pendias A and Pendias H 2001 *Trace Metals in Soils and Plants* 2nd edn (Boca Raton, FL: CRC Press)

[2] Duffus J H 2002 Heavy metals-a meaningless term *Pure Appl. Chem.* **74** 793–807

[3] Nagajyoti P C, Lee K D and Sreekanth T V M 2008 Heavy metals, occurrence and toxicity for plants: a review *Environ. Chem. Lett.* **8** 199–216

[4] Alloway B J and Jackson A P 1991 The behaviour of heavy metals in sewage sludge amended soils *Sci. Total Environ.* **100** 151–76

[5] Nies D H 1999 Microbial heavy metal resistance *Appl. Environ. Microbiol.* **51** 730–50

[6] Bruins M R, Kapil S and Oehme F W 2000 Microbial resistance to metals in the environment *Ecotoxicol. Environ. Saf.* **45** 198–207

[7] Jarup L 2003 Hazards of heavy metal contamination *British Med. Bull.* **68** 167–82

[8] Briffa J, Sinagra E and Blundell R 2020 Heavy metal pollution in the environment and their toxicological effects on humans *Heliyon* **6** E04691

[9] Moore F, Dehbandi R, Keshavarzi B and Amjadian K 2016 Potentially toxic elements in the soil and two indigenous plant species in Dashkasan epithermal gold mining area, West Iran *Environ. Earth Sci.* **75** 268

[10] Singh B R and Steinnes E 1994 Soil and water contamination by heavy metals. in soil processes and water quality *Advances in Soil Science* ed R Lal and B A Stewart (Boca Raton, FL: Lewis Publishers, CRC Press, Inc.) pp 233–70

[11] Taylor S R 1964 Abundance of chemical elements in the continental crust *Geochem. Cosmochem. Acta* **28** 1273–85

[12] Kabata-Penadias A 1993 Behavioural properties of trace metals in soil *Appl. Geochem Suppl.* **2** 3–9

[13] Kabata-Pendias A 2004 Soil-plant transfer of trace elements—an environmental issue *Geoderma* **122** 143–9

[14] Allaway B J 1990 *Heavy Metals in Soils* (Glasgow, Scotland/New York: Blackie/Wiley)

[15] Brady N C and Weil R R 2002 The nature and properties of soils 13th edn (Upper Saddle River, NJ: Prentice-Hall)

[16] Nriagu J O 1989 A global assessment of natural sources of atmospheric trace metals *Nature* **338** 47–9

[17] He Z L, Yang X E and Stoffella P J 2005 Trace elements in agroecosystems and impacts on the environment *J. Trace Elements Med. Biol.* **19** 125–40

[18] Wei B and Yang L 2010 A review of heavy metal contaminations in urban soils, urban road dusts and agricultural soils from China *Microchem. J.* **94** 99–107

[19] Elias M S, Wo Y M, Hamzah M S, Shukor S A, Rahman S A, Salim N A A, Azman A M and Hashim A 2017 Analysis of soil samples from Gebeng area using NAA technique *Advancing Nuclear Science and Engineering for Sustainable Nuclear Energy Knowledge AIP Conf. Proc.* **1799** 030008-1–10

[20] Interstate Technology; Regulatory Council 2003 *Characterization and remediation of soils at closed small arms firing ranges* (Washington, DC: Interstate Technology and Regulatory Council, Small Arms Firing Range Team) pp 1–2

[21] Stauffer M, Alvarado C and Pignolet, A. J 2017 Persistent mercury in shooting range soils: the legacy from former primers *Bull. Environ. Toxicol.* **98** 14–21

[22] Staszewski T, Malawska M, Studnik-Wójcikowska B, Galera H and Wilkomirski B 2015 Soil and plants contamination with selected heavy metals in the area of a railway junction *Arch. Environ. Protect.* **41** 35–42

[23] Newman M C and Jagoe R H 1998 Allozymes reflect the population-level effect of mercury: simulations of the mosquitofish (*Gambusia holbrooki* Girard) GPI-2 response *Ecotoxicology* **7** 141–50

[24] Adriano D C 1986 *Trace Elements in the Terrestrial Environment* (New York: Springer) 105–23 pp533

[25] Fernandes J C and Henriques F S 1991 Biochemical, physiological and structural effects of excess copper in plants *Bot. Rev.* **57** 246–73

[26] Hund-Rinke K and Korrdel W 2012 Ansaetzezur Bewertung der Verfuegbarkeit von Schadstoffenimnachsorgenden Bodenschutz—Teil III: Verfuegbarkeit/Bioverfuegbarkeit von Schadstoffenfuer Pflanzenund Bodenorganismen *Bodenchutz* **4** 113–9

[27] Templeton D M, Ariese F, Cornelis R, Danielsson L G, Muntau H, van Leeuwen H P and Lobinski R 2000 Guidelines for terms related to chemical speciation and fractionation of elements. Definitions, structural aspects, and methodological approaches (IUPAC Recommendations 2000) *Pure Appl. Chem.* **72** 1453–70

[28] Jones B and Turki A 1997 Distribution and speciation of heavy metals in surficial sediments from the Tees Estuary, North-East England *Mar. Pollut. Bull.* **34** 768–79

[29] Huang P M, Violante A, Bollag J M and Vityakon P 2005 *Soil Abiotic and Biotic Interactions and the Impact on the Ecosystem and Human Welfare* (Plymouth: Science Publishers) p 439

[30] Bennett B G 1982 Exposure of man to environmental nickel: an exposure commitment assessment *Sci. Total Environ.* **22** 203

[31] Rieuwerts J S, Thornton I, Farago M E and Ashmore M R 1998 Factors influencing metal bioavailability in soils: preliminary investigations for the development of a critical loads approach for metals *Chem. Spec. Bioavail.* **10** 61–75

[32] Sun Q, Xu D, Shen Z, Dou C, Dou Z, Li Y and Gao Y 2022 Effects of soil properties on heavy metal bioavailability and accumulation in crop grains under different farmland use patterns *Sci. Rep.* **12** 9211

[33] Bruemmer G W, Gerth J and Herms U 1986 Heavy metal species, mobility and availability in soils *J. Plant Nutr. Soil Sci.* **149** 382–98

[34] ISO 17402 2008 *Soil Quality—Requirements and Guidance for the Selection and Application of Methods for the Assessment of Bioavailability of Contaminants in Soil and Soil Materials* (Geneva, Switzerland: ISO) https://www.iso.org/standard/38349.html

[35] Kim R-Y, Yoon J-K, Kim T-S, Yang J E, Owens G and Kim K-R 2015 Bioavailability of heavy metals in soils: definitions and practical implementation—a critical review *Environ. Geochem. Health* **37** 1041–61

[36] ISO/FDIS 16198 2015 *Soil Quality—Plant-Based Test to Assess the Environmental Bioavailability of Trace Elements to Plants* (Geneva, Switzerland: ISO) https://www.iso.-org/standard/55834.html

[37] Wang C X, Mo Z, Wang H, Wang Z J and Cao Z H 2003 The transport, time-dependent distribution of heavy metals in paddy crops *Chemosphere* **50** 717–23

[38] Fischer G, Tubiello F N, van Velthuizen H and Wiberg D A 2007 Climate change impacts on irrigation water requirements: Effects of mitigation, 1990-2080 *Technol. Forecast. Soc. Change* **74** 1083–1107

[39] Blume H-P, Bruemmer G W, Horn R, Kandeler E, Koegel-Knabner I and Kretzschmar R *et al* 2010 *Scheffer/Schachtschabel—Lehrbuch der Bodenkunde* (Heidelberg, Berlin: Springer Spektrum) 16th edn

[40] Hodson M E, Vijver M G and Peijnenburg W J G M 2011 Bioavailability in soils *Dealing with Contaminated Sites: from Theory Toward Practical Application* ed F A Swartjes (Berlin: Springer) pp 721–47

[41] Tiller K G, Gerth J and Brummer G 1984 The relative affinities of Cd, Ni, and Zn for different soil clay fractions and goethite *Geoderma* **34** 17–35

[42] Herms U and Bruemmer G W 1984 Einflussgoressen der Schwermetallloeslichkeitund -bindung in Boeden *J. Plant Nutr. Soil Sci.* **147** 400–24

[43] Baes C F and Mesmer R E 1976 *The Hydrolysis of Cations* (New York: Wiley-Interscience)

[44] Huges M N and Poole R K 1991 Metal speciation and microbial growth-the hard (and soft) facts *Gen. Microbiol.* **137** 725–34

[45] Kabata-Pendias A and Mukherjee A B 2007 *Trace Elements from Soil to Human* (Berlin: Springer) 193–5

[46] Olaniran A, Balgobind A and Pillay B 2013 Bioavailability of heavy metals in soil: impact on microbial biodegradation of organic compounds and possible improvement strategies *Int. J. Mol. Sci.* **14** 10197–228

[47] Qian J, Shan X-Q, Wang Z-J and Tu Q 1996 Distribution and plant availability of heavy metals in different particle size fractions of soil *Sci. Total Environ.* **187** 131–41

[48] Zaky M H and Elwa A-S M 2020 Heavy metals content relating to soil physical properties *Egypt J. Appl. Sci.* **35** 50–62

[49] Sharma S, Sehkon N, Deswal S and John S 2009 Transport and fate of copper in soils *Int. J. Civil Environ. Eng.* **1** 19–39

[50] McCauley A, Jones C and Jacobsen J 2009 *Nutrient Management. Module No. 8* (Bozeman, MT: Montana State University Extension Publications) http://msuextension.org

[51] Pikuła D and Stępień W 2021 Effect of the degree of soil contamination with heavy metals on their mobility in the soil profile in a microplot experiment *Agronomy* **11** 878

[52] Xiao R, Zhang M X, Yao X Y, Ma Z W, Yu F H and Bai J H 2016 Heavy metal distribution in different soil aggregate size classes from restored brackish Marsh, oil exploitation zone, and tidal mud flat of the Yellow River Delta *J. Soils Sediments* **16** 821–30

[53] Sharma R K and Agrawal M 2005 Biological effects of heavy metals: an overview *J. Environ. Biol.* **26** 301–13

[54] Duruibe J O, Ogwuegbu M O C and Egwurugwu J N 2007 Heavy metal pollution and human biotoxic effects *Int. J. Phys. Sci.* **2** 112–8

[55] Valko M, Morris H and Cronin M 2005 Metals, toxicity and oxidative stress *Curr. Med. Chem.* **12** 1161–208

[56] 2008 *Heavy Metal Poisoning* (Quincy, MA: National Organization for Rare Disorders (NORD)) https://rarediseases.org/rare-diseases/heavy-metal-poisoning/

[57] Aldor I, Fourest E and Volesky B 1995 Desorption of cadmium from algal biosorbent *Can. J. Chem. Eng.* **73** 516

[58] Lansdown A B G 2010 A pharmacological and toxicological profile of silver as an antimicrobial agent in medical devices *Adv. Pharmacol. Sci.* **16** 910686

[59] Akpor O B, Ohiobor G O and Olaolu T D 2014 Heavy metal pollutants in wastewater effluents: sources, effects and remediation *Adv. Biosci. Bioeng.* **2** 37–43

[60] Wuana R A and Okieimen F E 2011 Heavy metals in contaminated soils: a review of sources, chemistry, risks and best available strategies for remediation *Int. Scholar. Res. Notices* 2011

[61] Reddy K and Adams J 2015 *Sustainable Remediation of Contaminated Sites* (New York, NY: Momentum Press) https://www.momentumpress.net/books/sustainable-remediation-contaminated-sites

[62] Cundy A B, Bardos R P, Church A, Puschenreiter M, Friesl-Hanl W, Müller I, Neu S, Mench M, Witters N and Vangronsveld J 2013 Developing principles of sustainability and stakeholder engagement for ‘gentle’ remediation approaches: the European context *J. Environ. Manage.* **129** 283–91

[63] Thevarajah B, Nishshanka G K S H, Premaratne M, Wasath W A J, Nimarshana P H V, Malik A and Ariyadasa T U 2022 Cyanobacterial pigment production in wastewaters treated for heavy metal removal: current status and perspectives *J. Environ. Chem. Eng.* **11** 108999

[64] Thomé A, Reginatto C, Vanzetto G and Braun A B 2018 Remediation technologies applied in polluted soils: new perspectives in this field *The Int. Congress on Environmental Geotechnics* (Singapore: Springer) 186–203

[65] Liu L, Li W, Song W and Guo M 2018 Remediation techniques for heavy metal-contaminated soils: principles and applicability *Sci. Total Environ.* **633** 206–19

[66] Khalid S, Shahid M, Niazi N K, Murtaza B, Bibi I and Dumat C 2017 A comparison of technologies for remediation of heavy metal contaminated soils *J. Geochem. Explor.* **182** 247–68

[67] Song B *et al* 2017 Evaluation methods for assessing effectiveness of *in situ* remediation of soil and sediment contaminated with organic pollutants and heavy metals *Environ. Int.* **105** 43–55

[68] Samaksaman U, Peng T H, Kuo J H, Lu C H and Wey M Y 2016 Thermal treatment of soil co-contaminated with lube oil and heavy metals in a low-temperature two-stage fluidized bed incinerator *Appl. Therm. Eng.* **93** 131–8

[69] Siegel L 2008 When is enough, enough? *Community Perspectives on Groundwater Treatment at Department of Defense Facilities* (Mountain View, CA: Centre for Public Environmental Oversight) http://-www.cpeo.org/pubs/Enough.pdf

[70] Park D, Yun Y S and Park J M 2010 The past, present, and future trends of biosorption *Biotechnol. Bioprocess Eng.* **15** 86–102

[71] Enenche E D, David O A, Inya A I and Abraham A K 2021 Adsorption of Pb (II) and Cd (II) Ions from their aqueous solutions unto carbonised orange peels *World* **6** 55–61

[72] Oklo D A, Daniel E E, Peter A A and Simeon E O 2021 Adsorption isotherms and kinetics of Pb (II) and Cd (II) ions onto carbonised sugarcane bagasse *Adv. Biochem.* **9** 44–9

[73] Hlihor R M, Diaconu M, Leon F, Curteanu S, Tavares T and Gavrilescu M 2015 Experimental analysis and mathematical prediction of Cd (II) removal by biosorption using support vector machines and genetic algorithms *New Biotechnol.* **32** 358–68

[74] Chojnacka K 2009 *Biosorption and Bioaccumulation in Practice* (Hauppauge, NY: Nova Science Publishers)

[75] 2016 *Bioremediation of Aquatic and Terrestrial Ecosystems* ed M Fingerman (Boca Raton, FL: CRC Press)

[76] Robalds A, Naja G M and Klavins M 2016 Highlighting inconsistencies regarding metal biosorption *J. Hazard. Mater.* **304** 553–6

[77] Wang J L and Chen C 2009 Biosorbents for heavy metals removal and their future a review *Biotechnol. Adv.* **27** 195–226

[78] Naja G and Volesky B 2011 The mechanism of metal cation and anion biosorption *Microbial Biosorption of Metals* (Dordrecht: Springer) 19–58 pp

[79] Paula E D, Simion A I, Grigoraş C G and Favier-Teodorescu L 2013 Optimization of barley husks acid hydrolysis process using the response surface methodology *Rev. Roum. Chim.* **58** 517–25

[80] Abed R M, Al-Kharusi S and Al-Hinai M 2015 Effect of biostimulation, temperature and salinity on respiration activities and bacterial community composition in an oil polluted desert soil *Int. Biodeterior. Biodegrad.* **98** 43–52

[81] Abdolali A, Guo W S, Ngo H H, Chen S S, Nguyen N C and Tung K L 2014 Typical lignocellulosic wastes and by-products for biosorption process in water and wastewater treatment: a critical review *Bioresour. Technol.* **160** 57–66

[82] Zhou X, Zhang J, Pan Z and Li D 2019 Review of methods for the detection and determination of malachite green and leuco-malachite green in aquaculture *Crit. Rev. Anal. Chem.* **49** 1–20

[83] Hlihor R M, Apostol L C and Gavrilescu M 2017 Environmental bioremediation by biosorption and bioaccumulation: principles and applications *Enhancing Cleanup of Environmental Pollutants* (Cham: Springer) 289–315

[84] Hlihor R M, Bulgariu L, Sobariu D L, Diaconu M, Tavares T and Gavrilescu M 2014 Recent advances in biosorption of heavy metals: support tools for biosorption equilibrium, kinetics and mechanism *Rev. Roum. Chim.* **59** 527–38

[85] Nguyen D T, Tran T V, Kumar P S, Din A T, Jalil A A and Vo D V 2022 Invasive plants as biosorbents for environmental remediation: a review *Environ. Chem. Lett.* **20** 1–31

[86] Bansal S, Pandey P K and Upadhayay S 2021 Methylene blue dye removal from wastewater using ailanthus excelsaroxb as adsorbent *Water Conserv. Sci. Eng.* **6** 1–9

[87] Dahou T, Defoort F, Khiari B, Labaki M, Dupont C and Jeguirim M 2021 Role of inorganics on the biomass char gasification reactivity: a review involving reaction mechanisms and kinetics models *Renew. Sustain. Energy Rev.* **135** 110136

[88] Johnson L R, Breger B and Drummond F 2019 Novel plant–insect interactions in an urban environment: enemies, protectors, and pollinators of invasive knotweeds *Ecosphere* **10** e02885

[89] Amrhar O, El Gana L and Mobarak M 2021 Calculation of adsorption isotherms by statistical physics models: a review *Environ. Chem. Lett.* **19** 4519–47

[90] Khamparia S and Jaspal D 2016 Investigation of adsorption of Rhodamine B onto a natural adsorbent *Argemonemexicana J. Environ. Manage.* **183** 786–93

[91] Nguyen D T, Dang H H, Vo D V, Bach L G, Nguyen T D and Van Tran T 2021 Biogenic synthesis of MgO nanoparticles from different extracts (flower, bark, leaf) of *Tecoma stans* (L.) and their utilization in selected organic dyes treatment *J. Hazard. Mater.* **404** 124146

[92] Rigueto C V, Piccin J S, Dettmer A, Rosseto M, Dotto G L, de Oliveira Schmitz A P, Perondi D, de Freitas T S, Loss R A and Geraldi C A 2020 Water hyacinth (*Eichhorniacrassipes*) roots, an amazon natural waste, as an alternative biosorbent to uptake a reactive textile dye from aqueous solutions *Ecol. Eng.* **150** 105817

[93] Tounsadi H, Khalidi A, Abdennouri M and Barka N 2016 Potential capability of natural biosorbents: *Diplotaxisharra* and *Glebioniscoronaria* L. on the removal efficiency of dyes from aqueous solutions *Desalin. Water Treat.* **57** 16633–42

[94] Bassey U, Suleiman M A, Ochigbo S S, Ndamitso M M, Daniel E D, Otolo S E and Chukwudi A 2015 Adsorption isotherm, kinetics and thermodynamics study of Cr (VI) ions onto modified activated carbon from endocarp of *Canariumschweinfurthii Int. Res. J. Pure Appl. Chem.* **6** 46

[95] Wang C, Xiong C, He Y, Yang C, Li X, Zheng J and Wang S 2021 Facile preparation of magnetic Zr-MOF for adsorption of Pb (II) and Cr (VI) from water: adsorption characteristics and mechanisms *Chem. Eng. J.* **415** 128923

[96] Shooto N D 2020 Removal of toxic hexavalent chromium (Cr (VI)) and divalent lead (Pb (II)) ions from aqueous solution by modified rhizomes of *Acoruscalamus Surf. Interfaces* **20** 100624

[97] Sheng Y, Tan X, Gu Y, Zhou X, Tu M and Xu Y 2021 Effect of ascorbic acid assisted dilute acid pretreatment on lignin removal and enzyme digestibility of agricultural residues *Renew. Energy* **163** 732–9

[98] Mallesh D, Anbarasan J, Kumar P M, Upendar K, Chandrashekar P, Rao B V and Lingaiah N 2020 Synthesis, characterization of carbon adsorbents derived from waste biomass and its application to CO_2 capture *Appl. Surf. Sci.* **530** 147226

[99] Jönsson L J and Martín C 2016 Pretreatment of lignocellulose: formation of inhibitory by-products and strategies for minimizing their effects *Bioresour. Technol.* **199** 103–2

[100] Saxena A, Bhardwaj M, Allen T, Kumar S and Sahney R 2017 Adsorption of heavy metals from wastewater using agricultural–industrial wastes as biosorbents *Water Sci.* **31** 189–97

[101] Kanamarlapudi S L R K, Chintalpudi V K and Muddada S 2018 Application of biosorption for removal of heavy metals from wastewater *Biosorption* **18** 70–116

[102] Torres E 2020 Biosorption: a review of the latest advances *Processes* **8** 1584

[103] Kiran and Kumar B 2022 Biosorbent: types, mechanism and its biosorption data *Environmental Chemistry and Its Applications for the Pollution Abatement* (Patna: Blue Hill Publications) 1–7 pp

[104] Gaur N, Flora G, Yadav M and Tiwari A 2014 A review with recent advancements on bioremediation-based abolition of heavy metals *Environ. Sci.: Process. Impacts* **16** 180–93

[105] Santaeufemia S, Torres E, Mera R and Abalde J 2016 Bioremediation of oxytetracycline in seawater by living and dead biomass of the microalga *Phaeodactylumtricornutum J. Hazard. Mater.* **320** 315–25

[106] Yu R *et al* 2020 Behavior and mechanism of cesium biosorption from aqueous solution by living *Synechococcus* PCC7002 *Microorganisms* **8** 491

[107] Joo J H, Hassan S H and Oh S E 2010 Comparative study of biosorption of Zn^{2+} by *Pseudomonas aeruginosa* and *Bacillus cereus Int. Biodeterior. Biodegrad.* **64** 734–41

[108] Shamim S 2018 Biosorption of heavy metals *Biosorption* **2** 21–49

[109] Mathew B B, Jaishankar M, Biju V G and Beeregowda K N 2016 Role of bioadsorbents in reducing toxic metals *J. Toxicol.* Article ID 4369604 13

[110] Rogowska A, Pomastowski P, Rafinska K, Railean-Plugaru V, Zloch M, Walczak J and Buszewski B 2019 A study of zearalenone biosorption and metabolisation by prokaryotic and eukaryotic cells *Toxicon.* **169** 81–90

[111] Peng H, Li D, Ye J, Xu H, Xie W, Zhang Y, Wu M, Xu L, Liang Y and Liu W 2019 Biosorption behavior of the *Ochrobactrum* MT180101 on ionic copper and chelate copper *J. Environ. Manag.* **235** 224–30

[112] Wang L, Xiao H, He N, Sun D and Duan S 2019 Biosorption and biodegradation of the environmental hormone nonylphenol by four marine microalgae *Sci. Rep.* **9** 5277

[113] Reddy S and Osborne J W 2020 Biodegradation and biosorption of Reactive Red 120 dye by immobilized Pseudomonas guariconensis: kinetic and toxicity study *Water Environ. Res.*

[114] Wang X, Li D, Gao P, Gu W, He X, Yang W and Tang W 2020 Analysis of biosorption and biotransformation mechanism of *Pseudomonas chengduensis* strain MBR under Cd(II) stress from genomic perspective *Ecotoxicol. Environ. Saf.* **198** 110655

[115] Tsezos M, Remoundaki E and Hatzikioseyian A 2006 Biosorption-principles and applications for metal immobilization from waste-water streams *Proc. EU-Asia Workshop on Clean Production and Nanotechnologies (Seoul)* pp 23–33

[116] Abbas S H, Ismail I M, Mostafa T M and Sulaymon A H 2014 Biosorption of heavy metals: a review *J. Chem. Sci. Technol* **3** 74–102

[117] Bashir A, Malik L A, Ahad S, Manzoor T, Bhat M A, Dar G N and Pandith A H 2019 Removal of heavy metal ions from aqueous system by ion-exchange and biosorption methods *Environ. Chem. Lett.* **17** 729–54

[118] Ali Redha A 2020 Removal of heavy metals from aqueous media by biosorption *Arab J. Basic Appl. Sci.* **27** 183–93

[119] Kuyucak N and Vole B 1988 Biosorbents for recovery of metals from industrial solutions *Biotechnol. Left* **10** 137–42

[120] Sharma S K and Mudhoo A 2010 *Green Chemistry for Environmental Sustainability* (London: Taylor and Francis)

[121] Perpetuo E A, Souza C B and Nascimento C A O 2011 Engineering bacteria for bioremediation *Progress in Molecular and Environmental Bioengineering* ed A Carpi (London: IntechOpen) 605–32

[122] Srivastava S and Goyal P 2010 *Novel Biomaterials: Decontamination of Toxic Metals from Wastewater* (Berlin: Springer)

[123] Cimá-Mukul C A, Abdellaoui Y, Abatal M, Vargas J, Arlette A, Santiago A A and Barrón Zambrano J A 2019 Eco-efficient biosorbent based on *Leucaena leucocephala* residues for the simultaneous removal of Pb(II) and Cd(II) ions from water system: sorption and mechanism *Bioinorg. Chem. Appl.* **2019** 2814047

[124] Ayers P W 2005 An elementary derivation of the hard/soft-acid/base principle *J. Chem. Phys.* **122** 141102

[125] Garcıa-Mendieta A, Olguın M T and Solache-Rıos M 2012 Biosorption properties of green tomato husk (*Physalis philadelphica* Lam) for iron, manganese and iron-manganese from aqueous systems *Desalination* **284** 167–74

[126] Nandi D, Ghosh S K, Ghosh A, Siengchin S, Roy A, Gupta K and Ghosh U C 2021 Arsenic removal from water by graphene nanoplatelets prepared from nail waste: a physicochemical study of adsorption based on process optimization, kinetics, isotherm and thermodynamics *Environ. Nanotechnol., Monitor. Manag.* **16** 100564

[127] Oyewole O A, Adamu B B, Oladoja E O, Balogun A N, Okunlola B M and Odiniya E E 2018 A review on heavy metals biosorption in the environment *Brazil J. Biol. Sci.* **5** 225–36

[128] Greene B and Darnall D W 1990 Microbial oxygenic photoautotrophs (cyanobacteria and algae) for metal-ion binding *Microbial Mineral Recovery* ed H L Ehrlich and C L Brierley (New York: McGraw-Hill) pp 227–302

[129] Deng and Wang P 2012 Isolation of marine bacteria highly resistant to mercury and their bioaccumulation process *Bioresour. Technol.* **121** 342–7

[130] Volesky B 2003 Biosorbent materials *Biotechnol. Bioeng. Symp.* **16** 121–6

[131] Hajahmadi Z, Younesi H, Bahramifar N, Khakpour H and Pirzadeh K 2015 Multicomponent isotherm for biosorption of Zn(II), CO(II) and Cd(II) from ternary mixture onto pretreated dried *Aspergillus niger* biomass *Water Resour. Ind.* **11** 71–80

[132] Zouboulis A L, Matis K A and Hancock I C 1997 Biosorption of metals from dilute aqueous solutions *Sep. Purif. Methods* **26** 255–95

[133] Okoya A A, Akinyele A B, Amuda O S and Ofoezie I E 2015 Chitosan grafted modified maize cob for removal of lead and chromium from wastewater *Ethiop. J. Environ. Studi. Manag.* **8** 881–92

[134] Ashraf M A, Wajid A, Mahmood K, Maah M J and Yusoff I 2011 Low cost biosorbent banana peel (*Musa sapientum*) for the removal of heavy metals *Sci. Res. Essays* **6** 4055–64

[135] Ahalya N, Ramachandra T V and Kanamadi R D 2003 Biosorption of heavy metals *Res. J. Chem. Environ.* **7** 71–9

[136] Goyal N, Jain S C and Banerjee U C 2003 Comparative studies on the microbial adsorption of heavy metals *Adv. Environ. Res.* **7** 311–9

[137] Ahmad A, Bhat A H and Buang A 2019 Enhanced biosorption of transition metals by living *Chlorella vulgaris* immobilized in Ca-alginate beads *Environ. Technol.* **40** 1793–809

[138] White C, Sayer J A and Gadd G M 1997 Microbial solubilization and immobilization of toxic metals: key biogeochemical processes for treatment of contamination *FEMS Microbiol. Rev.* **20** 503–16

[139] Dai Q H, Bian X Y, Li R, Jiang C B, Ge J M, Li B L and Ou J 2019 Biosorption of lead (II) from aqueous solution by lactic acid bacteria *Water Sci. Technol.* **79** 627–34

[140] Hu X, Cao J, Yang H, Li D, Qiao Y, Zhao J, Zhang Z and Huang L 2020 Pb^{2+} biosorption from aqueous solutions by live and dead biosorbents of the hydrocarbon-degrading strain *Rhodococcus sp.* HX-2 *PLoS One* **15** e0226557

[141] Barquilha C E R, Cossich E S, Tavares C R G and da Silva E A 2019 Biosorption of nickel (II) and Cu^{2+}, ions from synthetic and real effluents by alginate-based biosorbent produced from seaweed Sargassum sp *Environ. Sci. Pollut. Res. Int.* **26** 11100–12

[142] Calero C I, Salles A, Semelman M and Sigman M 2013 Front. Hum. Neurosci *Sec. Cognitive Neuroscience* **7**

[143] Sao K 2014 A review on heavy metals uptake by plants through biosorption *Int. Proc. Econ. Develop. Res.* **75** 78–83

[144] Wang J L and Chen C 2009 Biosorbents for heavy metals removal and their future a review *Biotechnol. Adv.* **27** 195–22

[145] Aksu Z, Sag Y and Kutsal T 1992 The biosorption of copper by *C. vulgaris* and *Z. ramigera* *Environ. Technol.* **13** 579–86

[146] Ashraf M 2011 Low cost biosorbent banana peel (*Musa sapientum*) for the removal of heavy metals *Sci. Res. Essays* **6** 4055–64 http://umexpert.um.edu.my/file/publication/0000409471909.pdf

[147] Selatnia A, Madani A, Bakhti M Z, Kertous L, Mansouri Y and Yous R 2004 Biosorption of Ni^{2+} from aqueous solution by a NaOH-treated bacterial dead *Streptomyces rimosus* *Biomass. Miner. Eng.* **17** 903–11

[148] Baysal Z, Çinar E, Bulut Y, Alkan H and Dogru M 2009 Equilibrium and thermodynamic studies on biosorption of Pb (II) onto *Candida albicans* biomass *J. Hazard. Mater.* **161** 62–7

[149] Norberg A B and Persson H 1984 Accumulation of heavy-metal ions by *Zoogloearamigera* *Biotechnol. Bioeng.* **26** 239–46

[150] Chang J S and Hong J 1994 Biosorption of mercury by the inactivated cells of *Pseudomonas aeruginosa* PU21 (Rip64) *Biotechnol. Bioeng.* **44** 999–1006

[151] Bassey U, Suleiman M A T, Ochigbo S S, Ndamitso M M, Daniel E D, Otolo S E and Chukwudi A 2015 Adsorption isotherm, kinetics and thermodynamics study of Cr (VI) ions onto modified activated carbon from endocarp of *Canarium schweinfurthii* *Int. Res. J. Pure Appl. Chem.* **6** 46

[152] Urrutia M M 1997 General bacterial sorption processes *Biosorbents for Metal Ions* ed J Wase and C Forster (London: CRC Press) 39–66 pp

[153] Vasudevan P, Padmavathy V, Tewari N and Dhingra S C 2001 Biosorption of heavy metal ions *J. Sci. Ind. Res.* **60** 112–20

[154] Kinoshita H, Sohma Y, Ohtake F, Ishida M, Kawai Y, Kitazawa H, Saito T and Kimura K 2013 Biosorption of heavy metals by lactic acid bacteria and identification of mercury binding protein *Res. Microbiol.* **164** 701–9

[155] Iihan S, Nourbakhsh M, Kilicarslan S and Ozdag H 2004 Removal of chromium, lead and copper ions from industrial waste waters by *Staphylococcus saprophyticus* *Turkish Electron. J. Biotechnol.* **2** 50–7

[156] Rincon J, Gonzalez F, Ballester A, Blazquez M L and Munoz J A 2005 Biosorption of heavy metals by chemically-activated alga *Fucus vesiculosus* *J. Chem. Technol. Biotechnol.* **80** 403–1407

[157] Brinza L, Dring M J and Gavrilescu M 2007 Marine micro- and macro-algal species as biosorbents for heavy metals *Environ. Eng. Manage. J.* **6** 237–51

[158] Kuyicak N and Volesky B 1990 Biosorption by fungal biomass ed B Volesky *Biosorption of Heavy Metals* (Boca Raton, FL: CRC Press) pp 173–98

[159] Khani M H 2011 Uranium biosorption by *Padina* sp. algae biomass: kinetics and thermodynamics *Environ. Sci. Pollut. Res. (Int.)* **18** 1593–605

[160] Flouty R and Estephane G 2012 Bioaccumulation and biosorption of copper and lead by a unicellular algae *Chlamydomonas reinhardtii* in single and binary metal systems: a comparative study *J. Environ. Manage.* **111** 106–14

[161] Trinelli M A, Areco M M and Afonso Mdos S 2013 Co-biosorption of copper and glyphosate by *Ulva lactuca* *Colloids Surf.,* B **105** 251–8

[162] Oyedepo T A 2011 Biosorption of lead (II) and copper (II) metal ions on *Calotropis procera* (Ait.) *Sci. J. Pure Appl. Chem.* 1–7

[163] Gadd G M and Griffiths A J 1980 Effect of copper on morphology of *Aureobasidium pullulans Trans Brit. Mycol. Soc.* **74** 387–92
[164] Gadd G M and Mowll J L 1985 Copper uptake by yeast-like cells, hyphae and chlamydospores of *Aureobasidium pullulans Exp. Mycol.* **9** 230–40
[165] Remacle J 1990 The cell wall and metal binding ed B Volesky *Biosorption of Heavy Metals* (Boca Raton, FL: CRC Press) pp 83–92
[166] Huang C and Huang C P 1996 Application of *Aspergillus oryzae* and *Rhizopus oryzae* for Cu (II) removal *Water Res.* **30** 1985–90
[167] Arica M Y, Bayramoglu G, Yilmaz M, Bektas S and Genc O 2004 Biosorption of Hg(II) Cd(II) and Zn(II) by Ca-alginate and immobilized wood-rotting fungus *Funaliatrogii J. Hazard. Mater.* B **109** 191–9
[168] Mohamad O A, Hao X, Xie P, Hatab S, Lin Y and Wei G 2012 *Microbes Environ.* **27** 234–41
[169] Freitas G R, da Silva M G C and Vieira M G A 2019 Biosorption technology for removal of toxic metals: a review of commercial biosorbents and patents *Environ. Sci. Pollut. Res. Int.* **26** 19097–118

Chapter 6

The photocatalytic degradation of dyes and metal ions using functionalized biosorbents

Benedict Okundaye, Kingsley Eghonghon Ukhurebor, Uyiosa Osagie Aigbe, Efosa Aigbe, Vincent Aizebeoje Balogun and Ahmed El Nemr

As a result of increased growth in urbanization and industrialization, environmental pollution is now rising globally. Organic substances which are produced by several industries (dyes and heavy metal ions) have led to significant environmental contamination. Recently, it has been discovered that metal-oxide-based semiconductors make effective photocatalysts for the degradation of organic contaminants in effluents. However, their large bandgaps and the length of time it takes for photogenerated electron–hole pairs to recombine restrict their photocatalytic performance. The main focus of this chapter is to provide a guide for effective, economic-based approaches that provide potential advantages to solutions for environmental pollution.

6.1 Introduction

Natural resource (especially water) protection is a serious issue, given today's economic expansion and population increase [1, 2]. According to Yan *et al* [3], around two million tons of waste pollutants are heedlessly dumped into the public water supply each day, severely harming the well-being of people and causing a large number of fatalities.

The most prominent contaminants include heavy metal ions (HMIs), antibiotics derived from industrial waste, monoaromatic hydrocarbons, residual toxic colorants, and toxic dyes. According to some reported studies, there are various concentrations of these contaminants in the environment [4–7]. Globally, water resources receive random releases of more than 54 000 tons of antibiotics and 175 000 tons of dyes per year [8, 9].

Concern has intensified as water contamination has become an unsolvable problem [10, 11]. According to some studies, several strategies for wastewater

doi:10.1088/978-0-7503-5598-8ch6

purification, such as ion exchange, adsorption, membrane separation, and electrochemical purification, have been devised and put into practice to manage water resources [12–16]. Despite significant efforts to address this issue, a number of challenges still exist, including inadequate capacity for adsorption, easy inactivation, secondary contaminants, and high energy utilization [17–21]. Photoinduced catalytic degradation technology has recently received a lot of interest as a solution to this difficult problem [22].

The most affordable, efficient, and even limitless source of energy is solar energy. Advances in nanoscience and nanotechnology have made it possible to systematically design and create novel semiconducting photocatalysts that employ active radicals to destroy organic pollutants [9, 23]. Additional advantages of the photocatalytic method include affordability, environmental friendliness, and low energy consumption [24–26]. Notwithstanding being a self-driving prospective method for gathering endless solar energy, the effectiveness of semiconducting photocatalysis technology in transforming solar energy is far lower than anticipated [27]. This is due to the difficulty of developing a single photocatalyst that is capable of absorbing all solar energy while preserving robust photostability and low rates of photoinduced carrier recombination [28, 29]. Co-photocatalysts are thought to partially meet the challenge of acquisition, since each part of a composite possesses certain mutually beneficial characteristics. Several substances are capable of being utilized in photocatalytic procedures; however, metal oxides (MOs), particularly titanium oxide (TiO_2), cerium oxide (CeO_2), zinc oxide (ZnO), zirconium oxide (ZrO_2), manganese oxide (MnO_2), and tin oxide (SnO_2), have been identified as outstanding photocatalysts in this category owing to their excellent stability over a wide pH range, improved effectiveness, inexpensiveness, easy availability, reduced toxic effects, environmental friendliness, and strongly oxidizing photogenerated holes [30–32].

In the future, it will be necessary to keep improving the environment through the promotion of a more environmentally friendly and sustainable ecosystem. Environmental pollution caused by industrial activities has a detrimental impact on the health of people, aquatic life, terrestrial animals, and vegetation, since industrial effluents contribute significantly to environmental contamination. Hence, this chapter focuses on merging contemporary and traditional ways to remove various prominent contaminants from industrial wastewater in order to create a sustainable environment. However, the chapter's goal is also emphasized, which is to examine how various prominent contaminants such as dyes and HMIs used in the industrial sector behave before photocatalysts are applied to reduce their effects.

6.2 The sources and effects of dyes and metal ions on the environment

Industrial development and advancements in technology, agriculture, and other anthropogenic activities are the primary causes of the release of hazardous substances such as HMIs and dyes into the natural environment [33–36].

HMIs that are naturally present in the environment are closely linked to the soil and are not harmful to animals or plants. When they exceed the recommended limits, the natural environment is contaminated by these non-degraded HMIs, dyes, and metalloids, which pose a major threat to the well-being of every living thing [37, 38]. Most HMIs last a long period in the natural environment due to their nondegradable characteristics, which lead to bioaccumulation and toxicity. According to Mani and Kumar [39], the toxicity of HMIs and dyes depends on the concentration at which they interact with organisms, the duration of exposure, the dosage absorbed, and the means of exposure.

A number of industries release noxious HMIs (such as Ag, As, Cd, Cr, Pb, Hg, Zn, etc.) into the environment. Public health is seriously endangered by these HMIs [6, 40]. Similarly, aromatic color formulations known as dyes (synthetic dyes such as [41] acidic dyes, azoic or naphthol dyes, basic dyes, chrome or mordant dyes, developed (or diazo dyes), direct dyes, disperse or acetate dyes, reactive or fiber-reactive dyes, sulfur dyes, vat dye) are utilized by the textile, paper, leather, food, cosmetics, tanneries, and pharmaceutical industries, as well as other sectors [42, 43]. These synthetic dyes can be purchased as powder, paste, liquid, or granules and are manufactured from petrochemicals and other chemicals [44]. Fabrics consume little energy, color rapidly and uniformly, have a variety of color hues, and maintain stability in the face of several environmental variables [45]. Therefore, when synthetic dyes are released into the biosphere without being thoroughly cleansed or even substantially cleaned, they are usually considered harmful [46, 47].

According to several studies, textile wastewater contains organic and inorganic contaminants such as HMIs, chlorinated compounds, chromium, pigments, etc. Such wastewater has a high pH and a high biological and chemical demand for oxygen [43, 48, 49]. Particularly in developing nations, these toxins have been blamed for polluting nearby landfills and agricultural land; they inhibit plant development by causing oxidative stress and reducing the rate of photosynthesis and CO_2 absorption [50]. Wastewater's dark hue and high turbidity, which reduces marine photosynthesis, also contributes to the damage caused [48].

As a result, these toxins endanger atmospheric, terrestrial, and aquatic species as well as human beings in both direct and indirect ways. Their well-documented effects on human life include damage to the digestive tract, vascular occlusion, fatigue, high blood pressure, memory loss, speech problems, increased allergies, irritability, osteoporosis, neurological disorders, obstructive lung disease, renal dysfunction, and aggressive behavior [51]. Figure 6.1 (as adapted from Onyancha *et al* [52]) illustrates some of the sources and effects of noxious HMIs on human bodies. Additionally, HMI contaminants quickly alter the chemical, physical, and biological characteristics of water due to their non-biodegradability. The hazards of these contaminants can still be mitigated by treating polluted water and reducing pollutant concentrations to allowable levels before polluted water is discharged into the environment [53]. Due to the potential harm these pollutants may cause to human life and the requirement for clean and safe water, researchers have been forced to create accurate detection methods that can measure these contaminants.

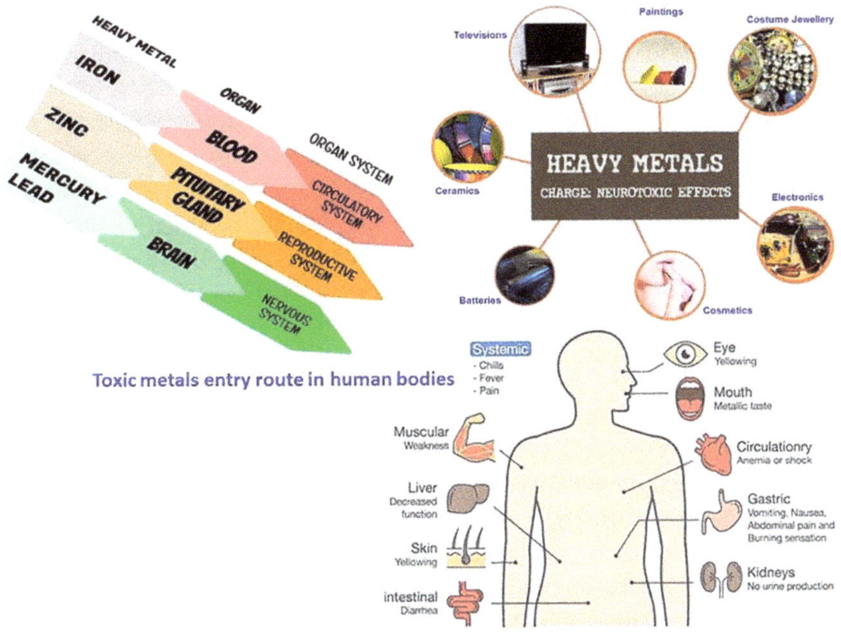

Figure 6.1. Some of the sources and effects of noxious HMIs in human bodies. Reprinted from [52], Copyright (2022), with permission from Elsevier.

Considering the fact that these noxious HMIs and dyes pose serious health risks to people and other living entities, removing them from industrial effluent is an important concern. To lessen the pollution burden on the environment, it is important to develop contemporary treatment and management technologies to eliminate dangerous HMIs and dyes from industrial waste.

6.3 Techniques for removing dyes and metal ions from the environment

A current key international fear is ecological pollution brought about by the disposal of dyes and HMIs, which imperils both human health and ecological organisms. The increased use of dyes and HMIs over the last few years has heightened the concentration of these substances in groundwater and surface waters, thereby posing serious threats to the environment and human health. Due to the increasing water demands and inadequate water resources, it is fundamental to avoid the release of these unsafe compounds into the water cycle and to eliminate them from wastewater [54]. Thus, it is a priority to eliminate such contaminants from poisonous water and create a contamination-free ecosystem [55, 56].

Industries encounter various challenges in decreasing the discharge of contaminants into the environment as well as water usage and energy consumption. Hence, to achieve environmental safety protection, various treatment techniques have been established and have grown into an important research area. The precise procedures control contaminants' removal from wastewater. Owing to their toxic properties, a superior sequestration implementation technique is needed to confiscate contaminants [57].

Various technologies have been developed for the management and remediation of contaminant-laden effluents. The remediation technologies generally utilized comprise chemical and electrochemical technologies, ion exchange, photocatalysis, adsorption, biological treatment techniques, etc. However, most of these techniques are constrained by the creation of noxious sludge, extreme operating and maintenance costs, and the complicated methods involved in the treatment [58]. Processes such adsorption are being looked into as a way to overcome these weaknesses because of their significant effect on the bioavailability and transport of unsafe contaminants. Adsorption is an efficient and inexpensive way of eliminating contaminants from effluent. Another advantage of the adsorption procedure is that it is repeatedly reversible, permitting the adsorbent to be retrieved. The efficacy of adsorbents is affected by a variety of variables, including the contact time, starting concentration, temperature, pH, and rotation speed [6]. Adsorbent procedures are utilized separately to treat wastewater depending largely on the quantity of the contaminants' load [59]. One of the critical issues facing regularly used effluent treatment processes is their efficiency in terms of complete contaminant sequestration and energy use [60]. Figure 6.2, which was adopted and modified from Farhan *et al* [61], depicts the various types of techniques employed for the sequestration of

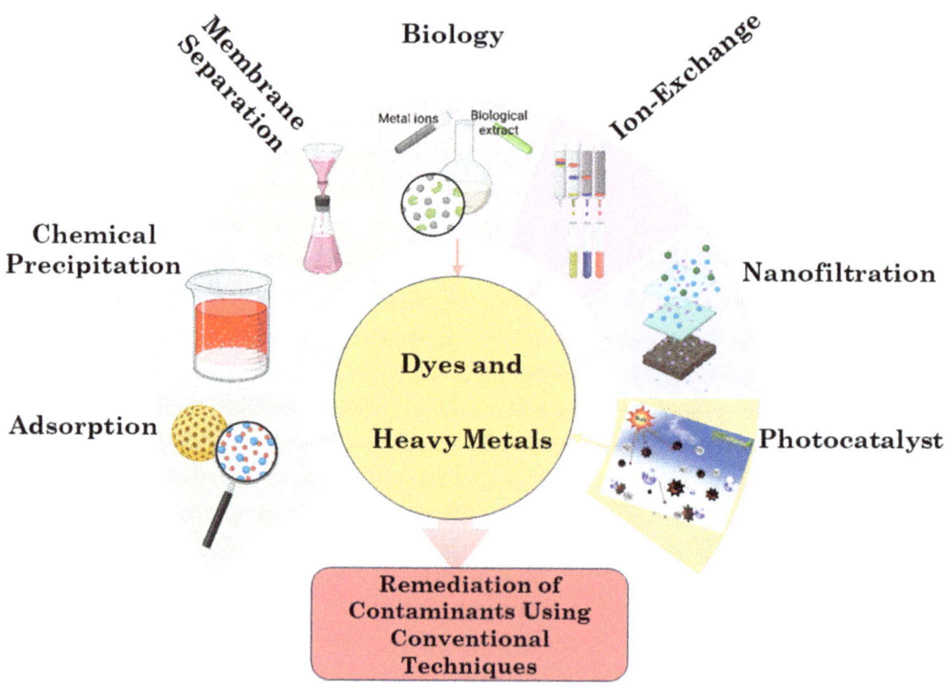

Figure 6.2. Various techniques for the treatment of different contaminants. Adapted by permission from Springer Nature [61], Copyright (2023).

Table 6.1. A comparison of the pro and cons of different treatment techniques used for contaminant sequestration from wastewater.

Procedures	Pro	Cons
Biological	This procedure is beneficial in HMI elimination.	The technique needs to be developed.
Oxidation	Electricity is not required.	The use of oxidation leads to corrosion in the system.
Photocatalyst	Simultaneous HMI and organic contaminant removal.	MIs take a long time to eliminate.
Coagulation	Inexpensive and has dewatering characteristics.	Sludge creation and chemical utilization are elevated.
Electrochemical treatment	Minimal chemical usage and is effective for critical MI elimination.	A large electrical supply is needed, and the treatment technique requires an extraordinary initial investment.
Adsorption	Reduced sludge creation, uses low-cost sorbents and a simple process.	Desorption.
Ion exchange	Extraordinary modification of parts.	MI elimination is limited by elevated operating costs.
Membrane filtration	Elevated elimination of HMIs with low space requirements	Very costly technique; suffers from membrane contamination and an intricate procedure

contaminants. Table 6.1 gives an overview of the pros and cons of various treatment procedures for contaminant elimination [57, 59].

6.4 Photocatalytic degradation of dyes and metal ions using functionalized biosorbents

Among the various techniques, photodegradation and adsorption are established techniques which are effective methods for the sequestration of noxious pollutants from water. For this reason, several sorbents/catalysts have been created for the elimination of such toxic chemicals from effluents. However, most of these techniques suffer from certain weaknesses, which explains the need for the urgent development of inexpensive materials and successful procedures for obtaining good drinking water free from pollutants. In addition, various semiconducting MOs have received great attention for use in the effective photodegradation of organic pollutants and the sorption of HMIs [62].

Energy generation and the control and elimination of ecological contamination, depend completely on the catalytic process. The success of the abovementioned objectives is determined by catalytic effectiveness, which is determined by the magnitude of active sites, increased surface area, and therefore reduced catalytic particle size. All these properties are present in organically derived nanoscale catalysts, which are also theoretically and financially feasible [63].

The term 'photocatalysis' first appeared in the scientific literature in 1911, when scientists learned about the light sorption capacity of uranyl ions, in which a catalyst was employed to accelerate the chemical reaction rate and any physical transformation. The pioneering work on photocatalysis attracted widespread attention from the scientific community in the early 1970s, owing to the disaster in the oil sector [5–7] and the severe environmental damage resulting from this emergency [64]. In photocatalysis, a catalyst is exploited to accelerate chemical reactions under the influence of ultraviolet (UV) or visible light or infrared radiation. 'Photo' and 'catalysis' are two words that originate in ancient Greek; 'photo' and 'catalysis' denote light and an enhanced reaction rate obtained by employing an exogenous material that is not consumed in the procedure, respectively. A catalyst intensifies the chemical reaction rate by lowering the activation energy needed. Photocatalyst can be homogeneous or heterogeneous. To remove contaminants, several reactions are involved in heterogeneous photocatalysis. Metal complexes are used in catalysts in homogeneous photocatalysis. Recently, a handful of heterogeneous semiconducting photocatalysts such as ZrO_2, TiO_2, SnO_2, CeO_2, ZrO, ZnO, etc. have been employed for the degradation of various contaminants due to their wide bandgaps (E_g), as depicted in table 6.2 [2, 65].

The various benefits of photocatalysis are the complete mineralization of refractory contaminants into safe components, an absence of resultant contamination, low cost, and a simple strategy [64]. Specifically, these materials have attracted huge interest owing to their extraordinary photosensitivity, elevated sorption capacity, excellent quantum effectiveness, nontoxic qualities, and large E_g values. In addition, these nanoparticles (NPs), which have extraordinary surface areas and porosities show better photocatalytic action than their bulk equivalents because they reduce the gap between the photon sorption sites and inhibit electron–hole (e^-–h^+) recombination [62, 66, 67]. The rate of recombination is changed by doped ions which act as trapping centers for charge carriers and alter the surface

Table 6.2. MOs and their corresponding E_g values.

MO semiconductors	E_g (eV)
TiO_2	3.0
MnO	3.8
FeO	2.4
SnO_2	3.7
CeO_2	3.2
ZnO	3.4
Fe_2O_3	2.0
WO_3	2.7
MnO_2	1.3
ZrO_2	5.0

features by improving the sorption capacity. Dopants act as sinks that accept photoexcited electrons, thereby accelerating the reduction procedure and slowing e^-–h^+ recombination [68]. However, under visible light, doped MOs are more effective, owing to surface defects that tend to be enhanced when two or more MOs are doped, which shifts the absorption towards the visible light region. The doping of MOs introduces a novel energy phase, and this results in electron blocking and optical E_g energy reduction [69].

The procedure involves illuminating semiconductor materials using the visible or ultraviolet band to stimulate electrons to migrate from the valence band to the conduction band, thus creating holes. These holes initiate the separation of water into protons and hydroxyl (OH^-) ions, which then oxidize and decrease contaminants. Owing to interactions between electrons in the conductive bands and the dissolved oxygen (O_2), superoxide anions are also created during the process. Organic compounds are increasingly degraded as a result of the redox reaction with holes and electrons. It has been proposed that the green production of MO-based NPs, such as silver, titanium, and zinc, may have implications for photocatalytic remediation. Cost-effectiveness, heightened stability, and reaction rate are some of their advantages [63, 70].

From the perspective of practical use, the stability and reusability of photocatalysts are critical parameters that impact the whole process of upscaling; in this regard, the reclamation of MO NPs after the reaction is tremendously problematic and ineffective, which therefore impedes their industrial-scale application. The most significant problem is their presence in definitively treated wastewater, which could possibly cause genotoxicity and cytotoxicity to the marine environment and human beings. To overcome this, the practicality and effectiveness of photocatalysts have been established by employing immobilization methods based on waste materials during photoreactions. These materials used for fixing MOs represent a fascinating way in which to resolve the severe problems faced in waste disposal and cost reduction [64].

The lowest unavailable and highest inactive energy bands in semiconductors are described by the E_g value. Photoelectrons have a huge E_g energy (ΔE), which permits valence band (VB) electrons to become excited [60]. The photocatalytic degradation mechanism is illustrated in figure 6.3 and the process is characterized by the following stages [64, 71–75]:

(a) On the illumination of the catalyst's surface by a light source with an energy corresponding to or greater than the E_g energy (E_g) of the semiconductor, electrons in the valence band (VB) are initiated, separated by diffusion and migration, and transferred to the conduction band (CB) (equation (6.1)). Hence, molecules of water would experience ionization (equation (6.2)).

$$\text{Photocatalyst} + h\upsilon\ (\text{photon}) \xrightarrow{h\upsilon \gg E_g} h^+_{VB} + e^-_{CB} \tag{6.1}$$

$$H_2O \rightarrow OH^- + H^+ \tag{6.2}$$

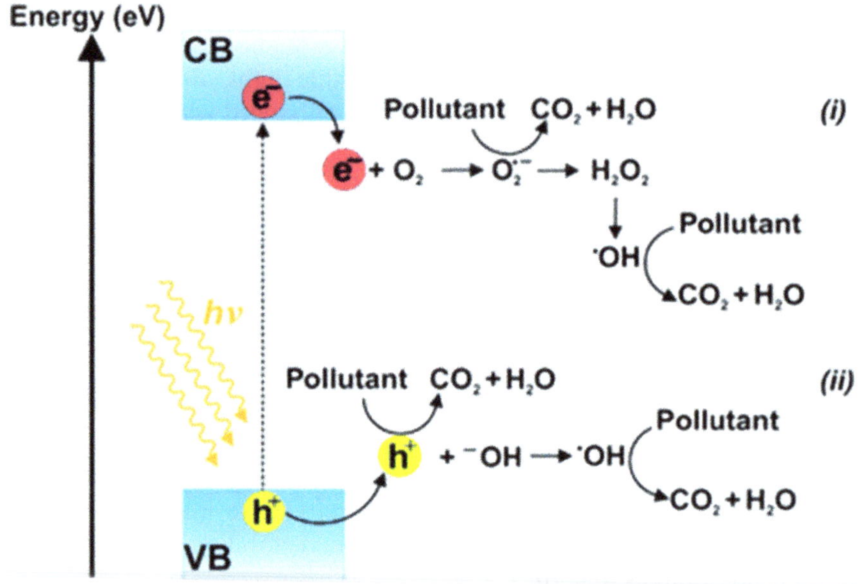

Figure 6.3. Photocatalytic interactions ($hv \geqslant E_g$) showing the oxidation–reduction reactions of electron donors/acceptors that take place in the CB and the VB. Adapted from [64] with permission from the Royal Society of Chemistry, Copyright (2021).

(b) The positive holes that are left in the semiconductor's VB react with molecules of water to create OH^- radicals (equation (6.3)). Due to their strongly oxidizing nature, the OH^- radicals degrade organic contaminants into intermediates (or, in some cases, water) and carbon dioxide (equation (6.4)).

$$O_2 + e_{CB}^- \rightarrow O_2 \tag{6.3}$$

$$O_2 + \text{contaminant} \rightarrow \text{Intermeiates} + H_2O + CO_2 \tag{6.4}$$

The immobilization of catalysts on supporting materials can be carried out using hydrothermal methods, precipitation, sol–gel methods, pyrolysis, hydrolysis, co-precipitation, dip coating, wet chemical fabrication, etc. These methods are dependent on the kind of photocatalyst used, the target contaminant(s), and the supporting material.

MOs functionalized with cellulose nanocrystals, copper oxide (CuO), chitosan, single-walled carbon nanotubes (SWCNTs) and magnetite (Fe_3O_4) to make various nanocomposites have been found to efficiently degrade thiazine dye, Rose Bengal, Cr(VI), As(V), Congo red, and methylene blue using light illumination [76–78].

6.5 Conclusions

Environmental contamination by noxious HMIs and organic pollutants such as dyes constitutes a worldwide threat on a progressively more substantial scale. However owing to the shortcomings of MOs, such as their large E_g values, e^-–h^+ pair recombination, and their working limits under UV radiation, their application in the photodegradation of contaminants has been restricted. The current advancements in nanotechnology offer a huge opportunity for the creation of desirable MOs and their composites with extraordinary surface areas, distinctive surface functionalities, and optical and photocatalytic features that improve their photocatalytic degradation of contaminants under the influence of illumination. Hence, this chapter has described the recent developments in photocatalytic activity for the degradation of pollutants in wastewater.

References

[1] Tripathi M, Singh P, Singh R, Bala S, Pathak N, Singh S, Chauhan R and Singh P.K 2023 Microbial biosorbent for remediation of dyes and heavy metals pollution: a green strategy for sustainable environment *Front. Microbiol.* **14** 1168954

[2] Kumari H *et al* 2023 A review on photocatalysis used for wastewater treatment: dye degradation *Water Air Soil Pollut.* **234** 349

[3] Yan X, Feng J, Li P, Li J, Ren B, Gao S and Cao R 2021 Fast and efficient removal of mercury ions using zirconium-based metal–organic framework filter membranes *Inorg. Chem. Commun.* **131** 108796

[4] Yang Y, Jiang K, Guo J, Li J, Peng X, Hong B, Wang X and Ge H 2020 Facile fabrication of Au/Fe_3O_4 nanocomposites as excellent nanocatalyst for ultrafast recyclable reduction of 4-nitropheol *Chem. Eng. J.* **381** 122596

[5] Aigbe U, Ukhurebor K, Onyancha R, Osibote O, Darmokoesoemo H and Kusuma H 2021 Fly ash-based adsorbent for adsorption of heavy metals and dyes from aqueous solution: a review *J. Mate. Res. Technol.* **14** 2751–74

[6] Aigbe U, Ukhurebor K, Onyancha R, Okundaye B, Pal K, Osibote O, Esiekpe E, Kusuma H and Darmokoesoemo H 2022 A facile review on the sorption of heavy metals and dyes using bionanocomposites *Adsorp. Sci. Technol.* **8030175** 1–36

[7] Devi S, Suman C S, Singh S, Kumar P, Kumar S, Kumar A and Kumar V 2023 Magnetic Fe_2O_3/CNT nanocomposites: characterization and photocatalytic application towards the degradation of rose bengal dye *Ceram. Int.* **49** 20071–9

[8] Zhang Q, Ying G, Pan C, Liu Y and Zhao J 2015 Comprehensive evaluation of antibiotics emission and fate in the river basins of China: source analysis, multimedia modeling, and linkage to bacterial resistance *Environ. Sci. Technol.* **49** 6772–82

[9] He X, Yang D, Zhang X, Liu M, Kang Z, Lin C, Jia N and Luque R 2019 Waste eggshell membrane-templated CuO–ZnO nanocomposites with enhanced adsorption, catalysis and antibacterial properties for water purification *Chem. Eng. J.* **369** 621–33

[10] Anani A, Adama K, Ukhurebor K, Habib A, Abanihi V and Pal K 2023 Application of nanofibrous protein for the purification of contaminated water as a next generational sorption technology: a review *Nanotechnol.* **34** 1–8

[11] Ukhurebor K, Hossain I, Pal K, Jokthan G, Osang F, Ebrima F and Katal D 2023 Applications and contemporary issues with adsorption for water monitoring and remediation: a facile review *Top. Catal.* **2023**
[12] Li J, Huang L, Jiang X, Zhang L and Sun X 2021 Preparation and characterization of ternary Cu/Cu_2O/C composite: an extraordinary adsorbent for removing anionic organic dyes from water *Chem. Eng. J.* **404** 127091
[13] Suman S, Chahal A and Kumar P 2020 Zn doped α-Fe_2O_3: an efficient material for UV driven photocatalysis and electrical conductivity *Crystals* **10** 273
[14] Suman S, Chahal A and Kumar P 2020 Annealing effect on photocatalytic and magnetic properties of Zn doped hematite nanoparticles *AIP Conf. Proc.* **2265** 030476
[15] Eldeeb T, Aigbe U, Ukhurebor K, Onyancha R, El-Nemr M, Hassaan M, Ragab S, Osibote O and El Nemr A 2022 Adsorption of methylene blue (MB) dye on ozone, purified and sonicated sawdust biochars *Biomass Convers. and Biorefin.* **2022** 1–23
[16] Eldeeb T *et al* 2022 Biosorption of acid brown 14 dye to mandarin-biochar-CO-TETA derived from mandarin peels *Biomass Convers. and Biorefin.* **2022** 1–21
[17] Neolaka Y, Riwu A, Aigbe U, Ukhurebor K, Onyancha R, Darmokoesoemo H and Kusuma H 2023 Potential of activated carbon from various sources as a low-cost adsorbent to remove heavy metals and synthetic dyes *Res. Chem.* **5** 100711
[18] Neolaka Y *et al* 2022 Synthesis of Zinc(II)-natural zeolite mordenite type as a drug carrier for ibuprofen: Drug release kinetic modeling and cytotoxicity study *Res. Chem.* **4** 100578
[19] Eleryan A, Aigbe U, Ukhurebor K, Onyancha R, Hassaan M, Elkatory M, Ragab S, Osibote O, Kusuma H and El Nemr A 2023 Adsorption of direct blue 106 dye using zinc oxide nanoparticles prepared via green synthesis technique *Environ. Sci. Pollut. Res.* **2023** 1–17
[20] Eleryan A, Hassaan M, Aigbe U, Ukhurebor K, Onyancha R, El Nemr M, Ragab S, Hossain I and El Nemr A 2023 Kinetic and isotherm studies of acid orange 7 dye absorption using sulphonated mandarin biochar treated with TETA *Biomass Convers. Biorefin.* **2023** 1–12
[21] El-Nemr M, Aigbe U, Hassaan M, Ukhurebor K, Ragab S, Onyancha R, Osibote O and El Nemr A 2022 The use of biochar-NH_2 produced from watermelon peels as a natural adsorbent for the removal of Cu(II) ion from water *Biomass Convers. Biorefin.* **2022**
[22] Feng Y, Su X, Chen Y, Liu Y, Zhao X, Lu C, Ma Y, Lu G and Ma M 2023 Research progress of graphene oxide-based magnetic composites in adsorption and photocatalytic degradation of pollutants. A review *Mater. Res. Bull.* **162** 112207
[23] Hu X, Zhang Y, Wang B, Li H and Dong W 2019 Novel g-C_3N_4/$BiOC_1xI_1$-x nanosheets with rich oxygen vacancies for enhanced photocatalytic degradation of organic contaminants under visible and simulated solar light *Appl. Catalysis B.* **256** 117789
[24] Huang D, Yan X, Yan M, Zeng G, Zhou C, Wan J, Cheng M and Xue W 2018 Graphitic carbon nitride-based heterojunction photoactive nanocomposites: applications and mechanism insight *ACS Appl. Mater. Interfaces.* **10** 21035–55
[25] Wei X, Ou C, Guan X X, Peng Z and Zheng X 2019 Facile assembly of CdS-reduced graphene oxide heterojunction with enhanced elimination performance for organic pollutants in wastewater *Appl. Surf. Sci.* **469** 666–73
[26] Wu H, Tan H, Toe C, Scott J, Wang L, Amal R and Ng Y 2020 Photocatalytic and photoelectrochemical systems: similarities and differences *Adv. Mater.* **32** 1–21

[27] Alam U, Khan A, Ali D, Bahnemann D and Muneer M 2018 Comparative photocatalytic activity of sol-gel derived rare earth metal (La, Nd, Sm and Dy)-doped ZnO photocatalysts for degradation of dyes *RSC Adv.* **8** 17582–94
[28] Bie C, Zhu B, Xu F, Zhang L and Yu J 2019 In situ grown monolayer N-doped graphene on CdS hollow spheres with seamless contact for photocatalytic CO_2 reduction *Adv. Mater.* **31** 1–6
[29] Xu C, Anusuyadevi P, Aymonier C, Luque R and Marre S 2019 Nanostructured materials for photocatalysis *Chem. Soc. Rev.* **48** 3868–902
[30] Keerthana S, Yuvakkumar R, Kumar P, Ravi G and Velauthapillai D 2021 Rare earth metal (Sm) doped zinc ferrite ($ZnFe_2O_4$) for improved photocatalytic elimination of toxic dye from aquatic system *Environ. Res.* **197** 111047
[31] Shaheen K *et al* 2020 Metal oxides nanomaterials for the photocatalytic mineralization of toxic water wastes under solar light illumination *J. Water Process. Eng.* **34** 101138
[32] Singh K, Harish S, Kristy A, Shivani V, Archana J, Navaneethan M, Shimomura M and Hayakawa Y 2018 Erbium doped TiO_2 interconnected mesoporous spheres as an efficient visible light catalyst for photocatalytic applications *Appl. Surf. Sci.* **449** 755–63
[33] Garg A and Chopra L 2022 Dye waste: a significant environmental hazard *Mater. Today Proc.* **48** 1310–5
[34] Kinigopoulou V, Pashalidis I, Kalderis D and Anastopoulos I 2022 Microplastics as carriers of inorganic and organic contaminants in the environment: a review of recent progress *J. Mol. Liq.* **350** 118580
[35] Pham V, Kim J, Chang S and Chung W 2022 Bacterial biosorbents, an efficient heavy metals green clean-up strategy: prospects, challenges, and opportunities *Microorganisms* **10** 610
[36] Saravanan A, Kumar P, Duc P and Rangasamy G 2023 Strategies for microbial bioremediation of environmental pollutants from industrial wastewater: a sustainable approach *Chemosphere* **313** 137323
[37] Yin K W Q L M a C L 2019 Microorganism remediation strategies towards heavy metals *Chem. Eng. J.* **360** 1553–63
[38] Elgarahy A, Elwakeel K, Mohammad S and Elshoubaky G 2021 A critical review of biosorption of dyes, heavy metals and metalloids from wastewater as an efficient and green process *Clean. Eng. Technol.* **4** 100209
[39] Mani D and Kumar C 2014 Biotechnological advances in bioremediation of heavy metals contaminated ecosystems: an overview with special reference to phytoremediation *Int. J. Environ. Sci. Technol.* **11** 843–72
[40] Tripathi M, Upadhyay S, Kaur M and Kaur K 2018 Toxicity concerns of hexavalent chromium from tannery waste *J. Biotechnol. Bioeng.* **2** 40–4
[41] Ali H 2010 Biodegradation of synthetic dyes-a review *Water Air Soil Pollut.* **213** 251–73
[42] Shindhal T, Rakholiya P, Varjani S, Pandey A, Ngo H and Guo W 2021 A critical review on advances in the practices and perspectives for the treatment of dye industry wastewater *Bioengineered.* **12** 70–87
[43] Singh P, Singh P, Singh R and Singh R L 2021 Bio-decolorization of azo dye acid blue 113 by soil bacterium Klebsiella variicola RMLP1 *J. Ecophysiol. Occup. Health* **21** 64–71
[44] Gita S, Hussan A and Choudhury T 2017 Impact of textile dyes waste on aquatic environments and its treatment *Environ. Ecol.* **35** 2349–53
[45] Hossen M, Hussain M, Hakim A, Islam K, Uddin M and Azad A 2019 Biodegradation of reactive textile dye Novacron super black G by free cells of newly isolated *Alcaligenes faecalis* AZ_{26} and *Bacillus* spp obtained from textile effluents *Heliyon* **5** E02068

[46] Ito T, Adachi Y, Yamanashi Y and Shimada Y 2016 Long-term natural remediation process in textile dye-polluted river sediment driven by bacterial community changes *Water Res.* **100** 458–65

[47] Singh L 2017 Biodegradation of synthetic dyes: a mycoremediation approach for degradation/decolourization of textile dyes and effluents *J. Appl. Biotechnol. Bioeng.* **3** 430–5

[48] Donkadokula N, Kola A N I and Saroj D A review on advanced physico-chemical and biological textile dye wastewater treatment techniques *Rev. Environ. Sci. Biotechnol.* **19** 543–60

[49] Singh P, Singh R, Singh P and Singh R 2021 Efficient decolorization of dye acid blue 113 by soil bacterium Bacillus subtilis RMLP2 *Toxicol. Int.* **28** 269–80

[50] Vikrant K, Giri B, Raza N, Roy K, Kim K and Rai B 2018 Recent advancements in bioremediation of dye: current status and challenges *Bioresour. Technol.* **253** 355–67

[51] Onyancha R, Aigbe U, Ukhurebor K and Muchiri P 2021 Facile synthesis and applications of carbon nanotubes in heavy-metal remediation and biomedical fields: a comprehensive review *J. Mol. Struct.* **1238**

[52] Onyancha R, Aigbe U, Ukhurebor K, Kusuma H, Darmokoesoemo H, Osibote O and Pal K 2022 Influence of magnetism-mediated potentialities of recyclable adsorbents of heavy metal ions from aqueous solutions - an organized review *Results Chem.* **4** 100452

[53] Ukhurebor K, Aigbe U, Onyancha R, Nwankwo W, Osibote O, Paumo H, Ama O, Adetunji C and Siloko I 2021 Effect of hexavalent chromium on the environment and removal techniques: a review *J. Environ. Manage.* **280** 111809

[54] Zhao X, Baharinikoo L, Farahani M, Mahdizadeh B and Farizhandi A 2022 Experimental modelling studies on the removal of dyes and heavy metal ions using $ZnFe_2O_4$ nanoparticles *Sci. Rep.* **12** 5987

[55] Mishra J, Saini R and Singh D 2021 Review paper on removal of heavy metal ions from industrial waste water effluent *IOP Conf. Ser.: Mater. Sci. Eng.* **1168** 012027

[56] Badawi A, Abd Elkodous M and Ali G 2021 Recent advances in dye and metal ion removal using efficient adsorbents and novel nano-based materials: an overview *RSC Adv.* **11** 36528–53

[57] Carolin C, Kumar P, Saravanan A, Joshiba G and Naushad M 2017 Efficient techniques for the removal of toxic heavy metals from aquatic environment: a review *J. Environ. Chem. Eng.* **5** 2782–99

[58] Aragaw T and Bogale F Biomass-based adsorbents for removal of dyes from wastewater: a review *Front. Environ. Sci.* **9** 558

[59] Velusamy S, Roy A, Sundaram S and K M T 2021 A review on heavy metal ions and containing dyes removal through graphene oxide-based adsorption strategies for textile wastewater treatment *Chem. Rec.* **21** 1570–610

[60] Pattnaik A, Sahu J, Poonia A and Ghosh P 2023 Current perspective of nano-engineered metal oxide based photocatalysts in advanced oxidation processes for degradation of organic pollutants in wastewater *Chem. Eng. Res. Des.* **2023** 667–86

[61] Farhan A, Zulfiqar M, Samiah R E, Nawaz S, Iqbal H, Jesionowski T, Bilal M and Zdarta J 2023 Removal of toxic metals from water by nanocomposites through advanced remediation processes and photocatalytic oxidation *Curr. Pollut. Rep.* **2023** 1–23

[62] Singh S, Barick K C and Bahadur D 2013 Functional oxide nanomaterials and nanocomposites for the removal of heavy metals and dyes *Nanomater. Nanotechnol.* **3** 3–20 2013

[63] Singh E, Osmani R and Banerjee R 2020 Nanobioremediation: an emerging approach for a cleaner environment *Microbial Bioremediat. Biodegrad.* 309–63

[64] Son B, Long N and Hang N 2021 Fly ash-, foundry sand-, clay-, and pumice-based metal oxide nanocomposites as green photocatalysts *RSC Adv.* **11** 30805–26

[65] Serpone N 2018 Heterogeneous photocatalysis and prospects of TiO_2-based photocatalytic DeNOxing the atmospheric environment *Catalysts* **8** 553

[66] Aigbe U, Ukhurebor K, Onyancha R, Ama O, Osibote O, Kusuma H, Okanigbuan P, Azi S and Osifo P 2022 Dendrimers for environmental remediation *Nanotechnol. Environ. Remediat.* (New York, NY: Wiley) 243–64

[67] Yadav S, Rani N and Saini K 2022 A review on transition metal oxides based nanocomposites, their synthesis techniques, different morphologies and potential applications *IOP Conf. Ser.: Mater. Sci. Eng.* **1225** 012004

[68] Ahuja P, Ujjain S, Kanojia R and Attri P 2021 Transition metal oxides and their composites for photocatalytic d.ye degradation *J. Compos. Sci.* **5** 82

[69] Oyewo O, Elemike E, Onwudiwe D and Onyango M 2020 Metal oxide-cellulose nanocomposites for the removal of toxic metals and dyes from wastewater *Int. J. Biol. Macromol.* **164** 2477–96

[70] Joshi N, Gururani P and Gairola S 2022 Metal oxide nanoparticles and their nanocomposite-based materials as photocatalysts in the degradation of dyes *Biointerface Res. Appl. Chem.* **12** 6557–79

[71] Fadillah G, Yudha S, Sagadevan S, Fatimah I and Muraza O 2020 Magnetic iron oxide/clay nanocomposites for adsorption and catalytic oxidation in water treatment applications *Open Chem.* **18** 1148–66

[72] Qahtan T, Owolabi T, Olubi E and Hezam A 2023 State-of-the-art, challenges and prospects of heterogeneous tandem photocatalysis *Coord. Chem. Rev.* **492** 215276

[73] Singh J, Dutta T, Kim K, Rawat M, Samddar P and Kumar P 2018 Green synthesis of metals and their oxide nanoparticles: applications for environmental remediation *J. Nanobiotechnol.* **16** 1–24

[74] Nandy S, Savant S and Haussener S 2021 Prospects and challenges in designing photocatalytic particle suspension reactors for solar fuel processing *Chem. Sci.* **12** 9866–84

[75] Li D, Xu H, Jiao L and Jiang H 2019 Metal-organic frameworks for catalysis: state of the art, challenges, and opportunities *EnergyChem* **1** 100005

[76] Cumba R, Ligalig C, Tingson J, Molina M, Alguno A, Deocaris C, Latayada F, Primadona I and Capangpangan R 2022 Photocatalytic activity of cellulose nanocrystals/zinc oxide nanocomposite against thiazine dye under UV and visible light irradiation *ASEAN J. Chem. Eng.* **22** 168–77

[77] Elfeky A, Salem S, Elzaref A, Owda M, Eladawy H, Saeed A, Awad M, Abou-Zeid R and Fouda A 2020 Multifunctional cellulose nanocrystal/metal oxide hybrid, photo-degradation, antibacterial and larvicidal activities *Carbohydrate Polym.* **230** 115711

[78] ZabihiSahebi A, Koushkbaghi S, Pishnamazi M, Askari A, Khosravi R and Irani M 2019 Synthesis of cellulose acetate/chitosan/SWCNT/Fe_3O_4/TiO_2 composite nanofibers for the removal of Cr (VI), As (V), Methylene blue and Congo red from aqueous solutions *Int. J. Biol. Macromol.* **140** 1296–304

Chapter 7

The effective sensing and sequestration of metal ions and dyes using functionalized biosorbents

Robert Birundu Onyancha, Uyiosa Osagie Aigbe, Kingsley Eghonghon Ukhurebor, Bill Clintone Oyomo, Stephen Chege and Michael O Atambo

The availability of excess metals and dyes beyond permitted limits in aqueous systems is harmful to humans and aquatic organisms. Metals and dyes are nonbiodegradable, hence their direct or indirect uptake accumulates in living organisms, leading to detrimental toxic substances. As a result of their toxicity and the urgent need for clean water, efforts to detect, mitigate, and remove these substances from aqueous solutions remain a priority for many researchers. Therefore, this chapter provides a brief introduction to this topic followed by a discussion of the sensing techniques used for metal ions and dyes, their removal strategies, and a summary that includes future perspectives.

7.1 Introduction

Toxic pollutants such as heavy metals (HMs) and dyes are mainly generated via natural occurrences and anthropogenic activities. These pollutants have been found to be highly non-degradable and extremely toxic, and they bioaccumulate in the body when they are present in the food chain. In the body, for instance, metal ions such as Cd^{2+}, As^{3+}, Pb^{2+}, Hg^{2+}, and Ag^{+} react with bioparticles to form noxious substances with attributes that depend on increased biomagnification and concentration [1]. In addition, dyes such as malachite green [2], methylene blue [3], and acid brown 14 [4], as well as many more azo dyes, are considered poisonous and thus threaten the environment.

Consequently, they directly and indirectly pose a danger to plants, animals, and humans, both on land and in aquatic systems. Their impact on human life is well documented; it comprises injury to the gastrointestinal tract, vascular occlusion, fatigue, elevated blood pressure, memory loss, speech problems, increased allergies,

doi:10.1088/978-0-7503-5598-8ch7

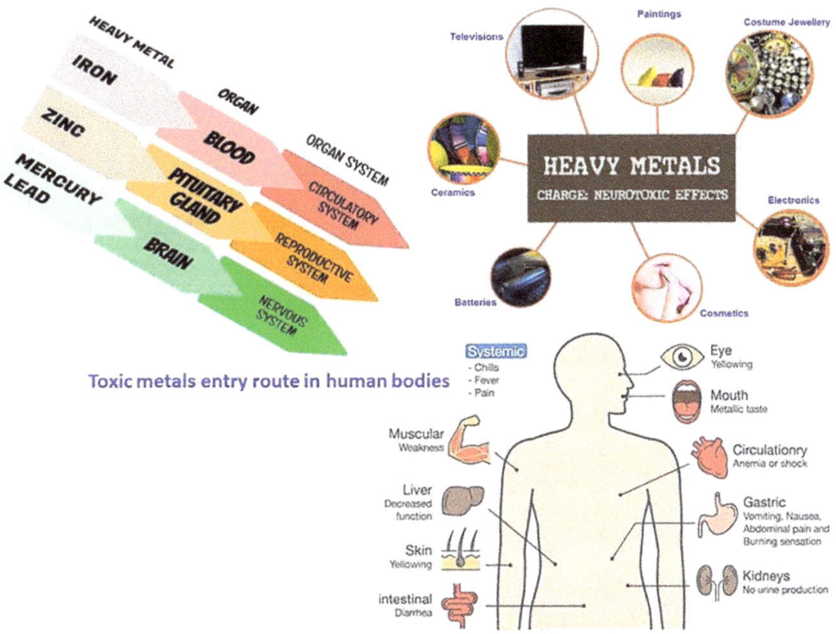

Figure 7.1. The sources and impacts of pollutants in human bodies. Reprinted from [1], Copyright (2022), with permission from Elsevier.

irritability, osteoporosis, neurological disorder, obstructive lung disease, renal dysfunction, and aggressive behavior [5] (figure 7.1 as adapted from [1]). Also, because metal pollutants are nonbiodegradable, they easily modify the physical, chemical, and biological properties of water.

The treatment of contaminated water and the reduction of pollutant concentrations to allowed limits before they are released into the environment remain the viable options for mitigating the dangers of these pollutants [1, 6].

Hence, the potential threat they pose to life coupled with the need for clean and safe water have compelled researchers to develop reliable detection techniques capable of quantifying these pollutants. Detection techniques for pollutants are classified into three types, namely optical, spectroscopic, and electrochemical. Spectroscopic methods are quite expensive and labor-intensive during the pretreatment period but extremely sensitive to pollutants. They include cold vapor atomic fluorescence spectroscopy, atomic emission spectroscopy, atomic absorption spectroscopy, and inductively coupled plasma mass spectroscopy. Conversely, electrochemical methods are cost-effective, reliable, time-saving, and easy to use. Examples include voltammetry, amperometry, impedance measurements, potentiometry, and coulometry. Pollutants can also be detected via absorption, luminescence, or reflection spectrometry; these optical measurements are categorized as conventional techniques [7].

In addition to detection, the sequestration of dyes and HMs ions from aqueous systems remains a priority. Consequently, several methods have been devised and

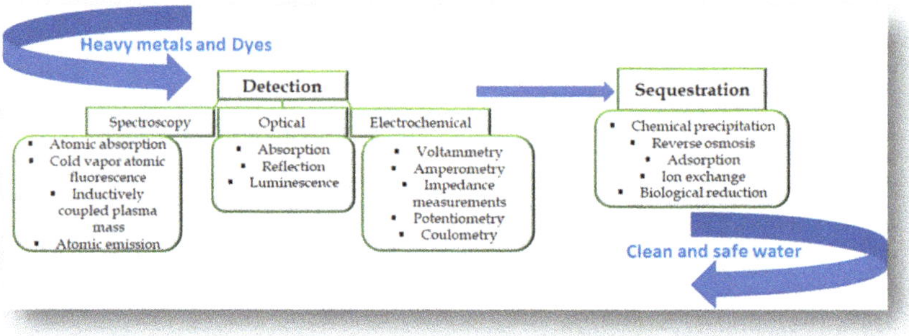

Figure 7.2. Detection techniques and sequestration strategies.

used in the removal process. These include ion exchange, adsorption, flocculation, chemical precipitation, electrochemical process, solvent extraction, evaporation, and membrane separation. However, except for adsorption, these possess inherent problems such as the production of toxic sledge, low efficiency, waste disposal cost, high operational cost, and incomplete sequestration of both metals and dyes [1].

The adsorption process provides excellent operability, high selectivity, easy usage, low operational cost, and great removal efficiency. For instance, adsorption studies of malachite green sequestration using activated potassium hydroxide leaf have shown that it is an efficient and less expensive process for treating malachite-green-contaminated industrial effluents [2]. The removal of methylene blue dye using biochars derived from sawdust revealed an adsorption dependence on the solution's pH; the maximum adsorption was recorded at a pH of 12 [3]. In addition, studies of inexpensive activated carbon (AC) synthesized from various sources have demonstrated that HMs and dyes can successfully be removed from aqueous solutions [8]. Agronomic waste materials have been extensively employed as bioadsorbents for dye and HM sequestration in an environmentally friendly process. For example, Mandarin-CO-TETA produced from mandarin peels has been employed in the removal of acid brown 14 dye and shown to yield a huge improvement in removal efficiency [4].

Therefore, in this chapter, we provide an overview of the detection and removal of metal ions and dyes using different methodologies, as shown in figure 7.2. This chapter briefly highlights a discussion of detection techniques and removal procedures for these contaminants.

7.2 The detection or sensing of heavy metals and dyes

An instrument or device deployed in the sensing of dyes and HMs in the environment is generally referred to as a detector. Detectors can precisely quantify the metal ions and dyes present, hence providing vital information necessary for removal processes. Importantly, a particular technique applicable to the detection of all metal ions and dyes is missing, even though various metal detection techniques have been developed and advanced. Metal ion detection techniques should ideally be environmentally friendly, cost-effective, easy to use, time-effective, and highly sensitive to ensure low-concentration detection. Sensing techniques can be broadly classified into spectroscopic, electrochemical, and optical [9].

7.2.1 Spectroscopic sensing techniques

Spectroscopic techniques can be applied to a wide range of elements, as they can concurrently establish the concentrations of pollutants in a sample even at low concentration limits. Nevertheless, they have inherent limitations, since they are quite expensive and required a skilled workforce to operate the equipment. Equally, they suffer from tedious sample preparation and immovable equipment due to its size, which limits on-site detection measurements [10]. Examples of these spectroscopic techniques include atomic absorption spectroscopy, inductively coupled plasma mass spectroscopy, x-ray fluorescence spectroscopy, neutron activation analysis, inductively coupled plasma optical emission spectroscopy, flame atomic absorption spectroscopy, high-resolution surface plasmon resonance spectroscopy, and anodic stripping voltammetry.

7.2.1.1 Atomic absorption spectroscopy (AAS)

Figure 7.3 shows a schematic representation of atomic absorption spectroscopy. The system comprises a light source, an atomizer, a monochromator, and a display system. In this technique, a particular wavelength is employed specifically to excite ground-state atoms to an excited state. The total absorbed energy measured during excitation necessarily corresponds to the concentration (quantity) of atoms in the analyzed sample.

The primary light is generally a hollow cathode lamp or an electrodeless discharge lamp made of identical elements in order to stimulate atoms/ions of the sample to be analyzed. It is important to note that only a gaseous medium with separated atoms or ions can be analyzed by this method. Through atomization, samples are decomposed to form a gaseous phase of atoms or ions which is subsequently used for analysis. Different types of detectors, such as photomultiplier tubes and

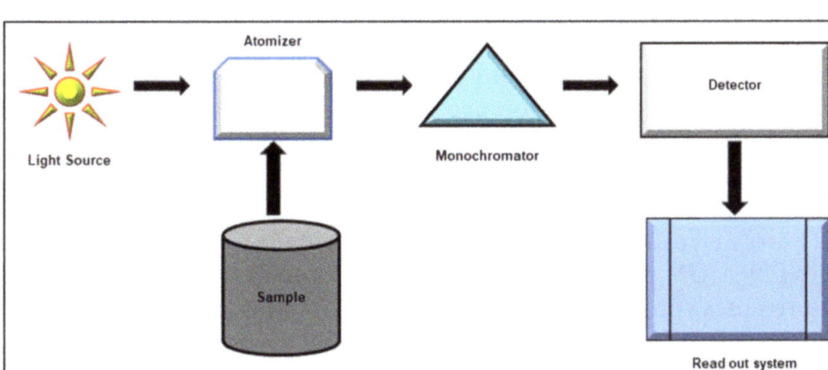

Figure 7.3. A typical atomic absorption spectrometer.

solid-state detectors, have successfully been used in this technique. A flow injection Hg system has been used for the detection of Hg due to its improved nature.

Cold vapor atomic absorption spectroscopy has been employed to evaluate the content of Hg^{2+} in a solution, and the results indicated a linear calibration curve between 0.05 and 85.6 ng ml^{-1} [11]. In addition, flame atomic absorption spectrometry studies have been reported for the analysis of Cd, Mn, Al, and Mg recovery using eight different digestion methods in vegetable samples [12]. The technique was found to be successful, and the quantification efficiency of metals was found to depend on the plant species, metallic ion, and digestion procedure.

7.2.1.2 Graphite furnace atomic absorption spectroscopy

The furnace in this setup comprises an open-ended graphite cylinder with a centralized hole meant for sample introduction. The direct introduction of the sample into the tube results in heating, thus removing the solvent and matrix systems. Sample atomization coupled with analysis then follows, in a way that is reminiscent of flame atomic absorption spectroscopy [13]. In this technique, more reproducible signals are obtained, which is attributed to the fact that sample is not in direct contact with the furnace wall; therefore, it is not exposed to a rapid change in temperature. The disadvantages of this technique are that it requires a longer analysis time compared to flame absorption spectroscopy [14], and it can only detect a few elements [15].

Graphite furnace atomic absorption spectroscopy (GFAAS) has recently been used to detect traces of cadmium in soil samples by employing a metal–organic framework (MOF, UIO-66-NH_2) to enhance accuracy [16]. Dispersive liquid–liquid microexcitation (DLLME) coupled with a graphite furnace was proposed as a way of detecting ultralow traces of copper in environmental water samples [17] The limit of detection for lead in blood was improved from 1 μg dl^{-1} to 0.2 μg dl^{-1} using GFAAS due to the use of newer instrumentation, different calibration strategies, and increased sample injection volume. This improved technique could be used in hospitals to quantify lower levels of lead in blood [18].

7.2.1.3 Atomic fluorescence spectroscopy (AFS)

In this technique, sample solution atomization is performed; then, the atoms are illuminated by a light source, resulting in excitation. Following excitation, the atoms experience radioactive deactivation accompanied by the emission of characteristic radiation, which is then detected. The cold vapor atomic fluorescence technique can be employed in Hg detection. Argon is used as a carrier gas to transport Hg atoms to the cell, where the atoms are excited by a collimated UV source. Consequently, the absorbed energy is re-emitted by the excited atoms in the form of fluorescence, which is measured by a UV photodiode detector or a photomultiplier tube detector. This technique requires simple instrumentation, has a small investment cost, and offers fast and convenient operation; thus, it is regarded as one of the most promising sensing strategies.

Cold vapor atomic fluorescence spectroscopy has recently been used to determine the amount of Hg in plankton samples [19]. Plankton accumulates Hg and transfers it to the aquatic biota. The consumption of mercury-contaminated seafood is clearly harmful to human health. The developed wet digestion strategies exhibited excellent recovery efficiencies of 94.1% ± 7.6% for the IAEA-450 method and 97.2% ± 4.6% for BCR-414.

Liquid chromatography atomic fluorescence spectrometry (LC-AFS) was recently used for the speciation analysis of ultralow trace levels of Hg in rice. A preconcentration technique was applied to real rice samples prior to LC-AFS analysis, since the sensitivity of LC-AFS is inferior to that of liquid chromatography inductively coupled plasma mass spectrometry (LC-ICP-MS). LC-AFS is, however, desirable for the speciation analysis of Hg in rice, due to its affordability and ease of operation. LC-ICP-MS is costly and complicated to operate [20].

In addition, a method was developed to determine the total content of HM ions in water via this technique using N, S-CQDs prepared through the graphene oxide-assisted production method. This fluorescence quenching technique displayed a wide range of linear responses for several water-soluble metal ions. The fluorescence quenching was attributed to integration resulting from coordination interactions between the N, Sulfur co-doped carbon quantum dots (S-CQDs) and the HM ions [21].

7.2.1.4 X-ray fluorescence spectroscopy (XFS)

The irradiation of the surface of a material by x-rays and gamma rays results in ionization. Notably, the ejection of electrons from the inner orbitals of the K shell or the L shell occurs due to bombardment of the material by high-energy radiation. This leads to the creation of electron vacancies which are filled by electrons transitioning from higher-energy shells, coupled with the release of photons, i.e. fluorescence. This photon energy can be measured and accordingly used to identify the composition of the sample because each element uniquely emits a particular energy spectrum. This technique can easily determine the elemental composition of a sample.

A research strategy was developed and optimized to quantify HM ions such as Pb (II) in vegetables at low limits of detection through wavelength-dispersive x-ray fluorescence spectroscopy (WD-XRF) or portable energy-dispersive x-ray fluorescence spectroscopy (ED-XRF) [22]. Despite significant compromises in the efficiency and precision of the results, the portability of this technology and the fact that it can be improved showed that it can be a promising strategy for HM detection.

A universal and independent pipeline for data analysis was recently added to a dynamical method of analysis which can rectify complex x-ray interactions. The pipeline was reported to yield more accuracy in processing XRF data for herbarium species compared to both the empirical calibration approach and the built-in instrument algorithm [23].

Furthermore, XRF has recently been used to perform a qualitative analysis of HM ions in fine jewelry. The number of samples was 143. Nickel (Ni) was reported to be present in 61 of the samples; this may have health effects, as it can cause contact dermatitis. Lead (Pb) was also reported to be present in 18 gemstone samples, but it was absent from all the metal alloy samples [24].

7.2.2 Electrochemical methods of sensing

The presence of HM ions in water generally leads to changes in its electrical parameters such as charge, impedance, and electroluminescence in an electrochemical setup [25]. Therefore, the presence of HM ions in any aqueous system can be detected by employing the following electrical signal detection techniques: potentiometric, volumetric, coulometric, and electrochemiluminescent techniques, impedance change measurements, and amperometry measurements.

The advantages of these techniques are that they are user-friendly, reliable, and economical. They only require simple procedures in order to monitor contaminated samples. Moreover, they have shorter analytical times than other spectroscopic techniques [26]. Conversely, these techniques provide lower sensitivities and poorer sensing limits compared to spectroscopic strategies and optical methods. To enhance the detection performance of these techniques, it is necessary to modify and re-engineer their designs [27].

7.2.2.1 Potentiometry

Potentiometry techniques are commonly employed for the qualitative analysis of ions in aqueous media; they are very effective in HM detection. These techniques are based on measuring the electromotive force (emf) in the absence of a current. An additional advantage of this technique and those highlighted in the section above is that they have a broad range of responses [28]. Their drawbacks are high detection limits and reduced sensitivity. To overcome these limitations, electrodes modified using carbon nanotubes (CNTs) and metallic nanoparticles have provided promising results because this modification offers enhanced sensitivity and lowers the limits of detection of HM ions [29, 30].

A potentiometric method that employs a highly selective erdosteine-based sensor (Cu-erdosteine sensor) has recently been used for the surveillance of Cu (II) ions in shredded canned tuna and food products. Negative health effects may result from long-term exposure to large amounts of Cu (II) ions [31].

The development and optimization of a potentiometric indicator electrode sensor using carbon paste electrodes (CPEs) has been reported [32]. The optimization of the CPE composition was performed by multivariate techniques such as D-optimal mixture design. Tap water and pharmaceutical product samples were tested for silver ions using the CPE sensor. The authors reported that the sensor had high sensitivity and selectivity for the detection of silver ions. Similarly, an approach was recently developed to demonstrate to students how to perform a potentiometric examination of the concentration of metal ions. This approach uses calcium-ion-selective electrodes (Ca-ISE) and was used to determine the zinc concentration in supplement tablets [33].

7.2.2.2 Amperometry

This technique involves an application of a fixed potential between working and reference electrodes in a solution that comprises electroactive species. Consequently, this leads to the establishment of extremely small currents owing to the reduction of

HMs ions, which are then measured. In this technique, the potential of the working electrode is fixed; thus, among various electroactive species, only a particular metal ion is detected. The analyte experiences a faradic reaction at an appropriate polarity and degree of applied potential. Nonetheless, the working electrode has a small surface area and only a small part of the analyte reacts. The electrode was modified to overcome this problem and enhance the sensitivity of this approach for metal ion recognition.

A sensor for the amperometric recognition of Hg ions (Hg^{2+}) was fabricated and used for Hg detection in water and milk samples [34]. The sensor was based on the immobilization of organic chelator ionophores and multiwalled carbon nanotubes (MWCNTs) on a glassy carbon electrode. The research literature also shows that the amperometric detection of Pb^{2+}, Cd^{2+}, and Cu^{2+} was successfully performed using a single carbon fiber electrode as the working electrode, which had the advantage of eliminating the toxic Hg electrode [35].

7.2.2.3 Voltammetry

Voltammetry is the most used technique among electrochemical methods due to its high sensitivity, high accuracy, and lower detection limits. Here, current measurements are performed at various potentials, thus generating current–voltage graphs for analysis. Examples of voltammetric modes include cyclic, pulse, stripping, and galvanostatic.

7.2.2.3.1 Cyclic voltammetry

In this method, the potential is scanned forward from a minimum value to a maximum, followed by a backward scan. Consequently, a cyclic voltammogram is produced by simply recording the current during the forward and backward scans and plotting it against the potential. Cyclic voltammograms can be used to measure crucial parameters, such as the cathodic and anodic peak potentials and related currents [36]. Cyclic voltammetry coupled with various modifiers has recently been used in the detection of lead, arsenic, and cadmium metal ions [37].

7.2.2.3.2 Pulse voltammetry

The method employs voltage pulses with diverse profiles and amplitudes. Pulse voltammetry can be classified further into normal pulse, differential pulse, and reverse pulse voltammetry. Due to its excellent sensitivity in the detection of metal ions, differential pulse voltammetry [38] is the most widely used pulse voltammetry technique.

Differential pulse voltammetry was recently used to observe the electrocatalytic activity of metals on Au-modified porous carbon threads which are employed as working electrodes for the simultaneous detection of Pb, Cu, Ca, and Hg [39]. Differential pulse stripping voltammetry exhibited an excellent linear correlation between the peak current and the Pb^{2+} concentration in an investigation that was recently carried out by Zhuang *et al* using a planar disk electrode (MWCNTs)/chitosan (CS)/lead (Pb^{2+}) ionophore IV nanomaterial [40].

7.2.2.3.3 ***Stripping voltammetry***

The two main steps in stripping voltammetry are as follows: the analyte solution is thoroughly agitated and deposited on the electrode in the first stage, which is called electrodeposition. The second stage, known as the voltammetric step, involves stripping the analyte off and analyzing it using any voltammetric technique. Stripping voltammetry has advantages, such as the ability to detect low concentrations, and its setup is simple and inexpensive [41]. This technique can be further categorized into two types, namely anodic stripping voltammetry (ASV) and cathodic stripping voltammetry (CSV).

The implementation of ASV has been shown to be challenging due to the pre-analysis optimization of parameters that depend on the metal ions to be detected. However, if proper optimization is done based on the application of interest, ASV is an effective and powerful voltammetric electrochemical technique [42].

7.2.2.4 Galvanostatic techniques

In these techniques, the measurement of electric potential is performed in conjunction with the application of electric current. Here, a galvanostat is employed to regulate the current between the two electrodes (working and reference). The advantage of these techniques compared to potentiostatic techniques is that they employ simple instrumentation. However, the huge inherent double-layer charging is disadvantageous. An example of the most commonly applied galvanostatic technique is galvanostatic stripping chronopotentiometry [43], which has been used to perform the simultaneous quantitative analysis of lead, copper, and cadmium ions in commercial butter and margarines [44].

7.2.3 Optical sensing methods

Common techniques such as luminescence spectrometry, reflection spectrometry, or absorption spectrometry can be employed in the sensing/recognition of the optical effects of a material. However, for the optical sensing of HMs integrated optics, ionophores, capillary-type devices, optical fibers, and certain indicator dyes are regularly used. Optical ion sensing has its limitations. There are optical indicators that can react with multiple metal ions, making it difficult to obtain accurate results. These indicators can also react with hydrogen ions, which further complicates the detection process. To address these issues, it is important to control the pH of the solution and apply an appropriate correction factor to compensate for any interference from other metal ions or hydrogen ions [45].

Optical chemical sensors (OCSs) are also used in sensing of HMs ions and have the advantage of being an on-site detection approach. A sample that is irradiated binds the OCS to the analyte, resulting in changes in optical properties such as absorption, emission, or transmission, which can be associated with analyte concentration. Ullah *et al* reviewed optical sensors, focusing on fluorescent sensors, surface plasmon resonance (SPR) sensors, and surface enhanced Raman scattering (SERS) [10].

An optical carbon-based nanosensor has been created that employs neural network architectures to analyze complex fluorescence signals from a carbon dot (CD) aqueous suspension containing HMs (Cu^{2+}, Ni^{2+}, Cr^{3+}) and anions (NO_3^-). The developed optical CD nanosensor was used for the simultaneous measurement of Cu^{2+}, Ni^{2+}, Cr^{3+}, and NO_3^- concentrations in liquid media [46].

SERS has been reported as a method for HM ion sensing in food and agricultural products. SERS offers advantages such as simple sampling, a quick response, non-invasiveness, and supersensitivity [47].

7.2.3.1 Indicator-dye-based sensors

This kind of sensor relies on an indicator dye that binds with HM ions, leading to a variation in the absorption or fluorescence properties of the dye. The direct optical determination of an HM ion (where the indicator acts as a transducer) is not possible using this type of HM ion sensor. Both static and dynamic quenching of the luminescence of an indicator dye occur in another group of indicators in which HMs act as 'quenchers.' Importantly, in static quenching, the interaction between the quencher and the fluorophore occurs in its ground state, while for dynamic quenching, interaction between the metal ions and the indicator occurs only in excited states. In dynamic quenching, the dye is not consumed, hence the process is reversible [48].

The advantage of fluorescent indicators is that they provide improved selectivity and sensitivity. The disadvantages of these types of sensors are: (1) several indicators used for metal ion detection are nonselective and can bind to multiple metal ions. (2) Some indicators are unsuitable for optical sensing of HM ions due to issues such as unfavorable analytical wavelengths, unfavorable stability, or a requirement for extra reagents. Some indicators also bind irreversibly or only under specific pH conditions. (3) A significant issue with these sensors is that a unique dye with an inimitable analytical wavelength must be employed for each metal ion, thus making the detection process more complex [49].

A quinoxaline–hydrazinobenzothiazole-based sensor was developed and used to detect Cu^{2+}, Co^{2+}, Ni^{2+} and Hg^{2+} metal ions via colorimetry and fluorescence. Naked-eye detection under normal daylight was also possible when the sensor was employed [50]. Also, the metal-ion sensing properties of a bispicolyl appended oxidized bisindole derivative (BIM-BPA) were investigated. The indicator-based sensor displayed a chromogenic response from red to light brown in the presence of Hg^{2+} and Cu^{2+} ions. The sensor was applied in biological media, such as plasma and albumin protein, which are competitive sites for hosting metal ions since they contain binding sites [51].

An anthraimidazoledione-based amphiphilic compound has been designed to probe for the dual sensing of Ca^{2+} ions. The compound is pink in its native state, but when Ca^{2+} is added, the color of the solution changes to orange with a concomitant appearance of blue fluorescence. This method has the advantage of providing a naked-eye response. The probe was tested in the presence of blood serum albumin protein and water samples for the detection of Ca^{2+} ions.

7.2.3.2 Ionophore-based sensors

Ionophores are well known to be organic molecules that form complexes with ions and have properties that make it possible to specify HM ions. These properties include the ability to bind to specific ions, ion carrier properties, and complexation properties. Ionophores are used for HM ion sensing by optical means. Ionophores can be transformed into effective HM ion sensors by the addition of a variety of sensing techniques such as chromogenic or fluorogenic moieties or a combination of both with appropriate dyes [52]. The removal of ions by membranes is another suitable way to detect HM ions. In optical sensors of this type, selective binding between HM ions and ionophores is responsible for the recognition of the metal ion, while a proton-selective chromatophore provides the optical signal. When HM ions bind to the ionophore, protons with a charge corresponding to that of the HM ion are expunged from the chromoionophore, leading to a change in the color or fluorescence of the chromoionophore. The ionophores contain heteroatoms such as nitrogen, oxygen, or sulfur, which coordinate with the HM ions via coordination bonds.

A new *o*-hydroxyazocompound **L**-bearing pyrrole residue was recently synthesized for use as a chemosensor [53]. Researchers tested the chemosensor for selectivity, sensitivity, and limit of detection, among other parameters, as an optical sensing material. The researchers concluded that the compound could be used for the detection of Cu (II) in samples of water or body fluids. When the compound interacts with Cu (II), it displays a color variation from yellow to pink. Equally, chromoionophore–ionophore chemistry can be used to aid in the optical detection of ionic species. A single-use lithium sensor has also been created by introducing ionophore-based chemistry into a microfluidic thread-based analytical device (μTAD). This was used for the optical evaluation of lithium ions in saliva [54].

7.3 Sequestration methods

Once the detection of dyes and HMs has been achieved in aqueous solutions, the next phase is the removal and mitigation process to ensure that their concentrations are within allowable limits. A plethora of strategies have been advanced over the years to remove pollutants and clean wastewater. Some of these strategies include ion exchange, adsorption, flocculation, chemical precipitation, electrochemical process, solvent extraction, evaporation, and membrane separation. Apart from adsorption, most of these methods have drawbacks, including the production of sludge which is unfriendly to the environment, the production of secondary toxic substances, and inefficiency at low concentrations [1].

7.3.1 Chemical precipitation (CP)

This technique is among the most efficient methods in removing pollutants from wastewater and involves the addition of chemical reagents to the wastewater, which then leads to a reaction with pollutants to form indissoluble residues which can be removed using sedimentation or filtering mechanisms to obtain water that is free of toxic metal ions. Hydroxide and sulfide precipitation are the two most prevalent mechanisms of chemical precipitation. Due to their low cost and ease of handling, several hydroxides are employed for metal precipitation [55].

A detailed study has demonstrated the use of magnesium hydroxy carbonate in the removal of VO_2^+, Cr^{3+}, and Fe^{3+} from aqueous solutions. The reaction process was completed within 20 min at a pH of 7.1. Removal efficiencies greater than 99.9% were achieved, and the precipitates were composed mainly of Fe_2O_3, V_2O_5, and Cr_2O_3. The treatment was deemed to be successful because of its reduction of HM concentrations [55].

A comparative analysis of the efficiency of dye removal was performed using $FeSO_4/Ca(OH)_2$, Al2(SO4)$_3$, $MgCl_2$/NaOH, and Polyaluminum chloride (PAC). It was established that $MgCl_2$ provided the greatest color removal efficiency due to its surface area and positive surface charge, which facilitated dye removal via charge neutralization. The efficiency was also found to depend on the pH, and optimum dye removal was achieved when the pH value was 12.0 or more [56].

Na_2CO_3, Na_2S, and $Ca(OH)_2$ precipitants were also employed in the sequestration of Zn (II), Pb (II), and Cu (II) HMs from wastewater. For Cu (II) and Zn (II) and at a concentration of 100 mg l^{-1}, three precipitants achieved a removal percentage of 99.99%. The removal percentages for lead were 99.75%, 97.78%, and 76.14% using Na_2S, $Ca(OH)_2$, and Na_2CO_3, respectively [57].

7.3.1.1 Hydroxide chemical precipitation (HCP)

HCP involves the use of metal hydroxides produced by reactions with hydroxide pollutants [58]. For instance, $Ca(OH)_2$ and (NaOH) were employed as precipitants to eliminate Cu(II) and Cr(VI) from aqueous systems [59]. Coagulants such as organic polymers and iron salts have often been used, and it has been found that they can also improve the removal of HM ions such as hydroxide precipitates from wastewater. Furthermore, several chemical reagents such as alum, ferric chloride, and lime chloride have been utilized in the sequestration of toxic HM ions from aqueous media [59].

The pH level influences the hydroxide precipitation process, which can cause problems when working with mixed metals, since the ideal pH value for one metal may result in the release of another metal into the solution [7]. Therefore, sulfide precipitation is an effective alternative technique for removing pollutants from wastewater.

7.3.1.2 Sulfide precipitation

Sulfide precipitation is another viable chemical precipitation technique that is employed to remove toxic and heavy metal ions from wastewater. This process yields metal sulfide precipitates that have significantly lower solubility than hydroxide precipitates. Moreover, this method enables the selective removal of metal ions at a faster reaction rate, and it also has superior settling properties coupled with the potential to recycle sulfide precipitates through smelting [58]. To remove Cd^{2+}, Cu^{2+}, and Pb^{2+}, iron sulfide and pyrite were used as precipitants in the sulfide precipitation technique [60]. It was established that adsorption was dependent on pH and that an elevation in temperature led to an improved adsorption capacity. The decreasing order of adsorption capacities of these metals was $Pb^{2+} > Cu^{2+} > Cd^{2+}$ and very low levels of Pb^{2+} and Cu^{2+} were desorbed in an acidic system.

A pyridine-based thiol ligand was utilized to extract copper and cadmium from a solution [61]. The authors of this paper found this approach to have numerous benefits, including the availability of multiple sites for the removal of HMs as well as stable metal–ligand precipitate formation. It was suggested that the ligand contains sulfur and binds to HMs, resulting in the formation of stable precipitates. The length of the ligand was also found to be sufficient to facilitate interactions between HMs and each terminal sulfur group. The researchers achieved a 99.99% removal rate for copper at a pH of 4.5 and a 99.88% removal rate for cadmium at an optimal pH of 6.

7.3.1.3 Chelating chemical precipitation (CCP)

It is necessary to remove not only poisonous HM ions but also chelating agents. This is achieved using chelating precipitation. Chelating precipitants such as trimercaptotriazine, potassium/sodium thiocarbonate, and sodium diethyldithiocarbamate can be utilized instead of sulfide or hydroxide precipitation to remove HMs from aqueous systems [7]. Novel chelating precipitants have been synthesized to resolve the problem that commercial HM precipitants lack essential binding sites. For instance, a thiol-based compound known as 1,3-benzenediamidoethanethiol was synthesized and successfully employed for the precipitation of Hg ions in wastewater [62].

7.3.2 Adsorption techniques

Due to their poor capacities and removal rates, chemical precipitation and other traditional techniques are not always desirable for removing HMs other than Hg. Therefore, adsorption is an appealing technique, since it is less expensive, simple, and easy to operate [63]. This method is popular in the sequestration of dyes and HMs because pollutants are attached to the adsorbent's surface, separated, and removed from the wastewater. The materials can be natural or manufactured, and they must possess a high affinity for the pollutant.

Adsorption has various advantages over the traditional techniques, including the ability to reduce sludge, its superior efficiency coupled with low cost, its ability to regenerate and use adsorbents, and the prospect of recovering the metal [1, 5, 6]. For the sequestration of pollutants, several adsorbents such as biomaterials, activated carbon (AC), carbon nanotube (CNT)-based materials, and layered double hydroxides have been utilized [6, 7]. Other adsorbents that have been employed in various studies include zeolite, fly ash, manganese oxides, agricultural waste products such as water hyacinth and hazelnut shells, sunflowers, orange peel pith, coconut coir pith, bacterial biosorbents, marine algae, fungal biosorbents, and microbial and plant-derived biomasses [58].

7.3.2.1 Activated carbon

AC has been shown to be very efficient in eliminating toxic HM ions and dyes from aquatic environments. This is mostly because of its large surface area, as a result of which, it a frequently used nanomaterial in wastewater treatment. Its exceptional ability in the sequestration of wastewater is due mostly to its intricate pore structure and the presence of several surface functional groups [7]. In addition, it can be produced from various sources that contain high levels of carbon, including industrial waste materials.

Researchers have reported the synthesis of AC from sugarcane bagasse using H_3PO_4, as well as its utilization in the elimination of Fe (II) and Mn (II) from contaminated water. A study revealed a high Brunauer-Emmett-Teller (BET) surface area of 671.54 $m^2\ g^{-1}$ and a maximum iodine adsorption value of 602.23 mg g^{-1}. At a temperature of 25°C, the adsorbent demonstrated a strong ability to remove both Mn (II) and Fe (II) from hydrated solutions, achieving absorption values of 7.01 mg g^{-1} and 5.40 mg g^{-1} for Fe (II) and Mn (II), respectively [64]. In another study, research was performed on AC and adsorption efficiencies of 78% for Cd^{2+} and 94% for Pb^{2+} were observed [65]. Al-Othman and his co-authors used peanut shells to prepare AC via chemical activation with potassium hydroxide (KOH) and used it to remove Cr (VI) from aqueous solutions with concentrations ranging from 10 to 100 mg l^{-1}. The Cr (VI) adsorption showed sensitivity to the pH value, and the best adsorption occurred between pH 2 and pH 4 [66].

Studies have been conducted to investigate the production of a new AC from fig sawdust and its application as a biosorbent for the elimination of Pb (II) from an aqueous medium [67]. The study revealed the possibility of employing low-cost and sustainable biomass as a precursor for the manufacture of AC. The authors evaluated the impact of various operational parameters on the Pb (II) adsorption capacity of the developed fig sawdust AC (FSAC), including time, pH, adsorbent dosage, and temperature. The study indicated that FSAC could be a less expensive alternative to more expensive materials for the removal of harmful Pb (II) in liquid waste [67]. Overall, the research study provided useful insights into the possibilities of employing biomass as a precursor for AC production and the application of the resulting product as a biosorbent for pollutant removal from wastewater.

Similarly, researchers have investigated the efficiency of AC generated from coconut shells (ACS) for the removal of Cu^{2+}, Fe^{2+}, Zn^{2+}, and Pb^{2+} from industrial wastewater produced by the electroplating method [68]. The researchers chemically activated the ACS using zinc chloride and ran batch adsorption tests to establish how the dose of adsorbent, contact time with pollutants, pH, and stirring rate affected the process. The results showed that metal ion adsorption was affected by the adsorbent's dosage, contact time, pH, and stirring rate. The optimal parameters for effective adsorption were determined to be a 1g dose of coconut shell AC, a 350 rpm stirring rate, and a pH of 6. Therefore, this study shows that employing coconut shell-based AC has considerable potential as a cost-effective option for metal ion removal in wastewater treatment.

7.3.2.2 Biosorption

Biosorption refers to a process which is non-metabolically mediated. It involves the passive or active attachment of a metal to live or dead biomass. The extraction of harmful HMs ions from industrial wastewater using dry biomass can be regarded as an alternative approach in industrial wastewater treatment. The biosorption of HMs involves physisorption and chemical adsorption, both of which are important in metal ion adsorption processes [7].

Biosorption is a sustainable and environmentally friendly technology for water treatment and is aligned with green chemistry principles. One of its key advantages is the use of biowastes as adsorbent materials, which can minimize waste and contribute to the circular economy. Biosorption can effectively remove different water contaminants. Compared to the usual wastewater treatment methods, such as chemical precipitation and membrane filtration, biosorption has lower operating and manufacturing costs and can achieve higher removal efficiency. Therefore, biosorption has great potential as a practical solution for sustainable water management and environmental protection [69].

The use of inexpensive biomass for pollution management has been investigated, particularly in eliminating toxic HM ions from wastewater. Numerous studies have reported that various agricultural materials such as rice bran, cottonseed crop mill waste, soybean, leaf-derived biosorbents, jute, sawdust, used mushroom substrate, corn cobs, coconut shells, cellulose-based sorbents, and groundnut husk have shown effectiveness in removing HM ions [7].

Moreover, phytoremediation, a plant-based approach, has been demonstrated to be an effective and environmentally acceptable way of removing HM ions [70]. In addition, biosorption utilizing diverse materials has attracted interest for metal ion removal due to its ease of availability, cost-effectiveness, selectivity, and capability to effectively remove metals even at low concentrations [7].

For instance, microalgae have been used in a study of Cu (II) biosorption. Particle diameters ranging from 14.13m to 123.6m with macropores of 1.868 ±0.337m were used. Based on the infrared spectra, the researchers also proposed a physical adsorption process, which was validated by the variation of the enthalpy (48.318 kJ mol^{-1}). They subsequently employed the Box-Behnken factorial design to further improve the biosorption process, allowing them to discover the ideal conditions for maximal Cu (II) removal. The greatest percentage of copper removal (83.14%) was obtained at an algae concentration of 1.28 g l^{-1}, a copper ion concentration of 5 mg l^{-1}, and a pH of 6.33 [71].

An investigation was conducted to examine the potential of ash and Fe-nanoparticle-loaded ash (nFe-A) as adsorbents for removing Pb (II) from aqueous solutions [72]. The study used batch experiments to determine the adsorption capacity, kinetics, and isotherm models. The findings indicated that both adsorbents reached equilibrium in 40 min. The research also demonstrated that these adsorbents can easily be regenerated and successfully employed in four cycles for the sequestration and retrieval of Pb (II) from effluent media. The breakthrough capabilities were 25 mg g^{-1} for ash and 30 mg g^{-1} for nFe-A. Overall, the study suggested that ash and nFe-A made from *Rosa canina* L. leaves are effective and economical adsorbents for the elimination of Pb (II) from aqueous systems and could be ideal substitutes for more expensive materials used to treat liquid wastes that contain toxic Pb(II).

7.3.2.3 Layered double hydroxides (LDHs)

LDHs are another type of adsorbent utilized in the adsorption process for the removal of HMs from effluents. They are rare layered anionic clay minerals, and their general formula is $(M^{2+}{}_{1-x}M^{3+}{}_{x}(OH)_2)^{X+}(A^{n-})_{x/n}{}^{*}\ mH_2O$. The divalent and trivalent metal cations are represented by M^{2+} and M^{3+}, respectively [7]. The interlayer anion is represented by A^{n-}, where x denotes the molar ratio of $M^{3+}/(M^{2+}/M^{3})$. LDHs are highly versatile materials due to their ability to substitute metal cations into the laminae and anions into the interlayers, resulting in LDHs with varied interlayer spacings.

The performance of LDHs for HM ion adsorption is affected by several factors including the kind of cations available in the laminae, the nature of the surface groups, and the type of interlayer anions [7]. LDHs with small anions, such as glutamates, citrates, and malates, are favored for efficient HM removal due to their potential to form chelates with HMs, accelerating their leaching. Moreover, combining LDHs with other materials can enhance their performance in terms of removing HMs. For example, the *in situ* oxidative polymerization of a polyaniline/LDH composite was shown to be extremely successful in the elimination of Cr (VI), and the adsorption was noted to be maximized at a value of 393.7 mg g^{-1} [73].

Rahmanian *et al* explored the utilization of ultrasound irradiation for the co-precipitation of citrate intercalated zinc iron layered double hydroxide (ZnFe/Cit-LDH) [74]. They characterized the synthesized LDH using various approaches and assessed its ability to remove Pb (II) from an effluent. The results revealed that the ZnFe/Cit-LDH produced had good crystallinity, a homogeneous and plate-like morphology, and particle sizes between 20 and 40 nm. With a high adsorption capacity of 94.3 mg g^{-1} at 30 min and at pH 6, the synthesized adsorbent showed outstanding prospects for the elimination of Pb (II) from an effluent. Based on these findings, ultrasonic irradiation can be a cost-effective and ecologically friendly approach to producing LDH-based products and could have applications in environmental remediation.

7.3.2.4 Carbon nanotubes

CNTs, which are essentially graphite sheets that have been rolled into a tube shape, can also be used as adsorbents in the adsorption process for the removal of HMs from wastewater. There are two distinct types of CNTs, namely single-walled CNTs (SWCNTs) and multiwalled CNTs (MWCNTs) [5]. CNTs have garnered significant attention over the last decade due to their inherent stability, excellent adsorbent properties, and exceptional mechanical and electrical characteristics [5]. In addition, CNTs possess a high degree of porosity and hollowness, a low mass density, and a significant specific surface area.

CNTs also have four interaction sites, including interstitial channels, internal sites, grooves, and outer surfaces, allowing them to interact with pollutant molecules easily and effectively. Because of these qualities, CNTs are well-suited for the removal of toxic contaminants from aqueous systems [7]. For instance, Hao *et al* [75] conducted a study, of the use of functionalized oxygen-containing CNTs treated with chitosan to form chitosan-coated CNTs in order to improve the adsorption of the CNTs. Functional oxygen-containing CNTs coated with metal-loaded chitosan were employed to produce supercapacitor electrodes using a simple HM adsorption and carbonization process, and the chitosan-modified CNTs displayed good adsorbent capacities of 123.7 mg g^{-1} for Cu(II) and 142.1 mg g^{-1} for Cr (VI).

The chemical vapor deposition technique was used to grow CNTs on iron-doped and nickel-doped activated alumina, creating nanoflower clusters for a water treatment study [76]. Batch-mode investigations demonstrated that at a pH value of 2, the nanoclusters (NCs) displayed optimum adsorption of Cr (VI), while the best pH range for Cd (II) was pH 7–9. The nanoflower clusters were found to have sorption capacities of 264.5 mg g^{-1} for Cr (VI) and 229.9 mg g^{-1} for Cd (II). Additionally, the sorption data suited the Freundlich and Langmuir models well, and thermodynamic factors suggested that the nanoflower clusters were suitable for the elimination of Cd (II) and Cr (VI).

An extensive investigation has examined the development and application of a nanocomposite material, MWCNTs/ThO_2, for the effective sequestration of Pb (II) from aqueous solutions [77]. The MWCNTs were chemically modified and mixed with thorium oxide (ThO_2) to generate the nanocomposite. The adsorption behavior of Pb^{+2} onto the nanocomposite was thoroughly investigated, and an equilibrium was established after 50 min. The optimal adsorption conditions were found to consist of a pH value of 5.5 and a temperature of 45°C. The adsorption process was discovered to be endothermic, implying that high temperatures should improve the nanocomposite's adsorption capacity. The study demonstrated the MWCNT/ThO_2 nanocomposite's promising potential for the elimination of Pb (II), highlighting its significance for water treatment and environmental remediation applications.

Similarly, research studies have reported the use of 8-hydroxyquinoline as a ligand for MWCNTs in the elimination of Cd^{2+}, Zn^{2+}, Pb^{2+}, and Cu^{2+} from wastewater [78]. To achieve efficient metal removal, optimal adsorption parameters such as the quantity of MWCNTs, the pH, temperature, ionic strength, and metal ion concentration were determined, and it was observed that the majority of the HMs desorbed when the pH was lower than 2.0. The study also established that the modified MWCNTs were better than pristine MWCNTs in terms of adsorption performance, as confirmed by real water samples obtained from the Red Sea and an effluent treatment plant. The results further demonstrated the recyclability of the modified MWCNTs across many adsorption/desorption cycles with no significant efficiency loss.

7.3.3 Ion exchange technique

Another promising approach for eliminating toxic HMs from wastewater is the ion exchange technique, owing to the fast kinetics of the ion exchanger materials and its high removal efficiency [7]. Ion exchange materials operate by exchanging metal ions in the material with ions of comparable charge. This approach is commonly employed in water treatment plants and is capable of handling massive amounts of wastewater. Additionally, ion exchange resins are easily regenerated and reused, hence making them cost-effective and alternative methods for wastewater treatment. HMs and dyes have always been removed from wastewater using synthetic and natural ion exchange resins. However, synthetic resins are often favored owing to their efficiency in extracting HMs from solutions [7].

Strongly acidic resins that have sulfonic acid groups ($–SO_3H$) and weakly acidic resins containing carboxylic acid groups constitute the two types of cation exchangers (–COOH). These groups exchange H^+ ions for HMs, making their removal easier [7]. The efficacy of the ion exchange technique is influenced by numerous parameters, including temperature, pH, starting metal concentration, contact duration, and percentage of metal ion charge. The removal of different HMs has been discovered to be impacted by pH, the original concentration of the metal ions, the adsorbent dosage, as well as the contact duration [7].

Two strongly acidic cation exchangers, Amberlite 252 and Amberjet 1200, were evaluated in a study of the removal of Cu (II) and Zn (II) from synthetic rinse water. The studies were carried out using a column system at starting metal concentrations of 350 mg l^{-1} for Cu (II) and 600 mg l^{-1} for Zn (II) at a flow rate of 2.5 ml min^{-1} and using 35 g of adsorbent (II). The Amberjet 1200 resin was shown to be successful in eliminating these metals from synthetic rinse water in the electroplating industry [79].

In another study, natural and modified clinoptilolite-K zeolites were used to confirm the effectiveness of the Autonomous Water Purifier (AWP) prototype, which uses an integrated zeolite filter to solve water-related problems in marginalized communities that lack access to clean and safe water and electricity and are prone to water-related diseases. Adsorption studies were carried out in aqueous solutions at various HM concentrations between 20 and 100 mg l^{-1} using varying amounts of both natural and altered zeolites. The data were a good fit for the Langmuir isotherm and revealed that Hg^{2+} and Cu^{2+} ions were adsorbed to a lower extent on the zeolites, whereas Pb^{2+} and Cd^{2+} ions were adsorbed to a higher extent onto the zeolites [80].

7.3.4 Membrane filtration methods

Membrane filtration is another method of separating pollutants from wastewater that employs a semipermeable membrane. This method works well for removing suspended solids and organic molecules. However, it can also be very useful in removing metal ions and colors from wastewater [7]. The membrane can be composed of a variety of materials, including cellulose acetate, polyamide, and polyethersulfone. Membrane filtration processes differ based on the type of membrane used. This method provides various benefits compared to conventional approaches, including environmental friendliness, superior separation efficiency, low energy consumption, energy savings, the absence of a phase change, and ease of scaling [58]. Several filtering techniques, including reverse osmosis, nanofiltration, and ultrafiltration, can be employed in the sequestration of HMs in wastewater [7].

7.3.4.1 Ultrafiltration

This is a low-pressure method that eliminates dissolved and colloidal materials [58]. The method uses a permeable membrane to separate HMs or large molecules depending on the size of the membrane's pores and the weight of the molecules. Because of its high packing density, this approach has advantages such as decreased area requirements and a reduced driving force [7]. Moreover, ultrafiltration can be enhanced to increase removal efficiency, for example, by polymer-enhanced ultrafiltration, which includes the binding of metal ions by water-soluble polymeric ligands to create macromolecular complexes [7]. Polymer-assisted ultrafiltration uses water-soluble polymers in contaminated water. The polymers remain in the membrane if their size is greater than the molecular cutoff weight of the membrane. Polymer-assisted ultrafiltration facilitates the elimination of tiny solute molecules that cannot be sequestrated by traditional ultrafiltration methods [58].

Another version of ultrafiltration is micelle-enhanced ultrafiltration, which is a physicochemical membrane separation technology in which an anionic surfactant is added to the contaminated water. As a result, the anionic surfactant monomers combine and form micelles. An electrostatic force traps HMs in the outer region of the micelle. The HMs that are not caught in the outer section of the micelle and the residual surfactant flow through the ultrafiltration membrane. Micelle-enhanced ultrafiltration offers various advantages which include a high flux, high elimination, and low cost of energy, but the overall cost of running the process is high [58].

7.3.4.2 Reverse osmosis

This is a process in which a pure solvent is forcefully pushed through a partially permeable membrane by the application of pressure, whereas the solute particles are left behind [58]. For example, Qdais *et al* successfully employed this technique in the sequestration of Cu^{2+} and Cd^{2+} from an aqueous medium; they removed more than 90% of the ions [81]. Mohsen-Nia *et al* [82] also employed this method to remove Cu^{2+} and Ni^{2+} from an aqueous solution using the disodium salt ethylenediaminetetraacetic acid (Na2EDTA). It was observed that adding Na2EDTA increased the chelated ion sizes of Cu^{2+} and Ni^{2+}. Furthermore, an increase in the rejection efficiency to 99.5% was also observed, and more than 95% of the HM ions were removed.

7.3.4.3 Nanofiltration

Nanofiltration is an efficient membrane filtration method that is applied to filtration processes whose pore sizes range between those of ultrafiltration and reverse osmosis filtration. Positively charged polyethylimine (PEI) cross-linked polyimide nanofiltration membranes possess excellent chemical, thermal, and mechanical characteristics [58]. In order for such a membrane to also remove anions, it can be changed by adsorbing a chelating polymer containing negatively charged functional groups onto a charged cross-linked P84 PEI hollow fiber substrate. This alteration enhances membrane performance, alters the surface charge and pore size of the membrane, and allows it to remove multivalent anions with a rejection effectiveness greater than 98% [58].

Another approach for making membranes suggested by Zhu and his co-authors is to simultaneously co-extrude polybenzimidazole (PBI) and polyethersulfone (PES)/polyvinylpyrrolidone (PVP) dopes via a triple-orifice spinneret using the dry-jet wet-phase inversion process [83]. PBI provides high resistance to chemicals and unique charge features, while PES/PVP is cheaper and characteristically hydrophilic; it has good mechanical properties and great spinnability and allows the simple construction of a porous membrane. This membrane exhibited high Cd (II) and Pb (II) rejection efficiencies of 95% and 98%, respectively. Furthermore, the research literature reports that a membrane has been created by conjoining it with nanoscale ferric oxides in the membrane matrix to increase the removal of Pb (II) from wastewater [84]. Furthermore, a polyamide nanofiltration membrane was used for lead removal; the results showed that increasing the pressure as well as the concentration of the initial feed increased lead sequestration to 97.5% [85].

7.3.5 The electrochemical method

Electrolytic recovery is another technique for removing metals from wastewater which involves passing electricity through a solution containing dissolved metals [7]. For a variety of reasons, the electrochemical treatment method looks to be a highly promising and suitable methodology for wastewater treatment since it requires few additives, is very effective, has low maintenance costs, achieves results quickly, and requires little effort [86]. Electrochemical wastewater treatment procedures include electrodeposition, electrocoagulation, electro-floatation, and electro-oxidation [7, 86]. This process uses a disposable electrode to introduce ions into the wastewater, which subsequently counterbalances the particles' charges in the solution and initiates the process of coagulation [86]. These ions introduced into the wastewater during the electrochemical treatment process help to eliminate unwanted contaminants through various mechanisms, such as chemical reactions, coagulation, or precipitation. In the electro-precipitation method, for example, pollutants present in wastewater are precipitated in the form of hydroxides through the application of an electric current.

In studies, electrocoagulation has been found to be effective in removing nickel, copper ions, and Cr (VI) from wastewater, achieving a removal efficiency of 100% within 20 min. Similarly, a sequestration efficacy of approximately 100% has been documented for Cd (II), Cu (II), and Ni (II) using the batch cylindrical iron reactor electrocoagulation method [87].

Electrocoagulation (EC) has some advantages over conventional coagulation and flotation techniques. For instance, it provides a better removal rate for small colloidal particles. In addition, the application of electricity to the system initiates the entire process, increasing the likelihood of coagulation [86]. Furthermore, the electrocoagulation method also offers reduced sludge production, elimination of the need for chemical use, and ease of operation. In contrast, conventional coagulation and flotation techniques require large amounts of chemicals to bring the metal concentration down to a level acceptable for discharge [7].

7.4 Conclusions

Metal ions and dyes pose severe environmental problems due to their noxious nature. They have been reported to create several health risks, which have resulted in deaths and adverse health impacts for both humans and other living things. As a result, researchers have been motivated to develop novel detection techniques coupled with reliable and efficient sequestration methods geared towards remedying the situation. Importantly, numerous methods and techniques have now been adopted and have shown excellent performance despite their inherent drawbacks. Their evolution and modification are an ongoing process. Equally, the study of techniques that offer ease of use, low operational costs, and high efficiency should be given prominence in research programs.

References

[1] Onyancha R, Aigbe U, Ukhurebor K, Kusuma H, Darmokoesoemo H, Osibote O and Pal K 2022 Influence of magnetism-mediated potentialities of recyclable adsorbents for heavy metal ions removal from aqueous solutions–an organized review *Results Chem.* **4** 100452

[2] Sudarni D, Aigbe U, Ukhurebor K, Onyancha R, Kusuma H, Darmokoesoemo H, Osibote O, Balogun V and Widyaningrum B 2021 Malachite green removal by activated potassium hydroxide clove leaf agrowaste biosorbent: characterization, kinetic, isotherm, and thermodynamic studies *Adsorpt. Sci. Technol.* **2021** 1145312

[3] Eldeeb T, Aigbe U, Ukhurebor K, Onyancha R, El-Nemr M, Hassaan M, Ragab S, Osibote O and El Nemr A 2022 Adsorption of methylene blue (MB) dye on ozone, purified and sonicated sawdust biochars *Biomass Convers. Biorefin.* 1–23

[4] Eldeeb T *et al* 2022 Biosorption of acid brown 14 dye to mandarin-CO-TETA derived from mandarin peels *Biomass Convers. Biorefin.* 1–21

[5] Onyancha R, Aigbe U, Ukhurebor K and Muchiri P 2021 Facile synthesis and applications of carbon nanotubes in heavy-metal remediation and biomedical fields: a comprehensive review *J. Mol. Struct.* **1238** 30462

[6] Ukhurebor K, Aigbe U, Onyancha R, Nwankwo W, Osibote O, Paumo H, Ama O, Adetunji C and Siloko I 2021 Effect of hexavalent chromium on the environment and removal techniques: a review *J. Environ. Manage.* **280** 111809

[7] Malik L, Bashir A, Qureashi A and Pandith A 2019 Detection and removal of heavy metal ions: a review *Environ. Chem. Lett.* **17** 1495–521

[8] Neolaka Y, Riwu A, Aigbe U, Ukhurebor K, Onyancha R, Darmokoesoemo H and Kusuma H 2022 Potential of activated carbon from various sources as a low-cost adsorbent to remove heavy metals and synthetic dyes *Results Chem.* **5** 100711

[9] Bashir A, Manzoor T, Malik L, Qureashi A and Pandith A 2020 Enhanced and selective adsorption of Zn (II), Pb (II), Cd (II), and Hg (II) ions by a dumbbell-and flower-shaped potato starch phosphate polymer: a combined experimental and DFT calculation stud *ACS Omega* **5** 4853–67

[10] Ullah N, Mansha M, Khan I and Qurashi A 2018 Nanomaterial-based optical chemical sensors for the detection of heavy metals in water: recent advances and challenges *TrAC Trends Anal. Chem.* **100** 155–66

[11] Ghaedi M, Reza Fathi M, Shokrollahi A and Shajarat F 2006 Highly selective and sensitive preconcentration of mercury ion and determination by cold vapor atomic absorption spectroscopy *Anal. Lett.* **39** 1171–85
[12] Bankaji I, Kouki R, Dridi N, Ferreira R, Hidouri S, Duarte B, Sleimi N and Caçador I 2023 Comparison of digestion methods using atomic absorption spectrometry for the determination of metal levels in plants *Separations* **10** 40
[13] Fouad H, Elrakaiby R and Hashim M 2015 The application of flame atomic absorption spectrometry for gold determination in some of its bearing rocks *Am. J. Anal.l Chem.* **6** 411
[14] Ieggli C, Bohrer D, Do Nascimento P and De Carvalho L 2011 Flame and graphite furnace atomic absorption spectrometry for trace element determination in vegetable oils, margarine and butter after sample emulsification *Food Addit. Contam.* **28** 640–8
[15] Nie L, Jin H, Wang G, Tian J and Lin R 2008 Study of determination method for heavy metals and harmful elements residues in four traditional Chinese medicine injections *Zhongguo Zhong yao za zhi= Zhongguo Zhongyao Zazhi= Chin. J. Chinese Materia Medica.* **33** 2764–7 (article in Chinese)
[16] Liu H, Cui H, Wang Y, Jiang Z, Lei L and Wei S 2023 Accurate determination of trace cadmium in soil samples with graphite furnace atomic absorption spectrometry using metal–organic frameworks as matrix modifiers *Appl. Spectrosc.* **77** 131–9
[17] Soylak M, Uzcan F and Goktas O 2023 Ultrasound-assisted quasi-hydrophobic deep eutectic solvent-based determination of trace Rhodamine B in water and food samples: a simple and green approach *J. Food Compos. Anal.* **120** 105287
[18] Pacer E, Palmer C and Parsons P 2022 Determination of lead in blood by graphite furnace atomic absorption spectrometry with Zeeman background correction: improving a well-established method to support a lower blood lead reference value for children *Spectrochim. Acta* B **190** 106324
[19] Santos J, Mehmeti L and Slaveykova V 2022 Simple acid digestion procedure for the determination of total mercury in plankton by cold vapor atomic fluorescence spectroscopy *Methods Protoc.* **5** 29
[20] Yao Z, Liu J, Mao X, Chen G, Ma Z and Li B 2022 Ultratrace mercury speciation analysis in rice by in-line solid phase extraction–liquid chromatography–atomic fluorescence spectrometry *Food Chem.* **379** 132116
[21] Shen Y *et al* 2022 Graphene oxide-assisted synthesis of N, S Co-doped carbon quantum dots for fluorescence detection of multiple heavy metal ions *Talanta* **241** 123224
[22] Byers H, McHenry L and Grundl T 2019 XRF techniques to quantify heavy metals in vegetables at low detection limits *Food Chem.* **1** 100001
[23] Purwadi I, Casey L, Ryan C, Erskine P and van der Ent A 2022 X-ray fluorescence spectroscopy (XRF) for metallome analysis of herbarium specimens *Plant Methods* **18** 139
[24] Lerma M, Cantu J, Banu K and Gardea-Torresdey J 2023 Environmental assessment in fine jewelry in the US-Mexico's Paso del Norte region: a qualitative study via x-ray fluorescence spectroscopy *Sci. Total Environ.* **863** 161004
[25] Cui L, Wu J and Ju H 2015 Electrochemical sensing of heavy metal ions with inorganic, organic and bio-materials *Biosens. Bioelectron.* **63** 276–86
[26] Pujol L, Evrard D, Groenen-Serrano K, Freyssinier M, Ruffien-Cizsak A and Gros P 2014 Electrochemical sensors and devices for heavy metals assay in water: the French groups' contribution *Front. Chem.* **2** 19

[27] Bansod B, Kumar T, Thakur R, Rana S and Singh I 2017 A review on various electrochemical techniques for heavy metal ions detection with different sensing platforms *Biosens. Bioelectron.* **94** 443–55

[28] Aragay G and Merkoçi A 2012 Nanomaterials application in electrochemical detection of heavy metals *Electrochim. Acta* **84** 49–61

[29] Düzgün A, Zelada-Guillén G, Crespo G, Macho S, Riu J and Rius F 2011 Nanostructured materials in potentiometry *Anal. Bioanal.Chem.* **399** 171–81

[30] Bakker E and Pretsch E 2008 Nanoscale potentiometry *TrAC Trends Anal. Chem.* **27** 612–8

[31] Saad A, Ismail N, Gaber N and Elzanfaly E 2023 Erdosteine-based potentiometric sensor for real-time surveillance of copper traces in food supplements and shredded canned tuna *J. Food Compos. Anal.* **115** 105026

[32] Adeli A, Khoshnood R, Beyramabadi S, Pordel M and Morsali A 2022 Multivariate optimization of a novel potentiometric sensor to determine silver ions in real water and pharmacological product samples *Monatsh. Chemie-Chem.* **153** 227–35

[33] Samayoa-Oviedo H, Halpern E, Mehnert S and Laskin J 2022 Potentiometric determination of Zinc in supplement tablets using a Ca-Ion selective electrode *J. Chem. Educ.* **99** 2661–6

[34] Elsebai B, Ghica M, Abbas M and Brett C 2023 Novel amperometric mercury-selective sensor based on organic chelator ionophore *Molecules* **28** 2809

[35] Li X Z D X J a C H 2007 In-channel indirect amperometric detection of heavy metal ions for electrophoresis on a poly (dimethylsiloxane) microchip *Talanta* **71** 1130–5

[36] Devi N, Sasidharan M and Sundramoorthy A 2018 Gold nanoparticles-thiol-functionalized reduced graphene oxide coated electrochemical sensor system for selective detection of mercury ion *J. Electrochem. Soc.* **165** B3046–53

[37] Rajendrachari S, Somashekharappa K, Mahale R, Vasanth S and Chikkegouda S 2022 A review on cyclic voltammetric investigation of toxic heavy metals *Frontiers in Voltammetry* (London: IntechOpen)

[38] Xia F, Zhang X, Zhou C, Sun D, Dong Y and Liu Z 2010 Simultaneous determination of copper, lead, and cadmium at hexagonal mesoporous silica immobilized quercetin modified carbon paste electrode *J. Anal. Methods Chem.* **2010** 824197

[39] Lahari S, Amreen K, Dubey S, Ponnalagu R and Goel S 2023 Optimized porous carbon-fibre microelectrode for multiplexed, highly reproducible and repeatable detection of heavy metals in real water samples *Environ. Res.* **220** 115192

[40] Zhuang Y, Wang C, Qu W, Yan Y, Wang P and Qiu C 2023 A planar disk electrode chip based on MWCNT/CS/Pb^{2+} Ionophore IV nanomaterial membrane for trace level Pb^{2+} detection *Mol.* **28** 4142

[41] Jothimuthu P, Wilson R, Herren J, Haynes E, Heineman W and Papautsky I 2011 Lab-on-a-chip sensor for detection of highly electronegative heavy metals by anodic stripping voltammetry *Biomed. Microdevices* **13** 695–703

[42] Borrill A, Reily N and Macpherson J 2019 Addressing the practicalities of anodic stripping voltammetry for heavy metal detection: a tutorial review *Analyst* **144** 6834–49

[43] Estela J, Tomás C, Cladera A and Cerda V 1995 Potentiometric stripping analysis: a review *Crit. Rev. Anal. Chem.* **25** 91–141

[44] Szłyk E and Szydłowska-Czerniak A 2004 Determination of cadmium, lead, and copper in margarines and butters by galvanostatic stripping chronopotentiometry *J. Agric. Food Chem.* **52** 4064–71

[45] Wolfbeis O 2006 Fiber-optic chemical sensors and biosensors *Anal. Chem.* **78** 3859–74

[46] Sarmanova O, Laptinskiy K, Burikov S, Chugreeva G and Dolenko T 2023 Implementing neural network approach to create carbon-based optical nanosensor of heavy metal ions in liquid media *Spectrochim. Acta* A **286** 122003

[47] Guo Z, Chen P, Yosri N, Chen Q, Elseedi H, Zou X and Yang H 2023 Detection of heavy metals in food and agricultural products by surface-enhanced Raman spectroscopy *Food Rev. Int.* **39** 1440–61

[48] Lakowicz J E 2006 *Principles of Fluorescence Spectroscopy* (Boston, MA: Springer)

[49] Oehme I and Wolfbeis O 1997 Optical sensors for determination of heavy metal ions *Microchim. Acta* **126** 177–92

[50] Leslee D, Venkatachalam U, Gunasekaran J, Karuppannan S and Kuppannan S 2023 Synthesis of a quinoxaline–hydrazinobenzothiazole based probe—single point detection of Cu^{2+}, Co^{2+}, Ni^{2+} and Hg^{2+} ions in real water samples *Org. Biomol. Chem.* **21** 4130–43

[51] Fernandes R and Dey N 2023 Metal ion responsive bifunctional bis (indolyl) methane derivative: excitation-triggered alteration in the sensing behavior *Mater. Chem. Phys.* **302** 127637

[52] Lerchi M, Bakker E, Rusterholz B and Simon W 1992 Lead-selective bulk optodes based on neutral ionophores with subnanomolar detection limits *Anal. Chem.* **64** 1534–40

[53] Galiński B, Chojnacki J and Wagner-Wysiecka E 2023 Simple colorimetric copper (II) sensor–Spectral characterization and possible applications *Spectrochim. Acta* A **293** 122472

[54] Lewińska I, Capitán-Vallvey L and Erenas M 2023 Thread-based microfluidic sensor for lithium monitoring in saliva *Talanta* **253** 124094

[55] Zhang Y and Duan X 2020 Chemical precipitation of heavy metals from wastewater by using the synthetical magnesium hydroxy carbonate *Water Sci. Technol.* **81** 1130–6

[56] Gao B, Yue Q, Wang Y and Zhou W 2007 Color removal from dye-containing wastewater by magnesium chloride *J. Environ. Manage.* **82** 167–72

[57] Chen Q, Yao Y, Li X, Lu J, Zhou J and Huang Z 2018 Comparison of heavy metal removals from aqueous solutions by chemical precipitation and characteristics of precipitates *J. Water Process Eng.* **26** 289–300

[58] Renu A M and Singh K 2017 Methodologies for removal of heavy metal ions from wastewater: an overview *Interdiscip. Environ. Rev.* **18** 124–42

[59] Mirbagheri S and Hosseini S 2005 Pilot plant investigation on petrochemical wastewater treatment for the removal of copper and chromium with the objective of reuse *Desalination* **171** 85–93

[60] Özverdi A and Erdem M 2006 Cu^{2+}, Cd^{2+} and Pb^{2+} adsorption from aqueous solutions by pyrite and synthetic iron sulphide *J. Hazard. Mater.* **137** 626–32

[61] Matlock M, Howerton B, Henke K and Atwood D 2001 A pyridine-thiol ligand with multiple bonding sites for heavy metal precipitation *J. Hazard. Mater.* **82** 55–63

[62] Matlock M, Howerton B, Van Aelstyn M, Nordstrom F and Atwood D 2002 Advanced mercury removal from gold leachate solutions prior to gold and silver extraction: a field study from an active gold mine in Peru *Environ. Sci. Technol.* **36** 1636–9

[63] Wang X, Deng W, Xie Y and Wang C 2013 S.elective removal of mercury ions using a chitosan–poly (vinyl alcohol) hydrogel adsorbent with three-dimensional network structure *Chem. Eng. J.* **228** 232–42

[64] Elwakeel K, El-Sayed G and Abo El-Nassr S 2015 Removal of ferrous and manganous from water by activated carbon obtained from sugarcane bagasse *Desalin. Water Treat.* **55** 471–83

[65] Al-Malack M, Al-Attas O and Basaleh A 2017 Competitive adsorption of Pb^{2+} and Cd^{2+} onto activated carbon produced from municipal organic solid waste *Desalin. Water Treat.* **60** 310–8

[66] Al-Othman Z, Ali R and Naushad M 2012 Hexavalent chromium removal from aqueous medium by activated carbon prepared from peanut shell: adsorption kinetics, equilibrium and thermodynamic studies *Chem. Eng. J.* **184** 238–47

[67] Ghasemi M, Naushad M, Ghasemi N and Khosravi-Fard Y 2014 A novel agricultural waste based adsorbent for the removal of Pb (II) from aqueous solution: kinetics, equilibrium and thermodynamic studies *J. Ind. Eng. Chem.* **20** 454–61

[68] Bernard E, Jimoh A and Odigure J 2013 Heavy metals removal from industrial wastewater by activated carbon prepared from coconut shell *Res. J Chem. Sci.* **3** 3–9 http://www.isca.me/rjcs/Archives/v3/i8/1.ISCA-RJCS-2013-087.pdf

[69] Elgarahy A, Elwakeel K, Mohammad S and Elshoubaky G 2021 A critical review of biosorption of dyes, heavy metals and metalloids from wastewater as an efficient and green process *Clean. Eng. Technol.* **4** 100209

[70] Muthusaravanan S, Sivarajasekar N, Vivek J, Paramasivan T, Naushad M, Prakashmaran J, Gayathri V and Al-Duaij O 2018 Phytoremediation of heavy metals: mechanisms, methods and enhancements *Environ. Chem. Lett.* **16** 1339–59

[71] Moreira V, Lebron Y, Freire S, Santos L, Palladino F and Jacob R 2019 Biosorption of copper ions from aqueous solution using *Chlorella pyrenoidosa*: optimization, equilibrium and kinetics studies *Microchem. J.* **145** 119–29

[72] Ghasemi M, Naushad M, Ghasemi N and Khosravi-Fard Y 2014 Adsorption of Pb (II) from aqueous solution using new adsorbents prepared from agricultural waste: adsorption isotherm and kinetic studies *J. Ind. Eng. Chem.* **20** 2193–9

[73] Zhu K, Gao Y, Tan X and Chen C 2016 Polyaniline-modified Mg/Al layered double hydroxide composites and their application in efficient removal of Cr (VI) *ACS Sustain. Chem. Eng.* **4** 4361–9

[74] Rahmanian O, Amini S and Dinari M 2018 Preparation of zinc/iron layered double hydroxide intercalated by citrate anion for capturing lead (II) from aqueous solution *J. Mol. Liq.* **256** 9–15

[75] Hao P, Ma X, Xie J, Lei F L L, Zhu W, Cheng X, Cui G and Tang B 2018 Removal of toxic metal ions using chitosan coated carbon nanotube composites for supercapacitors *Sci. China Chem.* **61** 797–805

[76] Sankararamakrishnan N, Jaiswal M and Verma N 2014 Composite nanofloral clusters of carbon nanotubes and activated alumina: an efficient sorbent for heavy metal removal *Chem. Eng. J.* **235** 1–9

[77] Mittal A, Naushad M, Sharma G, Alothman Z, Wabaidur S and Alam M 2016 Fabrication of MWCNTs/ThO_2 nanocomposite and its adsorption behavior for the removal of Pb (II) metal from aqueous medium *Desalin. Water Treat.* **57** 21863–9

[78] Kosa S, Al-Zhrani G and Salam M 2012 Removal of heavy metals from aqueous solutions by multi-walled carbon nanotubes modified with 8-hydroxyquinoline *Chem. Eng. J.* **181** 159–68

[79] Budak T 2013 Removal of heavy metals from wastewater using synthetic ion exchange resin *Asian J. Chem.* **25** 4207–10

[80] Becerra-Paniagua D, Hernández-Granados A, Ruiz A and Cedano K 2021 Ion exchange of heavy metals using a modified zeolite filter integrated into a prototype autonomous water purifier

(AWP) on a community scale *Handbook T-XVI CIERMMI Women in Science Engineering and Technology* (Lomas de Santa Fe: ECORFAN-México, S.C.) https://www.ecorfan.org/handbooks/Handbooks_Women_in_Science_TXVI/Handbooks_Women_in_Science_TXVI.pdf

[81] Qdais H and Moussa H 2004 Removal of heavy metals from wastewater by membrane processes: a comparative study *Desalination* **164** 105–10

[82] Mohsen-Nia M, Montazeri P and Modarress H 2007 Removal of Cu^{2+} and Ni^{2+} from wastewater with a chelating agent and reverse osmosis processes *Desalination* **217** 276–81

[83] Zhu W, Sun S, Gao J, Fu F and Chung T 2014 Dual-layer polybenzimidazole/polyethersulfone (PBI/PES) nanofiltration (NF) hollow fiber membranes for heavy metals removal from wastewater *J. Membr. Sci.* **456** 117–27

[84] Gholami A, Moghadassi A, Hosseini S, Shabani S and Gholami F 2014 Preparation and characterization of polyvinyl chloride based nanocomposite nanofiltration-membrane modified by iron oxide nanoparticles for lead removal from water *J. Ind. Eng. Chem.* **20** 1517–22

[85] Mehdipour S, Vatanpour V and Kariminia H 2015 Influence of ion interaction on lead removal by a polyamide nanofiltration membrane *Desalination* **362** 84–92

[86] Tran T, Leu H, Chiu K and Lin C 2017 Electrochemical treatment of heavy metal-containing wastewater with the removal of COD and heavy metal ions *J. Chin. Chem. Soc.* 493–502

[87] Un U and Ocal S 2015 Removal of heavy metals (Cd, Cu, Ni) by electrocoagulation *Int. J. Environ. Sci. Dev.* **6** 425

IOP Publishing

Adsorption Applications for Environmental Sustainability

Kingsley Eghonghon Ukhurebor, Uyiosa Osagie Aigbe and Robert Birundu Onyancha

Chapter 8

The application of microorganism-derived biosorbents in the removal of heavy metals and dyes

Joan Nyika and Megersa Olumana Dinka

This chapter explores the use of biosorbents in decontaminating polluted environments by retaining their heavy metals and dyes, which are refractory. Using reported empirical studies, it is shown that various bacterial, algal, and fungal species have different efficacies for pollutant removal. Microorganisms have a ubiquitous nature, and they are easy to modify and regenerate as pollutant biosorbents. Environmental factors, the physicochemical characteristics of the microbes used, and the target environments influence the resultant efficacy rates. To improve the efficacy rates, dead microbes are preferred. Biosorbents can be immobilized with carriers or genetically modified to enhance their stability.

8.1 Introduction

Environmental pollution is a growing challenge to modern economies due to its ability to disrupt ecosystems. Industrialization is a major driver of pollution and introduces contaminants to environmental compartments such as air, soil, and water through the disposal of resultant wastes in such environs [1]. Other anthropic agents of pollution include chemical matters, biological agents, and agricultural wastes. Once in the environment, pollutants bioaugment, bioaccumulate, resist degradation, and enter trophic chains, thereby causing health concerns for all living things. For instance, in humans, pollutants cause dysfunctions of the nervous system, kidney, liver, and stomach, muscular weakness, cancer, skin infections, and diseases such as dysentery [2, 3]. The effects of such environmental pollutants, including dyes and heavy metals, are shown in table 8.1 [2, 3]. A variety of chemical pollutants are introduced to the environment in the form of spills, leakages, emissions, effluent,

doi:10.1088/978-0-7503-5598-8ch8

Table 8.1. Effects of environmental pollutants on living systems.

Pollutants	Effects
	Dyes
Azo dye	Dermatitis, skin irritations, hypertension, blindness, and vomiting
Congo red dye	Carcinogenic and mutagenic effects
Malachite green dye	Tissue injury, respiratory problems, carcinogenesis, and mutagenesis
Methylene blue dye	Vomiting, sore throat, nausea, abdominal pain, skin rash, headache, breathing problems, diarrhea, and dark urine
Azure B	Mutagenic effects, cytotoxic effects, disruption of the nervous system, inhibition of glutathione reductase enzyme, which regulates the redox homeostasis of cells.
Disperse red 1 dye and disperse orange 1 dye	Mutagenic effects; hepatoxicity causes DNA adducts that can result in cancer through damage to genetic material, causing cell toxicity and death.
Sudan 1 dye	Forms aromatic amines that are carcinogens; interferes with liver function.
Basic red 9 dye	A carcinogen in humans; results in liver dysfunction, mutations, skin irritations, and allergic dermatitis.
Crystal violet dye	Causes chromosomal damage, cell toxicity and damage, liver cancer, skin irritation, and renal and respiratory failure.
	Heavy metals
Mercury (Hg)	Birth defects, tiredness, skin rashes, headaches, chromosomal damage, brain damage, nervous system dysfunctions
Arsenic (As)	Liver and lung cancers, brain damage, lung damage, skin irritations, and stomach problems
Zinc (Zn)	Arteriosclerosis, stomach cramps, vomiting, nausea, anemia, skin irritations
Copper (Cu)	Liver and kidney damage, stomach pains, dizziness, vomiting, diarrhea, headaches
Cadmium (Cd)	Damage to the immune system, cancers, bone fractures, vomiting, stomach pains, and central nervous system damage
Nickel (Ni)	Lung and respiratory tract cancers, cardiovascular and kidney diseases, lung fibrosis
Chromium (Cr)	Damage to the liver, kidney, and skin; ulcers, nose irritations, allergic reactions, and immune deficiency
Lead (Pb)	Brain and kidney damage, high blood pressure, nervous system dysfunction

and/or solid waste, which induces environmental toxicity, and their management at excess levels and/or due to overexposure is challenging [4]. These chemicals are introduced into land, soil, and water systems, where they enter food chains and negatively affect the health and survival of flora and fauna. In this chapter, heavy metals and dyes are the pollutants of interest.

Trace metals such as As, Cr, Co, Cu, Hg, Ni, Pb, Se, and Zn are some of the environmental pollutants that are hazardous even at minute levels and often result from industrial activities and their resultant wastes [3, 5]. These metals do not transform into nontoxic forms and are persistent in the environment, which makes their presence devastating to ecosystems, especially in this technological era [6]. Ayangbenro *et al* [6] noted that heavy metals have accumulated in water, soils, sediments, and the atmosphere at concentrations that have surpassed allowable limits. Other authors have noted that the nonbiodegradable nature of these pollutants allows their biomagnification, subsequent speciation, and bioavailability in environmental compartments in addition to enhancing remediation difficulties [2, 3, 6]. Soils and water polluted by heavy metals have a detrimental effect on crop development and production and negatively affect the function of soil microorganisms [7, 8]. In animal systems, some metals such as Pb, Hg, Cd, and Cr have no known biological role but are associated with carcinogenicity, mutagenicity, and toxicity in aquatic and human ecosystems [9].

Dyes are organic in nature and are used in industrial processes such as printing, cosmetics, and rubber, leather, paper, and textile making [10, 11]. They can be classified as acidic, basic, direct, or reactive based on their physicochemical characteristics [12]. With the growing population and the rise of industrialization, from 0.7 million to more than 1.5 million tons of dye are manufactured and used annually; out of this total, 10%–15% is disposed of as wastewater, which makes dyes significant environmental pollutants [12–14]. Dyes are highly soluble in water and soils, which makes their removal using conventional methods difficult to achieve. Exposure to excess concentrations of dyes is associated with increased cancer etiology, skin irritations, and respiratory disorders in humans, among other effects, as shown in table 8.1 [15]. In aquatic animals, dyes reduce dissolved chemical and biochemical oxygen levels and threaten their survival, while in plants, these contaminants reduce the rate of photosynthesis by blocking the penetration of light [16]. In soils, dyes interfere with the function of microorganisms [16]. Dyes pollute soil and water resources in addition to causing them aesthetic damage [16]. Just like heavy metals, dyes persist in the environment and enter food chains where they are biomagnified before their toxicity becomes apparent.

Given the known toxicity associated with the presence of heavy metals and dyes in the environment, a variety of techniques have been introduced to handle and treat wastes and wastewater laden with pollutants to curtail their contamination. Some of the commonest methods include electrolysis, electrodialysis, ion exchange, reverse osmosis, chemical and electrochemical techniques, membrane processes, biological treatment, and adsorption techniques [13]. Some of these methods are summarized in figure 8.1, based on the environments in which they are applied [2, 17]. The techniques, though varied, have limitations such as high maintenance costs, complex mechanisms involved, high operational costs, and the production of large quantities of toxic sludge [18]. Of the methods, adsorption has been found to be a more superior approach due to its design simplicity, ease of operation, cost-effectiveness, and ability to regenerate the adsorbents used [17, 19], though the challenge in its use is the production of large quantities of biological and chemical sludge [17]. Activated

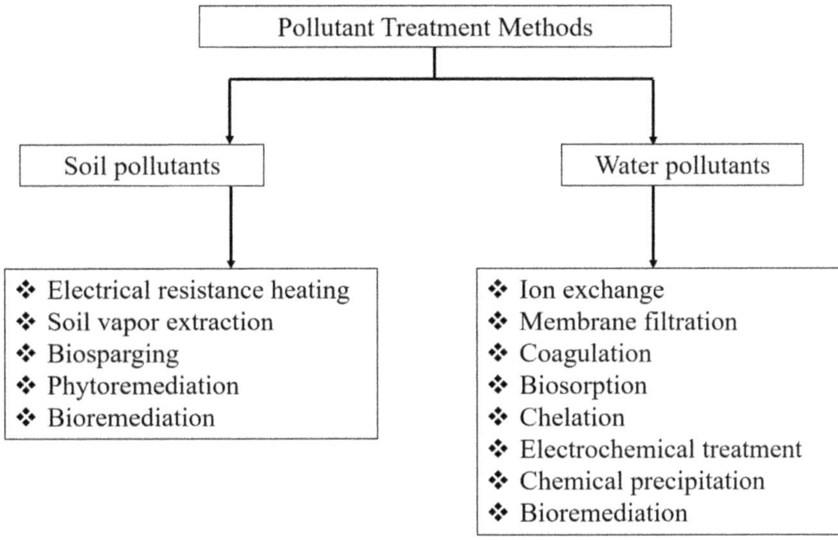

Figure 8.1. Methodologies used in the treatment of soil and water pollutants. Reprinted by permission from Springer Nature [9], copyright (2020).

carbon, which is considered to be a universal adsorbent, is also used in pollutant sorption, but its use is limited due to its high cost [20, 21]. Nonconventional sorbents are also used in contaminant remediation due to their low cost, but their sorption potential is limited [13]. To overcome these challenges, the use of bioadsorbents made from wood, microbial biomass, chitin, peat, bark, agricultural and industrial wastes, and charcoal is growing [22]. This trend is associated with their ubiquitous nature, low cost, eco-friendliness, and good surface features that allow pollutant removal. In this chapter, biosorbents derived from microorganisms and their potential to remediate trace metals and dyes in the environment are discussed in detail.

8.2 Mechanisms of biosorption

'Biosorption' is used to denote a pollutant removal technique that occurs via the passive binding of heavy metal and/or dye ions into nonliving biomass (biosorbents). The process can be comparable to bioaccumulation, in which living organisms actively metabolize pollutants [23]. Most research literature, including this chapter, defines biosorption as the removal of pollutant ions following their binding to either living or nonliving biomass. A number of factors determine the rate of biosorption, including the size and type of the biosorbent, ionic strength, biomass dose, temperature, original pH, and solvent concentration [3]. These factors influence the solubility of contaminants in solutions, the active binding sites used, and the protein structures and functional groups of biosorbents. In biosorption, techniques such as surface complexation, chelation, ion exchange, and biologic adsorption help to recover trace metals and dyes from polluted environments. The environment in

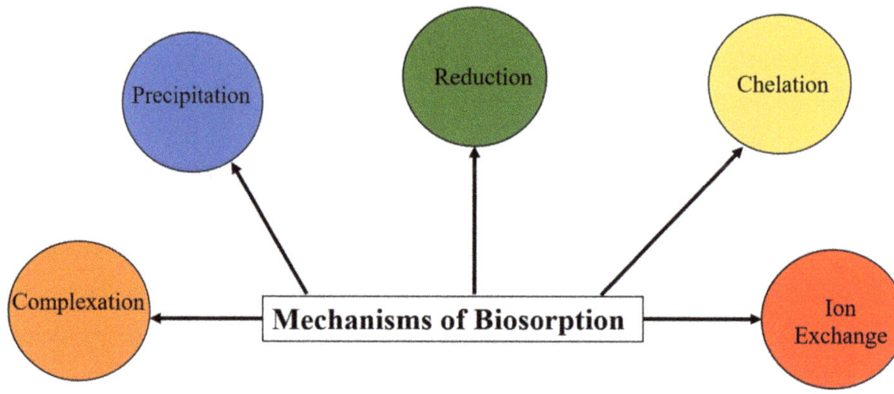

Figure 8.2. Mechanisms of biosorption. Reprinted from [3], Copyright (2022), with permission from Elsevier.

which the pollutants are located and the specific microbial cell metabolism mechanism used also influence biosorption. The biosorption of heavy metals and dyes, especially their removal from polluted water, is widely known to be effective due to the high efficiency of the technique, the ability to reuse biosorbents, the eco-friendly nature of the technique, the ability to recover the pollutants, and the minimal production of secondary waste [3]. Biosorbents commonly used include dead or living microbial cells such as fungi, algae, and bacteria. Microbial biosorbents have functional groups such as the carbonyl, hydroxyl, carboxyl, and amino moieties located on their cell walls, which form complexes with ions from the contaminants, thereby recovering them from polluted solutions [24]. Depending on the location of the pollutant, the removal of toxic trace metals and dyes can occur via cell surface precipitation (sorption) or via extracellular accumulation and precipitation. The mechanisms of biosorption are summarized in figure 8.2, based on the arrangement of sorbed pollutants and the presence or absence of metabolic activity [3].

Heavy metals and dyes interact with active groups on the cell walls of biosorbents to form complexes. These complexes consist of negatively charged or neutral polyatomic molecules surrounding an atom (usually the cations of dyes and heavy metals). The active groups include hydroxy–carboxy, phosphate, hydroxide, thiol, amino, and carboxy groups that interact with the contaminant ions. The bacteria *Pesudomonas syringae* has been observed to complex Zn, Mg, Hg, Cd, Cu, and Ca ions, thereby remediating wastewater [25]. The resultant complexes can be solubilized using organic acids such as fumaric, lactic, oxalic, malic, gluconic, and citric acids that are produced by microorganisms to enable their removal. Through sorption precipitation, pollutants can be deposited in solid form from solutions in cells and/or cell walls when the pH is high. This process may be either independent or cell metabolism dependent. In the cellular metabolism of microbes, contaminant removal is part of an active defense mechanism. As a result of this defense mechanism, the metal and dye ions react with the cell surface functional groups to induce precipitation [26]. Independent precipitation occurs during chemical interactions between a cell surface and a pollutant [27]. In the ion exchange mechanism, polysaccharide ions found on the cell walls of microbes are exchanged with cations from heavy metals and dyes. This mechanism can occur physically, using London-wheeled or electrostatic ionization, or chemically, via the covalent bonding of ions [3]. According to Selvi *et al* [28], Pb and Cd ions were chemically exchanged on the cell walls of the fungal species *Lactarius scrobiculatus* and *Amanita rubescens*.

8.3 Types of biosorbents

Many low-cost materials have been developed as adsorbents for the remediation of pollutants in polluted soil and water. They are classified into four categories:

(i) Natural biosorbents
(ii) Agricultural-waste-based biosorbents
(iii) Industrial-waste-based biosorbents
(iv) Microbe-based biomass biosorbents

Table 8.2 shows specific examples of such biosorbents, their sources, the adsorbates they take up, the functional groups they attract, and the biosorption mechanisms they use.

Natural materials such as zeolites, alginate, chitosan, and clays, other wastes from industrial activities such as slags, bagasse, and fly ash, and agricultural biomasses such as barks, peels, seeds, and straw are useful in the removal of dyes and heavy metals from wastewater [48]. Chitosan, for instance, is useful in the removal of reactive dyes via adsorption [49], while alginate sourced from seaweed and clays removes heavy metals [2]. The negative charges on natural biosorbents give them a high cation exchange capacity for the cations of dyes and heavy metals.

Agricultural wastes such as groundnut shells, corn cob, pomegranate, orange, banana peels, vegetable peels, sugar beet pulp, dry leaves, bark, wheat bran, and rice husks are useful biosorbents [50]. These wastes have active polymers such as lignin, pectin, hemicellulose, tannin, and cellulose that react with ions from dyes and heavy metals, aiding in their removal via adsorption [51]. The preference for the use of agricultural wastes is due to their chemical stability, ease of availability, eco-friendliness, and cost-effective nature in addition to their structural features, such as a variety of moiety groups, high porosity, and large surface area [52].

Industrial wastes can also serve as sorbents for environmental pollutants if their surfaces can be chemically modified to improve their adsorption capacity [2]. A case example is the use of sawdust, which is a low-cost, easily available by-product of the timber industry. This material contains cellulose, lignin, and hemicellulose as the active groups that assist in the removal of reactive dyes from polluted water [53]. Other industrial waste materials that serve as biosorbents include sugarcane bagasse, sludge, furnace slag, and fly ash.

Biological adsorbents sourced from yeasts, algae, fungi, and bacterial species are also key to the adsorption-based remediation of polluted environs.

Table 8.2. Different biosorbents used in the removal of various environmental pollutants, their functional moieties, and their target adsorbates.

Type of biosorbent	Source	Target adsorbate	Functional group	Mechanism used	Reference
Microorganism	Zeolite 13X-algal-alginate beads	Cu (II)	Methylene and amino	Surface binding	[39]
Microbes	*Pseudevernia furfuracea* (L.) Zopf (Nonliving lichen)	Methylene blue dye	Ester, ketone and aldehyde	Chemisorption and electrostatic interaction	[40]
Microbial	*Codium vermilara*	Cu (II)	Ether and carbonyl	Physical sorption	[41]
Algal-based	*Sargassum tenerrimum*	Cr (VI)	Hydroxyl, methyl and carboxyl	Oxidative reaction	[42]
Plant-based waste	Date palm empty fruit bunch	Cr (VI)	Alkene, carbonyl and methylene	Cationic adsorption	[43]
Natural	Leaves of *Solanum incanum*	U (IV)	Methylene and hydroxyl	Complexation and ion exchange	[44]
Agricultural	Wastes from date palm coir	Cr (VI)	Hydroxyl	Surface attribution	[43]
Biomaterial	Aerobic granular sludge containing bacteria and algae	Cr (VI)	Amino and hydroxyl	Surface complexation and electrostatic interaction	[45]
Bio-based	Living biofilm	Cu (II)	Amino, hydroxyl and carboxyl	Surface layer diffusion	[46]
Agricultural	Wastes of *Zea mays*	Oxybenzene, Cu (II)	Hydroxyl	Heterogenous adsorption	[47]
Microbial-based	*Bacillus subtilis*	Methylene blue	Carboxylic and hydroxyl	Ion exchange and surface interaction	[48]
Natural	Taro (*Colocasiaesculenta* (L.) Schott	Pb (II)	Amide and hydroxyl	Ion exchange	[49]
Microbial-based	*Sargassum filipendula*	Cd (II)	Hydroxyl, amide, carboxyl	Precipitation and ion exchange	[50]
Microalgal	*Chlorella vulgaris*	Hg (II)	Amides, amines, hydroxyl and carboxyl	Physical sorption	[51]

(*Continued*)

Table 8.2. (*Continued*)

Type of biosorbent	Source	Target adsorbate	Functional group	Mechanism used	Reference
Industrial	*Carica papaya* wood	Methylene blue dye and malachite green	Carboxyl and hydroxyl	Surface complexation	[52]
Industrial	Sugarcane bagasse	Eosin yellow dye			[53]
		Pb (II)	Carbonyl and carboxyl	Heterogenous adsorption	[54]
Industrial	Malt bagasse	Reactive blue BF-5G dye	Hydroxyl	Surface interaction and complexation	[55]
Agricultural	Lemon peel	Ni (II)	Amine, hydroxyl and carboxyl	Free binding sorption	[56]
	Cocoa pod husk	Cd, Cu, Pb, Zn (II)	Amine and hydroxyl	Ion exchange	[57]

Microorganisms have a high affinity for pollutants in water resources and soils [54]. Microbes also possess enzymes and antibiotics that enhance bioremediation. Their large-scale synthesis results in the production of by-products [55, 56]. Fungal species are key in dye decolorization, which takes place via the thiol, amino, and carboxyl functional moieties found on their cell surfaces [57]. Microorganisms can also be immobilized and used for the adsorption of reactive dyes under specified conditions [56]. Bacterial and algal species are also common in the cleansing of pollutants from polluted land and water resources [1–3]. The use of microbes as biosorbents is preferable due to their ease of preparation and the demand for a low-cost growth medium, which makes them economical to apply. The focus of this chapter is the use of microbe-based biosorbents for the bioremediation of heavy metals and dyes.

The principle behind biosorption is for the contaminants in aqueous solutions to be bound to (fixed on) a biological material (the biosorbent). As such, the biosorbent choice is key in determining the efficacy of the process. All biosorbents have chemical structures that allow the sequestration of contaminants via physical interactions or chemical bonding, which mostly occurs at the cell wall. Depending on the type of biosorbent, a pretreatment can be used to activate or functionalize it. For instance, agricultural-waste-derived biosorbents can undergo microwave functionalization, pyrolysis, or the addition of acid or alkaline solutions, while microbial biosorbents can be ground, immobilized, carbonized, or chemically modified before they are introduced to the target site [58]. To be highly effective, a biosorbent should be low cost, ubiquitous, and reusable; it should also have a high affinity for contaminants and offer easy desorption of the pollutants taken up [58, 59].

8.3.1 Microbial biosorption

Microbial biosorbents sourced from fungi, bacteria, and algae bind to pollutants such as toxic metals, antibiotics, and dyes via ionic interactions between the functional groups found in inactive or living microbes and those of the pollutants; the result is decontamination of the environment [58]. Decontamination occurs via the biodegradation and biosorption of the pollutants. In the use of microorganism-sourced sorbents to remove heavy metals and dyes from polluted environs, dead or inactive microbes were and are preferred to living ones due to their low toxicity, easy regeneration, and their ability to operate over wide pH and temperature ranges [2]. In addition, the use of dead microbial biomasses as sorbents does not require a nutrient medium, is suitable for maintaining pure cultures, and has high sorption and desorption rates [60]. Overall, the success rate of microbe-based biosorption is influenced by physicochemical factors including pH, pollutant concentration, temperature, structure, and the characteristics of the contaminant as well as the chemical composition of the microbial biomass used. For microbes, the mechanisms of sorption are precipitation, biodegradation, ion exchange, electrostatic interaction, chelation, and adsorption. Microbes applied as biosorbents can occur as either free or immobilized forms and can be used individually or as a consortium of several

Table 8.3. The differences between free and immobilized microbial biosorbents.

Biomass	Free	Immobilized
Advantages	- No cost is incurred to immobilize them - Functional groups are easily available, leading to high biosorption capability - High interfacial area	- Results in an absence of cell washouts - Easy to separate the effluent from the biomass - Higher productivity - Fluidized or fixed bed columns can be incorporated - High resistance, mechanical stability, and multiple uses - Easy to regenerate
Disadvantages	- Difficulty in regeneration - Difficult to separate the effluent from the biomass - Low mechanical stability and density due to their small size	- The carrier can interact with the biosorbent's active sites, leading to low removal efficacy - High resistance to mechanical diffusion - Reduced biosorbent capacity - Additional cost of the immobilization

microbes. In the former, movement is unrestricted, while in the latter, mobility is restricted using carriers. Immobilization aims to overcome the challenges of small size, low density, low elasticity, and insufficient mechanical stability associated with free microbial sorbents that cause their clogging, sorbent swelling, and low regeneration potential during liquid-phase separation from pollutants [60]. The differences between free and immobilized microbial biosorbents are summarized in table 8.3 [58].

Bacteria occur as gram-negative or gram-positive unicellular species. The nucleus, cytoplasm, cell membrane, and cell wall are their main components. The positive charge of gram-positive bacterial cell walls is a result of the phosphoryl and carboxyl moieties of the secondary polymer and the peptide chain, respectively [2]. The functional groups aid in metal and dye sequestration once pollutants' ions bind to them. The biosorption process is extracellular, and takes place on the surface of microbial cell walls. Fungal species such as mold and yeast have rigid cell walls comprising polysaccharides in their structure. The walls have surface functional groups, which facilitate biosorption. To improve the efficacy of contaminant biosorption, fungi such as yeast are immobilized using methods such as crosslinking and entrapment [61]. Algal cells contain alginate, cellulose, glycan, and chitin, which are important polysaccharide molecules that bind to pollutants to remove them [62]. Algal species also have hydroxyl, amino, sulfate, and carboxyl functional groups that enhance pollutant binding. To enhance their biosorption properties and stability, algal cells are pretreated through the addition of organic compounds, salts, alkali solutions, or acids.

8.3.2 The removal of heavy metals via microorganism-based biosorption

The removal of heavy metals from aqueous solutions using microbial adsorbents is growing and is becoming widely documented. Highly harmful metals including As, Cd, Cr, Hg, and Pb have been neutralized using this approach [7, 63]. Bacterial species, for instance, use the hydroxyl, amine, phosphonate, and carboxyl groups on their cell walls to bind to the metal cations and detoxify them. The gram-positive bacterial species have higher efficacy in heavy metal removal due to their thicker peptidoglycan layer, which provides better sorption compared to those of gram-negative bacteria. Some of the useful bacteria species, their metal targets, and their biosorption efficiencies are summarized in table 8.4. Fungal species are also key in

Table 8.4. Microorganisms used as heavy metal biosorbents and their target pollutants.

Microbial species	Targeted heavy metal	Biosorption efficacy (%)	References
Pseudomonas alcaliphila NEWG-2	Cr	96.6	[7]
Pseudomonas sp. *strain* DC-B3		41	
Pseudomonas aeruginosa G12		93	
Cellulosimicrobium funkei AR6		82.33	
Stenotrophomonas maltophilia		99.2	
Acinetobacter sp. WB-1		68.17	
Cellulosimicrobium sp.		96.98	
Stenotrophomonas sp.		90	
Cellulosimicrobium sp.	Pb	84.62	
Methylobacterium sp.		62.28	
Aeribacillus pallidus MRP280		96.78	
Bacillus sp. PZ-1		>90	
Arthrobacter viscosum		97	
Arthrobacter sp. 25		86.25	
Pseudomonas sp. I3		98.96	
Bacillus badius AK		60	
Klebsiella pneumonia	Cd	40.18	
Rhodotorula sp.		80	
Bacillus megaterium sp.		79	
Bacillus sp. Q3		93.76	
P. aeruginosa FZ-2	Hg	99.7	
Vibrio parahaemolyticus PG02		90	
P. aeruginosa	Pb, Cd	87 for Cd 98.5 for Pb	
Saccharomyces cerevisiae		76.2 for Cd 70.3 for Pb	
Desulfovibrio desulfuricans	Zn	100	
Micrococcus luteus DE2008	Cu, Pb	36.07 for Pb 25.42 for Cu	

(Continued)

Table 8.4. (*Continued*)

Microbial species	Targeted heavy metal	Biosorption efficacy (%)	References
Oceanbacillus profundus	Pb, Zn	97 for Pb 54 for Zn	
Staphylococcus epidermidis	Zn, Cr	56 for Cr 59 for Zn	
Streptomyces sp.	Cd, Pb, Cu	68.4 for Cd 74.5 for Cu 83.4 for Pb	
Klebsiella sp. USL2S	Cd, Hg, Ni, Pb	73.33 for Cd 85 for Hg 86.06 for Ni 97.13 for Pb	
Pseudomonas azotoformans JAW1	Cd, Cu, Pb	98.57 for Cd 69.76 for Cu 78.23 for Pb	
Chlorella vulgaris	Cd, Cu, Pb	95.5 for Cd 97.7 for Cu 99.4 for PB	[64]
Aspergillus sp. *Geobacillus* *Bacillus* sp. *P. aeruginosa* *Amanita muscaria* *S. cerevisiae* *Flavobacterium* sp. *Spirulina platensis Chlamydomonas* sp. *toebii* *Micrococcus* sp.	Cu	—	[65]
Trichoderma sp. *Rhizopus* sp. *Acinetobacter* sp. *Aspergillus niger* *Mycelial* *Arthrobacter* sp. *Bacillus* sp.	Cr	—	[66]
Myriophyllum spicatum Mycelial *Chlamydomonas* sp. *C. reinhardtii*	Pb	—	[67]

Aspergillus sp. *Pseudomonas* sp. *P. Trichoderma* sp. *Cyanobacteria* *Micrococcus* sp. *Nostoc* sp.			
Rhodotorula rubra *Enterobacteria* sp. *S. cerevisiae* *Cyanobacteria* *Bacillus* sp. *Klebsiella pneumoniae* *Vibrio parahaemolyticus* *Candida parapsilosis*	Hg	—	[68]
Pseudomonas sp. *Flavobacterium* sp. *Micrococcus* sp. *Aspergillus* sp.	As	—	[66]
S. cerevisiae *Aspergillus* sp. *Trichoderma* sp. *Pseudomonas* sp. *Bacillus* sp.	Ni	—	[69]
Bacillus sp. *S. cerevisiae* *Aspergillus* spp. *Streptomyces rimosus* *Pseudomonas* sp. *Botrytis cinerea* *Acinetobacter* sp.	Zn	—	[70]
Diplotaxis harra *Pseudomonas* sp. *S. cerevisiae* *Glebionis coronaria* *Trichoderma* sp. *Penicillium* *Geobacillus thermoleovorans* *Cyanobacteria* *Nostoc* sp. *Aspergillus* sp. *Bacillus* sp.	Cd	—	[65, 66]

heavy metal removal using their cell walls, which consist of an amorphous layer and a microfibrillar layer [60]. The wall has positive charges due to phosphate and carboxyl functional groups as well as negative charges due to the presence of amino groups. At different pH levels, depending on the target environment and pollutant, these species can bind to heavy metals to bioremediate them, as summarized in table 8.4. Using its chitosan, chitin, lipids, proteins, and mannan and glucan polymers, *Saccharomyces cerevisiae* in free or immobilized form results in the biosorption of heavy metals and their eventual removal from polluted sites. Yeast has been used to remove Co, Cr, Cs, Cu, Pb, Sr, Th, and U from aqueous solutions with high efficacy [60]. Algal species such as *Chlamydomonas* sp., cyanobacteria, and *Spirulina platensis* are also used in heavy metal biosorption but to a lesser extent than bacterial and fungal species (table 8.4).

8.3.3 The removal of dyes through microorganism-based biosorption

Synthetic dyes have a diverse nature and are classified as sulfur, vat, reactive, dispersive, direct, basic, or acidic based on their uses and physicochemical characteristics [71]. Dyes pose an environmental threat which is complicated by their diversity and assorted uses. Effluents containing dyes are hazardous to humans and the environment (table 8.1). Microorganisms such as bacteria, which can survive in harsh environmental conditions consisting of extreme temperatures, limited oxygen, and elevated pH conditions, are suitable for dye removal.

In microbial-based adsorption, bacterial species degrade dyes at a faster rate after adsorption compared to fungal species. Under anaerobic conditions, bacterial species can convert azo dyes that are considered recalcitrant to less toxic aromatic amines [71]. Under aerobic conditions, some bacterial species, such as *Micrococcus* sp., can also decolorize dyes from aqueous solutions using their azoreductase enzyme [72]. Some of the bacterial species with the ability to remove dyes from polluted solutions are summarized in table 8.5. Through biosorption and biodegradation, fungal species and yeasts can remove dyes from polluted environments. Microbes are chemically stable, cost-effective, and have the required mechanical features to allow the biosorption of dyes but unlike bacteria are limited by temperature increases that interfere with their surface adsorption properties and reduce the number of active binding sites [73]. Yeasts also survive in demanding environmental conditions, in which they adsorb and degrade dyes to decontaminate such environs at different efficiency rates. Some of the fungal species used in biosorption, the dyes they target, and their associated efficacy rates are summarized in table 8.5. Algal species have also been explored for dye removal (table 8.5) [17, 71] because their growth is not influenced by the presence of such contaminants [71]. The algal dye-removal mechanism relies on biosorption and enzymatic degradation and occurs at different efficacy rates depending on the targeted dye and the specific characteristics of the microbe used.

Table 8.5. Microbe-based adsorbents used in the removal of dyes from polluted environments.

Microbial species	Targeted dye	Biosorption efficacy (%)
Sphingomonas paucimoboilis	Methyl red	98
Nesterenkonia lacusekhoensis		97
Lysinibacillus fusiformis W1B6		96
Bjerkandera adusta	Methyl violet	94
Bacillus lentus BI377	Reactive red 141	99
Staphylococcus aureus	Orange II	76
Alishewanella sp.	Reactive blue 59	95
Acinetobacter sp. and *Klebsiella* sp.	Reactive orange 16	80
Halomonas sp. *strain* A55	Reactive red 184	96
Shewanella oneidensis MR-1	Napthol green B	95
Pseudomonas aeruginosa	Reactive red 21	81
Pseudomonas sp. and *Aspergillus ochraceus*	Reactive navy blue	80
Sterigmatomyces halophilus SSA1575	Reactive black 5	98
Aphanocapsa elachista	Reactive yellow 3 RN	49
Chlorella vulgaris	Reactive black 5	80
Armillaria sp. F022		80
Trichosporon akiyoshidainum HP2023		100
Bacillus pumilus HKG212	Remazol navy blue	>95
Bacillus halodurans MTCC 865	Acid black 24	90
Staphylococcus hominis RMLRT03	Acid orange	94
Myrothecium roridum		80
Pseudomonas sp., *Enterococcus* sp., *Escherichia* sp., *Staphylococcus* sp., and *Bacillus* sp.	Remazol brilliant violet 5R	100
Paraconiothyrium variabile	Acid red 18	97
Alcaligenes aquatilis 3c	Synazol red 6HBN	82
Pseudomonas extremorientalis BU118	Congo red	75
Geobacillus thermocatenulatus MS5		99
Geotrichum candidum		85
Serratia liquefaciens	Azure B	90
Lactococcus and *Dysgonomonas*	Metanil yellow	96
Cyathus bulle	Kiton blue A	88
Staphylococcus arlettae, *Klebsiella* sp., *Leucobacter albus*, and *Microbacterium* sp.	Disperse red 1	80
C. vulgaris		84
C. vulgaris	Disperse orange 2RL	55

(*Continued*)

Table 8.5. (*Continued*)

Microbial species	Targeted dye	Biosorption efficacy (%)
Brevibacillus laterosporus and *Galactomyces geotrichum*	Golden yellow HER	100
Ulva lactuca	Methylene blue	91
C. vulgaris	Direct blue 71	78
Peyronellaea prosopidis	Scarlet RR dye	85
Bjerkandera adusta	Crystal violet	91
B. adusta	Malachite green	96
Trichoderma harzianum	RR3	95
Aspergillus flavus	Methyl orange	53.62
Pleurotus ostreatus	Allura red AC	53
Lentinus concinnus	RY-86	97.6
Neonectria radicicola	Direct blue 86	99.6
Phoma sp.	Crystal violet	87
Spirulina platensis	Naphtol green B	94.4–99
Chlorella pyrenoidosa	Rhodamine B	97
Sargassum muticum	Methylene blue	90
Sargassum homeri	Acridine orange	>90
Phormidium animale	Acid red P-2BX	99.7
Bacillus cetenulatus	Basic blue 3	93
Acidithiobaccillus thiooxidans	Sulfur blue 15	87.5
Rhodopseudomonas palustris	Fast black K salt	72
Gluconacetobacter hansenii	Remazol	92

8.4 Prospects for the application of biosorbents for heavy metal and dye remediation

The use of microbial biosorbents in the removal of environmental pollutants is growing widely, as shown in this chapter. The efficacy of using biosorbents is dependent on the preparation, activation, and modification processes used to improve their physicochemical characteristics. A good microbial biosorbent should be readily available, should have good surface characteristics and adequate active sites, should be easily regenerated, and should have activation processes that are easily reproducible in order to maximize pollutant removal [74]. Biosorbents made from microorganisms are versatile, cost-effective, easy to operate, have high binding capacity to heavy metals and dyes, have good surface chemistry (surface area, many functional groups and pores), and do not produce large quantities of hazardous by-products, which justifies their preference in the bioremediation of heavy metals and dyes [75].

The efficacy of microbial biosorbents is also dependent on environmental factors such as nutrient availability, ionic strength, pH, and temperature, which affect the chemical and structural characteristics that determine biosorption capacity. Therefore, it is essential to focus on the characterization of such microorganisms via advanced research in order to understand the dynamics of their adsorption mechanisms and their ability to work in multipollutant systems such as polluted water and soils. According to Zhou *et al* [76], most actual polluted sites have diversified contaminants (a variety of dyes and heavy metals), and the application of stable and practical microbial biosorbents is key for successful remediation. The stability and efficacy of biosorbents can be advanced for better dye and heavy metal removal from polluted sites through the use of dead microbial biomass, the immobilization of sorbents on polysaccharide carriers or nanoparticles, and genetic engineering, [2, 17, 58, 71]. During these advancements, the cost and upscaling of the microbial-based biosorbents should not be compromised compared to existing conventional approaches for pollutant treatment. Aragaw and Bogale [17] noted that the processing, modification, and activation of biosorbents uses significant amounts of solvents and energy depending on the approach used; therefore, it is crucial to optimize the regeneration and reuse of such biosorbents. The safety of the modifications also needs to be investigated to avoid uncertainties associated with their application. In addition, it is essential to handle the adsorbent carefully and dispose of the eluent after the biosorption processes to prevent secondary pollution [77].

Microbial biosorbents have varied efficacy rates in heavy metal cleansing (table 8.4) and decolorizing dyes from the environment (table 8.5). The optimal conditions under which different microbes operate are also different, which makes it difficult to achieve complete pollutant removal. It is therefore imperative for research to focus more on optimizing the operational conditions of microbial biosorbents at the field scale. Most of the evaluated studies in this chapter sought to establish the efficiency of microorganisms in dye and heavy metal removal while neglecting the degradation and/or mineralization of the pollutants after their adsorption. Advanced research can also shed light on the ecotoxicity of metabolites produced following microbial biosorption. While some of the metabolites could be innocuous to microbial adsorbents, they could be harmful to other organisms in the ecosystem; therefore, this uncertainty should be clarified [71]. It is essential to consider the advantages and demerits associated with laboratory and field-scale applications of microbial biosorbents in order to optimize their efficacy while avoiding secondary environmental and health concerns.

8.5 Conclusions

In this chapter, biosorption technology has been shown to provide diverse combinations and mechanisms for the elimination of environmental pollutants. In particular, this approach has demonstrated great flexibility in dye and heavy metal removal. Various fungal, algal, and bacterial species have been shown to remove heavy metals and decolorize dyes with different efficacies. The use of microbial sorption for these purposes is preferred due to its cost efficiency, high effectiveness, and ability to regenerate and reuse such biomass. However, there are limited practical and field applications of biosorbents due to their operational complexities that prevent complete pollutant elimination. Contaminated environments are multiply polluted in nature and require the use of stable and highly efficient microbes which can withstand extreme environmental conditions. If such preconditions were met, it would enable the large-scale application of microbes in actual rather than laboratory settings. To realize such advances, this chapter recommended the need for extensive research through biotechnology to customize and produce microbial biosorbents that are compatible with targeted pollutants and the environments in which the contaminants are located; such biosorbents must target multiple contaminants and must not cause secondary pollution that would harm other ecosystems. The use of dead microbial biomass, its immobilization on polysaccharide carriers or nanoparticles, and genetic engineering could improve the efficacy of such biosorbents in the removal of environmental pollutants.

References

[1] Joseph L, Jun B, Flora J, Park C and Yoon Y 2019 Removal of heavy metals from water sources in the developing world using low-cost materials: a review *Chemosphere.* **229** 142–59

[2] Yaashikaa P, Kumar S, Saravanan A and Vo D 2020 Advances in biosorbents for removal of environmental pollutants: a review on pretreatment, removal mechanism and future outlook *J. Hazard. Mater.* **420** 126596

[3] Priya A, Gnanasekaran L, Dutta K, Rajendran S, Balakrishnan D and Moscoso M 2022 Biosorption of heavy metals by microorganisms: evaluation of different underlying mechanisms *Chemosphere.* **307** 135957

[4] Saravanan A, Kumar S, Jeevanantham S, Karishma S, Tajsabreen B, Yaashikaa P and Reshma B 2021 Effective water/wastewater treatment methodologies for toxic pollutants removal: processes and applications towards sustainable development *Chemosphere* **280** 130595

[5] Sedlakova A, Gresakova L, Godany A and Pristas P 2019 Bioaccumulation and biosorption of zinc by a novel Streptomyces K11 strain isolated from highly alkaline aluminum brown mud disposal site *Ecotoxicol. Environ. Saf.* **167** 204–11

[6] Ayangbenro S, Babalola O and Aremu O 2019 Bioflocculant production and heavy metal sorption by metal resistant bacterial isolates from gold mining soil *Chemosphere.* **231** 113–20

[7] Pham V, Kim J, Chang S and Chung W 2022 Bacterial biosorbents, an efficient heavy metals green clean-up strategy: prospects, challenges, and opportunities *Microorganisms.* **10** 610

[8] Oyewole O, Zobeashia S, Oladoja E, Raji R, Odiyina E and Musa A 2019 Biosorption of heavy metal polluted soil using bacteria and fungi isolated from soil *SN Appl. Sci.* **1** 857

[9] Ahmed S, Aktar S, Zaman S, Jahan R and Bari L 2020 Use of natural bio-sorbent in removing dye, heavy metal and antibiotic resistant bacteria from industrial wastewater *Appl. Water Sci.* **10** 107

[10] Piaskowski K, Swiderska R and Zarzycki P 2018 Dye removal from water and wastewater using various physical, chemical, and biological processes *J. AOAC Int.* **101** 1371–84

[11] Katheresan V, Kansedo J and Lau S 2018 Efficiency of various recent wastewater dye removal methods: a review *J. Environ. Chem. Eng.* **6** 4676–97

[12] Bhatia D, Sharma R, Singh J and Kanwar R 2017 Biological methods for textile dye removal from wastewater: a review *Crit. Rev. Environ. Sci. Technol.* **47** 1836–76

[13] Varghese A, Paul S and Latha M 2019 Remediation of heavy metals and dyes from wastewater using cellulose-based adsorbents *Environ. Chem. Lett.* **17** 867–77

[14] Syafiuddin A and Fulazzaky M 2021 Decolorization kinetics and mass transfer mechanisms of Remazol Brilliant Blue R dye mediated by different fungi *Biotechnol. Rep.* **29** e00573

[15] Amirza M, Adib M and Hamdan R 2017 Application of agricultural wastes activated carbon for dye removal—an overview *MATEC Web Conf.* **103** 06013–34

[16] Lellis B, Polonio C, Pamphile J and Polonio J 2019 Effects of textile dyes on health and the environment and bioremediation potential of living organisms *Biotechnol. Res. Innov* **3** 275–90

[17] Aragaw A and Bogale F 2021 Biomass-based adsorbents for removal of dyes from wastewater: a review *Front. Environ. Sci.* **9** 764958

[18] Guo G, Li X, Tian F, Liu T, Yang F and Ding K 2020 Azo dye decolorization by a halotolerant consortium under microaerophilic conditions *Chemosphere.* **244** 125510–37

[19] Aragaw A 2020 Recovery of iron hydroxides from electro-coagulated sludge for adsorption removals of dye wastewater: adsorption capacity and adsorbent characteristics *Surf. Interfaces.* **18** 100439

[20] Sivaranjanee R and Kumar P 2021 Treatment of textile wastewater using biochar produced from agricultural waste *Sustainable Technologies for Textile Wastewater Treatments* (Amsterdam: Elsevier) 187–208 pp

[21] Nyika J and Dinka M 2022 Activated bamboo charcoal in water treatment: a mini-review *Mater. Today Proc.* **56** 1904–7

[22] Kumar A, Singh R, Kumar S, Charaya M and Charaya M 2021 Biosorption: the removal of toxic dyes from industrial effluent using phytobiomass—a review *Plant Archives* **21** 1320–5

[23] Kumar G, Rao K, Yadav A, Kumar M and Sarathi T 2018 Biosorption of copper (II) and manganese (II) from waste water using low-cost bio adsorbents *J. Indian Chem. Soc.* **95** 1–8

[24] Cheng S, Show P, Lau B, Chang J and Ling T 2019 New prospects for modified algae in heavy metal adsorption *Trends Biotechnol.* **37** 1255–68

[25] Mark Y, Philip A, Nelson A, Muspratt A and Aikins S 2019 Safety assessment on microbial and heavy metal concentration in *Clarias gariepinus* (African catfish) cultured in treated wastewater pond in Kumasi, Ghana *Environ. Technol.* **40** 302–11

[26] Nanda M, Kumar V and Sharma D 2019 Multi metal tolerance mechanisms in bacteria: the resistance strategies acquired by bacteria that can be exploited to 'clean up' heavy metal contaminants from water *Aquat. Toxicol.* **212** 1–10

[27] Wang X, Ding S, Song W, Li H, Zhang Y, Ren W, Li M and Ding J 2021 A collaborative effect of algae-bacteria symbiotic and biological activated carbon system on black odorous water pretreated by UV photolysis *Biochem. Eng. J.* **169** 107983

[28] Selvi A, Rajasekar A, Theerthagiri J, Ananthaselvam A, Sathishkumar K, Madhavan J and Rahman P 2019 Integrated remediation processes toward heavy metal removal/recovery from various environments-a review *Front. Environ. Sci.* **7** 66

[29] Moghaddam S, Harun R, Mokhtar M and Zakaria R 2020 Kinetic and equilibrium modeling for the biosorption of metal ion by Zeolite 13X-Algal-Alginate Beads (ZABs) *J. Water Process. Eng* **33** 101057

[30] Koyuncu H and Kul A 2020 Biosorption study for removal of methylene blue dye from aqueous solution using a novel activated carbon obtained from nonliving lichen (*Pseudevernia furfuracea* (L.) Zopf.) *Surf. Interfaces.* **19** 100527

[31] Fawzy M 2020 Biosorption of copper ions from aqueous solution by *Codium vermilara*: optimization, kinetic, isotherm and thermodynamic studies *Adv. Powder Technol.* **31** 3724–35

[32] Bazzazzadeh R, Soudi M, Valinassab T and Moradlou O 2020 Kinetics and equilibrium studies on biosorption of hexavalent chromium from leather tanning wastewater by *Sargassum tenerrimum* from Chabahar-Bay Iran *Algal Res.* **48** 101896

[33] Rambabu K, Thanigaivelan A, Bharath G, Sivarajasekar N, Banat F and Show P 2021 Biosorption potential of *Phoenix dactylifera* coir wastes for toxic hexavalent chromium sequestration *Chemosphere.* **268** 128809

[34] Bakather O, Zouli N, Abutaleb A, Mahmoud M, Daher A, Hassan M, Eldoma M, Alasweda S and Fowad A 2020 Uranium (VI) ions uptake from liquid wastes by *Solanum incanum* leaves: biosorption, desorption and recovery *Alex. Eng. J.* **59** 1495–504

[35] Yang X, Zhao Z, Yu Y, Shimizu K, Zhang Z, Lei Z and Lee D 2020 Enhanced biosorption of Cr (VI) from synthetic wastewater using algal-bacterial aerobic granular sludge: batch experiments, kinetics and mechanisms *Sep. Purif. Technol.* **251** 117323

[36] Fathollahi A, Coupe S, El-Sheikh A and Nnadi E 2020 Cu (II) biosorption by living biofilms: isothermal, chemical, physical and biological evaluation *J. Environ. Manage.* **282** 111950

[37] Lakshmi S *et al* 2021 Biosorption of oxybenzene using biosorbent prepared by raw wastes of *Zea mays* and comparative study by using commercially available activated carbon *Saudi J. Biol. Sci.* **28** 3469–76

[38] Tural B, Ertas E, Enez B, Fincan S and Tural S 2017 Preparation and characterization of a novel magnetic biosorbent functionalized with biomass of *Bacillus subtilis*: kinetic and isotherm studies of biosorption processes in the removal of Methylene Blue *J. Environ. Chem. Eng* **5** 4795–802

[39] Saha G, Hoque M, Miah M, Holze R, Chowdhury D, Khandaker S and Chowdhury S 2017 Biosorptive removal of lead from aqueous solutions onto Taro (*Colocasiaesculenta* (L.) Schott) as a low cost bioadsorbent: characterization, equilibria, kinetics and biosorption-mechanism studies *J. Environ. Chem. Eng* **5** 2151–62

[40] Verma A, Kumar S and Kumar S 2017 Statistical modeling, equilibrium and kinetic studies of cadmium ions biosorption from aqueous solution using *S. filipendula J. Environ. Chem. Eng.* **5** 2290–304

[41] Kumar M, Singh A and Sikandar M 2020 Biosorption of Hg (II) from aqueous solution using algal biomass: kinetics and isotherm studies *Heliyon* **6** 03321

[42] Rangabhashiyam S, Lata S and Balasubramanian P 2018 Biosorption characteristics of methylene blue and malachite green from simulated wastewater onto *Carica papaya* wood biosorbent *Surf. Interfaces* **10** 197–215

[43] Bai M, Shaik O, Kavitha J, Varma M and Chittibabu N 2020 Biosorption of eosin yellow dye from aqueous solution using sugarcane bagasse: equilibrium, kinetics and thermodynamics. *Mater. Today Proc* **26** 842–9

[44] Ezeonuegbu B, Machido D, Whong C, Japhet W, Alexiou A, Elazab S, Qusty N, Yaro C and Batiha G 2021 Agricultural waste of sugarcane bagasse as efficient adsorbent for lead and nickel removal from untreated wastewater: biosorption, equilibrium isotherms, kinetics and desorption studies *Biotechnol. Rep.* **30** 00614

[45] Juchen P, Piffer H, Veit M, Goncalves G, Palacio S and Zanette J 2018 Biosorption of reactive blue BF-5G dye by malt bagasse: kinetic and equilibrium studies *J. Environ. Chem. Eng* **6** 7111–8

[46] Villen-Guzman M, Pinilla D, Lahoz C, Alonso C, Maroto J and Arhoun B 2019 Optimization of Ni (II) biosorption from aqueous solution on modified lemon peel *Environ. Res.* **179** 108849

[47] Njoku V 2014 Biosorption potential of cocoa pod husk for the removal of Zn (II) from aqueous phase *J. Environ. Chem. Eng* **2** 881–7

[48] Azimi A, Azari A, Rezakazemi M and Ansarpour M 2019 Removal of heavy metals from industrial wastewaters: a review *Chem Bio Eng Reviews* **4** 37–59

[49] Abhinaya M, Parthiban R, Kumar P and Vo N 2021 A review on cleaner strategies for extraction of chitosan and its application in toxic pollutant removal *Environ. Res.* **196** 110996

[50] Basri W *et al* 2019 A sugarcane-bagasse-based adsorbent employed for mitigating eutrophication threats and producing biodiesel simultaneously *Processes* **7** 572

[51] Saravanan A, Kumar P, Yaashikaa P, Karishma S, Jeevanantham S and Swetha S 2021 Mixed biosorbent of agro-waste and bacterial biomass for the separation of Pb (II) ions from water system *Chemosphere* **277** 130236

[52] Mo J, Yang Q, Zhang N, Zhang W, Zheng Y and Zhang Z 2018 A review on agro-industrial waste (AIW) derived adsorbents for water and wastewater treatment *J. Environ. Manage.* **227** 395–405

[53] Ahmaruzzaman M 2011 Industrial wastes as low-cost potential adsorbents for the treatment of wastewater laden with heavy metals *Adv. Colloid Interface Sci.* **166** 36–59

[54] Dai Q, Bian X, Li R, Jiang C, Ge J, Li B and Ou J 2019 Biosorption of lead (II) from aqueous solution by lactic acid bacteria *Water Sci. Technol.* **79** 627–34

[55] Huang H, Jia Q, Jing W, Dahms H and Wang L 2020 Screening strains for microbial biosorption technology of cadmium *Chemosphere* **251** 126428

[56] Show P, Ooi C, Song C, Chai W, Lin G, Liu B and Chang Y 2021 Purification of lysozyme from chicken egg white by high-density cation exchange adsorbents in stirred fluidized bed adsorption system *Food Chem.* **343** 128543

[57] Saravanan A, Karishma S, Kumar P, Varjani S, Yaashikaa P, Jeevanantham S, Ramamurthy R and Reshma B 2021 Simultaneous removal of Cu (II) and reactive green 6 dye from wastewater using immobilized mixed fungal biomass and its recovery *Chemosphere* **271** 129519

[58] Blaga A, Zaharia C and Suteu D 2021 Polysaccharides as a support for microbial-based adsorbents with application in removal of heavy metals and dyes *Polymers* **13** 2893

[59] Crini G, Lichtfouse E, Wilson L and Morin-Crini N 2019 Conventional and non-conventional adsorbents for wastewater treatment *Environ. Chem. Lett.* **17** 195–213

[60] Velkova Z, Kirova G, Stoytcheva M, Kostadinova S, Todorova K and Gochev V 2018 Immobilized microbial biosorbents for heavy metals removal *Eng. Life Sci.* **18** 871–81

[61] Qiao W, Zhang Y, Xia H, Luo Y, Liu S, Wang S and Wang W 2019 Bio-immobilization of lead by *Bacillus subtilis* X3 biomass isolated from lead mine soil under promotion of multiple adsorption mechanisms *R. Soc. Open Sci.* **6** 181701

[62] Lin Z, Li J, Luan Y and Dai W 2020 Application of algae for heavy metal adsorption: a 20-year-meta-analysis *Ecotoxicol. Environ. Saf* **190** 110089

[63] Aryal M and Liakopoulou-Kyriakides M 2015 Bioremoval of heavy metals by bacterial biomass *Environ. Monit. Assess.* **187** 4173

[64] Bilal M, Rasheed T, Sosa-Hernandez J, Raza A, Nabeel F and Iqbal H 2018 Biosorption: an interplay between marine algae and potentially toxic elements—a review *Mar. Drugs* **16** 1–16

[65] Papirio S, Frunzo L, Mattei M, Ferraro A, Race M, Acunto B, Pirozzi F and Esposito G 2017 *Heavy Metal Removal from Wastewaters by Biosorption: Mechanisms and Modeling* (Berlin: Springer Nature)

[66] Khan I, Aftab M, Shakir S, Ali M, Qayyum S, Rehman M, Haleem K and Touseef I 2019 Mycoremediation of heavy metal (Cd and Cr)–polluted soil through indigenous metal-lotolerant fungal isolates *Environ. Monit. Assess* **191** 1–11

[67] Deng Z, Cao L, Huang H, Jiang X, Wang W, Shi Y and Zhang R 2011 Characterization of Cd- and Pb-resistant fungal endophyte *Mucor* sp. CBRF59 isolated from rapes (*Brassica chinensis*) in a metal-contaminated soil *J. Hazard. Mater.* **185** 717–24

[68] Todorova K, Velkova Z, Stoytcheva M, Kirova G, Kostadinova S and Gochev V 2019 Novel composite biosorbent from *Bacillus cereus* for heavy metals removal from aqueous solutions *Biotechnol. Biotechnol. Equip.* **33** 730–8

[69] Jiang D, Yang Y, Huang C, Huang M, Chen J, Rao T and Ran X 2019 Removal of the heavy metal ion nickel (II) via an adsorption method using flower globular magnesium hydroxide *J. Hazard. Mater.* **373** 131–40

[70] Concordio-Reis P, Reis M and Freitas F 2020 Biosorption of heavy metals by the bacterial exopolysaccharide fucopol *App. Sci.* **10** 1–9

[71] Tang K, Darwish N, Alkahtani A, Ragab M, Gawwad A and Karacsony P 2022 Biological removal of dyes from wastewater: a review of its efficiency and advances *Tropical AquaticSoil Pollut.* **2** 59–75

[72] Olukanni O, Osuntoki A and Gbenle G 2009 Decolorization of azo dyes by a strain of *Micrococcus* isolated from a refuse dump soil *Biotechnology* **8** 442–8

[73] Erden E, Kaymaz Y and Pazarlioglu N 2011 Biosorption kinetics of a direct azo dye Sirius Blue K-CFN by *Trametes versicolor Electron. J. Biotechnol.* **14** 3

[74] Leng L, Xiong Q, Yang L, Li H, Zhou Y and Zhang W 2021 An overview on engineering the surface area and porosity of biochar *Sci. Total Environ.* **763** 144204

[75] Sherugar P, Padaki M, Naik N, George S and Murthy D 2022 Biomass-derived versatile activated carbon removes both heavy metals and dye molecules from wastewater with near-unity efficiency: mechanism and kinetics *Chemosphere* **287** 132085

[76] Zhou Y, Lu J, Zhou Y and Liu Y 2019 Recent advances for dyes removal using novel adsorbents: a review *Environ. Pollut.* **252** 352–65

[77] Patel H 2021 Review on solvent desorption study from exhausted adsorbent. *J. Saudi Chem. Soc.* **25** 101302

Chapter 9

The use of biosorbents derived from invasive plants for environmental remediation

W O Egboduku, T Egboduku, J O Emegha and O Imarhiagbe

Unlike organic pollutants, heavy metals are persistent environmental contaminants that cannot be biodegraded. While chemical/physical methods can facilitate the removal of these substances, there is an increasing demand to embrace eco-friendly alternatives for remediation purposes. There is evidence to suggest that invasive-plant-derived biosorbents can effectively remove contaminants, including toxic metals, from the environment. Biosorbents can be either natural or altered. The utilization of common invasive plants as biosorbent sources for bioremediation is explored in this chapter. This approach could prove crucial in mitigating the adverse impact of these species on the environment, thereby promoting a waste-to-wealth strategy.

9.1 Introduction

Environmental remediation involves the removal of contaminants or pollutants from the environment, which includes water and soil [1]. The purpose of such remediation is more than just eliminating pollutants, as it includes protecting people and the environment from harm. Studies of exposure routes have highlighted the hazardous effects of contaminants including fine particles, heavy metals, and emissions from decommissioned petrochemical facilities. Moreover, there is a growing understanding of the degradation and reuptake of pollutants as well as intricate phase transformation procedures; for instance, those that take place between the soil–water or air–surface water interfaces [1]. In the process of environmental remediation, an essential step is to analyze and improve the quality of the three main ecological compartments. The goal is to minimize pollution levels by establishing acceptable limits for specific land uses. This can be achieved through risk characterization that includes evaluating hazard, exposure, and dose–response

doi:10.1088/978-0-7503-5598-8ch9

factors [2]. The impact of industrialization and technological progress on human civilization is significant, yet their effects have contributed to a decline in water quality [3]. Numerous published studies have illustrated that approximately one billion individuals globally lack access to potable drinking water [4]. The improper disposal of a vast array of organic and inorganic substances such as pharmaceuticals, hydrocarbons, heavy metals, and textile dyes is the primary cause of this situation. The contamination caused by heavy metals in water sources is particularly concerning because even low levels of these metals can induce reactive oxygen species, resulting in cytotoxicity [1, 5]. One example of the significant negative effects that can occur due to prolonged exposure to a particular substance is arsenic. Consumption of this element at a level of more than 0.05 mg l^{-1} over an extended period can lead to acute symptoms such as chronic respiratory disorders, sensory loss, skin discoloration, and even skin cancer, as documented by [5]. Similarly, the growth in industries involved in dye production has led to elevated quantities of wastewater containing more than 10 000 dyes, which has intensified the concentration levels of colored effluents present within it [1]. The presence of pollutants in water bodies can lead to reduced aesthetic quality and hinder light penetration, causing disruption within the aquatic ecosystem. As part of an attempt to address this issue sustainably, biosorbents are gaining popularity as a means of removing toxic organic compounds or heavy metals [5]. These sorbent materials are derived from various sources, such as bacterial biomass, yeasts, and fungi with fibrous structures along with by-products generated by the food and pharmaceutical production industries, which include agricultural wastes comprised mainly of polysaccharides—all of which contribute towards environmental remediation efforts [1]. Every biomaterial should display exceptional biosorption proficiency and attraction towards all organic compounds and inorganic ions. Examples of some essential bioadsorbents include fungi, notably the filamentous species of *Alternaria*, *Aspergillus*, *Rhizopus*, and *Penicillium* along with yeast strains such as *Saccharomyces cerevisiae* and *Saccharomyces carlsbergensis* [1].

Microorganisms, which are commonly utilized in the food and pharmaceutical sectors, generate a surplus of waste that can be accessed at no or low cost. In addition, algae serve as an effective biosorbent source. Red, green, and brown algae have potential for use as biosorbents; among them, brown algae such as *Chlorella vulgaris* have proven to be particularly proficient [1, 5].

9.2 Categories of biosorbents

Several biomaterials derived from bryophytes, fungi, bacteria, or environmental waste are known to adsorb heavy metals and other pollutants. This chapter uses two significant categories of biosorbents for environmental remediation [6].

1. **Natural biosorbents:** natural biosorbents are mainly obtained from living organisms. The cells of living plants and other organisms take up toxic substances such as heavy metals by bioaccumulation. Toxic substances are also removed by dead cells by biosorption (table 9.1).

Table 9.1. Summary of the types of biosorbents and their sources.

Categories of biosorbents	Examples
1 Natural Fungi	- *Aspergillus fumigates* - *Aspergillus niger* - *Aspergillus flavus* - *Rhizopus oryzae* - *Rhizopus nigricans* - *Rhizopus stolonifer*
Algae	- *Pithophora* sp. - *Scenedesmus spinosus* - *Spirogyra hyaline* - *Nannochloropsis oculata* - *Chlorella vulgaris* - *Auxenochlorella protothecoides* - *Sargassum polycystum*
Bacterial	- *Micrococcus luteus* - *Enterobacter cloacae* - *Shewanella oneidensis* - *Bacillus cereus* - *Geobacter sulfurreducens* - *Bacillus subtilis*
Plants	- Stems of *Musa acuminata* and *Musa paradisiaca* (banana) - Pectins from *Beta vulgaris* (sugar beet) - The stem bark of *Acacia leucocephala* (acacia) - Leaves, seeds, and roots of *Nympheae lotus* (water lily) - Leaves, seeds, and roots of *Eichhornia crassipes* (water hyacinth) - Leaves, seeds, and roots of *Santalum album* (sandalwood) - Powdered leaf of *Utricularia aurea* (bladderwort) - Biomass of *Momordica charantia* (bitter melon) - Leaves of *Azadirachta indica* (neem) - Leaves of *Magnifera indica* (mango) - Leaves of *Cajanus cajan* (pigeon pea)
Agricultural wastes	- Husk of *Oriza sativa* (rice) - Peels of *Punica granatum* (pomegranate) - Charcoal husk of *Arachis hypogaea* (peanut)
2 Modified	- Fly ash or coal ash - Biochar produced from the husk of *O. sativa* (rice) - Biochar from the peduncle of *Musa paradisiaca* (banana) - Biochar produced from wheat straw pellets - Husk of *A. hypogaea* (peanut) treated with NaOH - Biochar obtained from the peels of *Cucumis melo* (dew melon) - Biochar obtained from *Phoenix dactylifera* (date palm)

Table 9.2. The production process of invasive plant biochars and their operating conditions.

IPB production process	Conditions	Temperature (°C)	Time	Pressure	Biochar yield
Slow pyrolysis	- Absence of oxygen - Moisture content < 15%–20% - Heating rate < 10 °C min^{-1}	300–800	>60 min	1 atm	30%–55%
Fast pyrolysis	- Absence of oxygen - Moisture content < 15%–20% - Heating rate ⩾ 200 °C min^{-1}	450–600	~0.02	1 atm	10–25
Gasification	- Low oxygen supply - Moisture content 10%–20% - Heating rate ~100 °C min^{-1}	750–1000	0.2–0.4	1–3 atm	14–25
Hydrothermal carbonization (HTC)	Moisture content 75%–90%	180–300	5–240	1–200	50–80
Torrefaction	- Absence of oxygen - Moisture content < 10% - Heating rate < 50 °C min^{-1}	200–300	15–60	1	70–80

2. **Modified biosorbents:** modified biosorbents are adsorbents that are chemically characterized. They are usually synthesized from other living organisms (see table 9.1).

9.3 Biosorbents derived from invasive plants

Invasive plants are part of the global ecosystem [7]. Although they are known to cause ecological or economic harm in new environments, i.e. those that they are not indigenous to, several scientists have reported the use of invasive plants for environmental remediation [7]. Invasive plants possess remarkable adaptability and an exceptional capability to spread swiftly. As such, it is crucial to find sustainable approaches that can alleviate their adverse effects on the indigenous ecosystem. Several methods have been suggested to check the invasion of alien plants; some of these include chemical, biological, and alternative controls [8]. Although these approaches have enduring outcomes when put into practice, they still demand significant labor and material resources. Consequently, there is a need to identify a more economical, effective, and sustainable way to address the issues of invasive plants. One plausible suggestion is to use them as biosorbents—emerging materials for environmental protection [9]. According to research

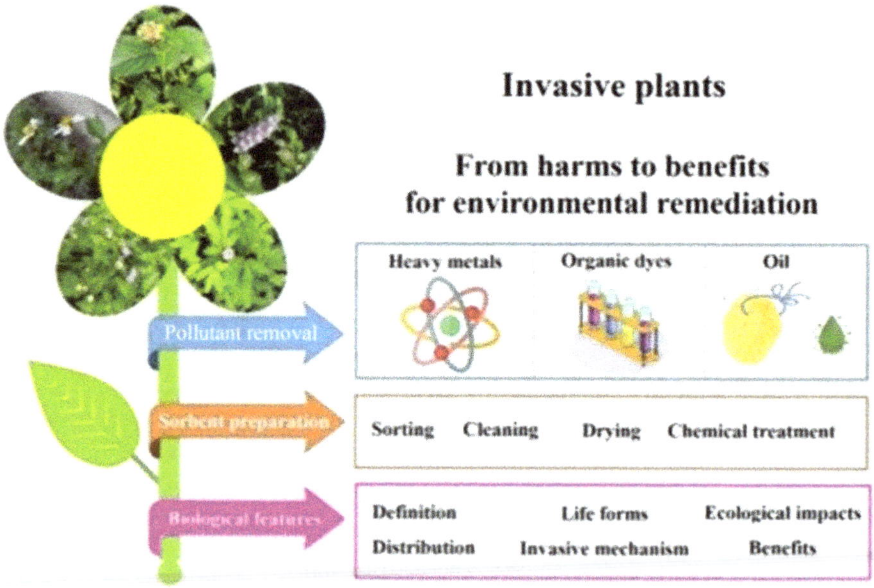

Figure 9.1. The conversion of invasive plants into biosorbents for environmental remediation. Reprinted by permission from Springer Nature [3], Copyright (2021).

findings, the creation of biosorbents from plants that have invaded an ecosystem can aid in reducing the ecological harm caused by biological invasions. The process of producing biochar using invasive plant species as a source material presents a viable approach for managing waste and curtailing the spread of such invasive vegetation [10]. Consequently, this technique offers promise in addressing the problems associated with invasive plants. Yang *et al* [11] reported that *Eichhornia crassipes*, *Solidago canadensis*, and *Alternanthera philoxeroides* have been identified as effective feedstocks for producing biochar and have potential applications as biosorbents in environmental remediation efforts. The invasive plants most currently used as biosorbents and biochars are *Eichhornia crassipes* (water hyacinth), *Eupatorium adenophorum* (cotton weed), *Spartina alterniflora* (smooth cord grass), *Alternanthera philoxeroides* (alligator weed), and *Solidago canadensis* L. (Canadian goldenrod) [3, 12]. The herbaceous nature of these invasive plants makes them readily available and abundant. Their stems possess numerous vascular bundles of small size, and adjacent to those are multiple thin-walled cells. This composition is favorable for manufacturing biochars which have increased porosity and a substantial specific surface area [13]. Table 9.2 shows biosorbents derived from some invasive plants and locations where they are used, while figure 9.1 shows the conversion of invasive plants into biosorbents for environmental remediation as adapted from [3].

9.4 The production of biosorbents/biochars from invasive plants

Pyrolysis is a chemical process of biomass breakdown that happens in the absence of oxygen. The term pyrolysis comes from two Greek words: 'pyro' meaning fire and 'lysis' which refers to decay or disintegration into fundamental components [14]. The conditions involved in pyrolysis, such as temperature, heating rate, etc. significantly affect the physicochemical properties of biochar. In the process of thermal decomposition, carbohydrates such as hemicellulose, cellulose, and lignin are subjected to cross-linking, depolymerization, and cleavage at distinct temperatures, which results in the formation of three forms of output: solid, liquid, and gas. The preparation of biochar through pyrolysis offers cost-effectiveness and the inherent benefits of a straightforward approach [11]. The process of pyrolysis, which involves heating organic matter in a nonreactive atmosphere, is intricate and comprises both simultaneous and successive reactions. The breakdown of the long chains of carbon, hydrogen, and oxygen compounds present in biomass occurs between 350 °C–550 ° C and 700 °C–800 °C in the absence of air/oxygen. This results in the formation of smaller molecules such as gases, condensable vapors (tars and oils), and solid charcoal under pyrolytic conditions. Pyrolysis can mainly be categorized into three different types: conventional, fast, or flash, depending on various operational factors. This indicates that this procedure offers multifaceted operations with diverse product yields based on their different characteristics due to the various setups used for optimization purposes during its application [11]. These classifications depend on factors such as the biomass particle size, solid residence time, heating rate, and temperature (table 9.2).

1. **Conventional pyrolysis:** pyrolysis has been utilized for millennia to improve the production of charcoal by employing low temperatures and heating rates. The traditional method features a long vapor residence time (5–30 min), resulting in continuous reactions among the vapor-phase components, leading to solid char and other liquids [15]. However, this conventional pyrolysis technique is not ideal for producing biochar because it suffers from technical deficiencies such as high residence times that cause primary product cracking during slow pyrolysis, which can negatively impact both the yield and the quality of the resulting biochar [16, 17].
2. **Fast pyrolysis:** in the fast pyrolysis process, biomass belonging to a particular plant is heated rapidly without oxygen at extremely high temperatures. According to research based on weight, this method produces oily products or other forms of liquids that amount to approximately 60%–75%. This output consists of biochar solids which make up about 15%–25% and a gaseous phase accounting for around 10%–20%. The ratio varies depending on the type of feedstock used in production. Specifically, when it comes to liquid yield from biomass sources, optimal results are achieved through low temperature settings with high heating rates and short residence times. The fast pyrolysis method is characterized by an accelerated heating rate and efficient heat transfer, resulting in prompt vapor cooling with minimal time for vapor to remain. This technology can be comparatively inexpensive to

implement and has high energy efficacy, particularly when applied on a small scale, surpassing those of other processes [18].

3. **Flash pyrolysis:** A potentially advantageous approach to generating biochar from plant biomass is through the rapid pyrolysis technique. This method involves quick devolatilization in an inert atmosphere, speedy heating of particles, high reaction temperatures ranging between 450 °C and 1000 °C, and a gas residency time that does not exceed one second. However, this process presents some technological challenges which include oil solids produced as well as an increase in viscosity over time due to the char's catalytic effect; moreover, the thermal stability of the oil and its propensity for corrosiveness remain disappointing [14].

In addition to pyrolysis, alternative techniques for obtaining biochar from various sources include gasification, hydrothermal carbonization, and torrefaction. The process of gasification differs from pyrolysis in that it involves an oxidizing agent and is typically used to generate syngas consisting of H_2, CO, CO_2, and CH_4. Biochar is often regarded as a secondary output, which has led to minimal exploration of its production optimization. Torrefaction has emerged as a promising method for generating biochar. This strategy involves removing the moisture content, carbon dioxide, and oxygen from biomass at temperatures ranging from 200 °C to 300 °C in anaerobic conditions. Polymerized saccharides are broken down in this process, resulting in an insoluble product with a low O/C ratio that repels water effectively. The gradual temperature increase during torrefaction makes it a form of gentle pyrolysis treatment commonly applied to achieve torrefied products in operational processes.

However, in academic discourse, torrefaction is not considered a highly favorable approach toward biochar manufacturing. This holds true even though it achieves comparatively greater product yields of 70–80 wt.%. Its inadequacy stems from the presence of substantial volatile components retained by the biomass during the torrefaction process and also from its failure to meet the desired physical and chemical attributes necessary for producing high-quality biochar (e.g. $O/C > 0.4$). As such, torrefaction has more often been utilized as an initial step for removing moisture content from feedstock, for densification purposes, or for enhancing a material's fragility [19]. Table 9.2 shows and compares the described biochar production processes' operating conditions and biochar yields.

The use of invasive plant-derived biosorbents has provided a beneficial pathway that can be used to produce high-quality biochar for various applications [19]. The utilization of invasive plants as raw materials in the production of biochar/biosorbent presents a promising solution that can reinforce both environmental and economic sustainability. In view of their unique morphology, invasive plant species possess inherent advantages in terms of biomass structure and texture, which can be harnessed to enhance the structural properties of biochar products. For instance, water hyacinth is characterized by its hollow architecture that confers excellent initial structural features to biochar preparations. Table 9.3 shows a list of

Table 9.3. Invasive plants and their uses in environmental remediation.

Invasive plant	Production method	Modification	Characteristics	Uses	References
Ambrosia artemisiifolia L.	Pyrolysis/350 °C–550 °C/120 min		High pH value (alkaline)	Adsorption of Cd(II) and Pb (II)	[20]
Conyza canadensis	Pyrolysis/350 °C–550 °C/120 min		High pH value (alkaline)	Adsorption of Cd(II) and Pb (II)	[20]
Acacia auriculiformis	Pyrolysis/500 °C/120 min	—	—	Removal of dyes from water	[3]
Ailanthus excelsa				Removal of dyes from water	[3]
Ambrosia trifida L.	Pyrolysis/700 °C/180 min		Aromatic, low polarity	Adsorption of trichloroethylene in water	[11]
Solidago canadensis L.	Pyrolysis/700 °C/240 min	Ca/Al hydrotalcite or hydroxylapatite modification		Adsorption of Eu(III) in water	[21]
Eichhornia crassipes	Pyrolysis/550 °C/60 min	Impregnated with phase-change materials (PCMs)	High porosity, high carbon content, good thermal conductivity	Storage of energy	[22]
E. crassipes	Pyrolysis/700 °C/120 min	Modified with metal oxide (Fe)	Large apertures, nanoparticles of Fe_3O_4 aggregate to form a cluster	Gasification of coal wastewater, degradation of 2,4,6-trichlorophenol	[23]
E. crassipes	Pyrolysis/<700 °C	Modification with metal oxide (Mn)	Rich in Mn-OH groups	Adsorption of heavy metals in water	[25].
E. crassipes	Pyrolysis/500 °C/180 min	Modification with KOH	Large number of heteroatoms, developed carbon nanonetwork	Energy storage	[24].
E. crassipes	Pyrolysis/250 °C/60 min	Modification with metal oxide (Fe)	Magnetic, more surface activity sites	Adsorption of As(V) in water	[25].
E. crassipes	Pyrolysis/400 °C/180 min	Modification with metal oxide (Fe)	Magnetic, presence of more OH groups	Adsorption of Cr (VI) in water	[26]
Prosopis juliflora	Pyrolysis/400 °C/60 min	Modification with KOH	Large number of heteroatoms	Energy storage	[27]

Eupatorium adenophorum	HTC/220 °C/60 min	Modification with HNO_3	More functional groups, more pores	Adsorption of Pb(II) in water	[28]
E. adenophorum	Pyrolysis/600 °C/120 min	Modification with Fe and Ni	Increased surface functional groups	Removal of 2,4,6-trichlorophenol from water	[29]
Spartina alterniflora	Pyrolysis/350 °C/120 min	—	Many oxygen functional groups	Adsorption of Cd in soil	[30]
Alternanthera philoxeroides	Pyrolysis/450 °C/120 min	Modification with H_2O_2	Oxygen-containing functional groups	Adsorption of metformin hydrochloride in water	[31]
A. philoxeroides	Pyrolysis		Rich microporous structure	Adsorption of rhodamine B in water	[32]
Sicyos angulatus	Pyrolysis/700 °C/120 min	Activated with steam for 45 min	Increased pore volume	Adsorption of sulfamethoazine (SMT) in water	[11].
Lantana camara	Pyrolysis/500 °C/240 min	—	High ash content	Used to reduce soil acidity	[11]
Pistia stratioles	Pyrolysis/400 °C–700 °C/180 min	Nitrogen doped	Large pore structure, graphene structure, rich functional groups	Adsorption of phthalate esters in water	[33]
E. crassipes		Treated with NaOH		Removal of heavy metals from water	[3]
Acorus calamus		Treated with HCl and NaOH		Adsorption of ions	[3]
Undaria pinnatifda	Pyrolysis	Treated with $CaCl_2$		Adsorption of ions	[3]
Imperata cylindrica		Treated with NaOH	Large particle size	Adsorption of Cu(II) in water	[3]
E. crassipes		Treated with 10% EDTA		Adsorption of Co(II)	[3]

invasive plants that can be used to produce biosorbents for environmental remediation, including their production methods.

9.5 Preparing biosorbents from invasive plants

Although little work has been done on the preparation of biosorbents from invasive plants, a few scientists have been able to describe methods that might be efficient for this task. Anandrao and Rhesma [34] prepared biosorbents from *Ailanthus excelsa* by first drying the stalk and washing it with distilled water to remove impurities. The dried particles were then crushed and soaked with concentrated H_2SO_4 and kept for 6 h in a fume cupboard. After 6 h, the mixture was thoroughly washed with cold water to completely remove the H_2SO_4, producing a sulphonated carbon which was then used for the adsorption of methylene blue. Generally, biosorbents can be prepared from plants through the following process (figure 9.2):

- **Collecting invasive plants:** in this stage, the invasive plants are contained, and their parts (stems, flowers, rhizomes, leaves, bark, etc.) are harvested, which are then referred to as 'explants.'
- **Processing stage:** the explants undergo sorting and washing with distilled water or ethanol to eliminate impurities. Through this process, contaminants and pathogens are removed, while some endogenous inorganic matter that can hinder adsorption is discarded along with them.
- **Drying:** the explants are subsequently dehydrated to eliminate moisture content through exposure to solar light or by using an oven at a temperature range of 50 °C–105 °C.

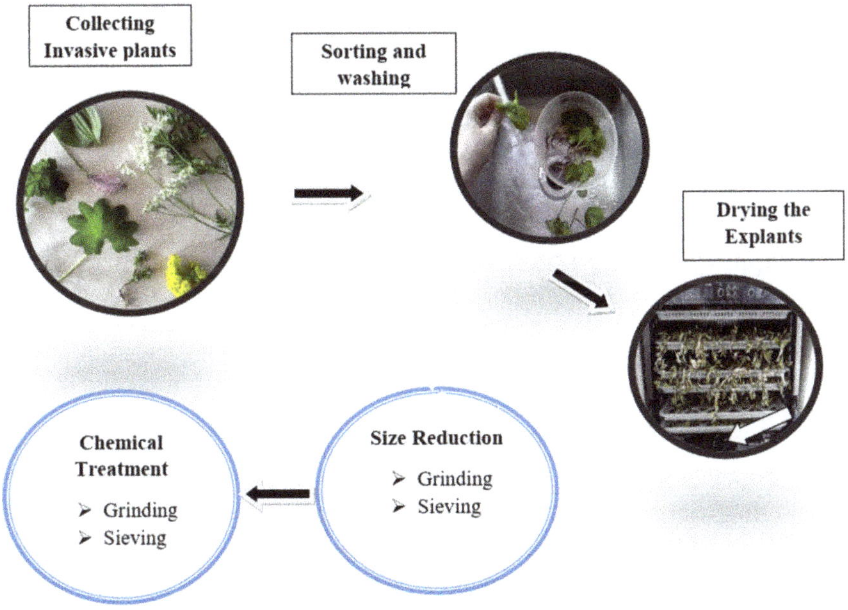

Figure 9.2. Preparation of biosorbents from invasive plants.

- **Reduction in size**: before a sieve is used to collect small particles, dried explants undergo size reduction through chopping or grinding. The main objective of this process is to increase the surface areas of adsorbents and reduce the diffusion path length for biosorbents. The result is an increase in the interactions between materials and pollutants, which enhances effectiveness [3]. The impact of explants' size reduction has been examined by scientists. In a study conducted by Khamparia and Jaspal [35], it was observed that the efficacy of rhodamine B elimination through adsorption on *Argemone mexicana* seed-based adsorbents increased considerably upon reducing the particle size from 710 to 300 μ. This stage is crucial in order to efficiently remove pollutants via the process of adsorption.
- **Chemical treatment:** chemical pretreatments play a significant role in the breakdown of lignin and cellulose within biomass or explants. This process facilitates an increase in surface area and porosity, thereby enhancing the adsorption of contaminants by biosorbents. Various chemicals, such as acids, bases, and oxidizers, are frequently utilized for this purpose [3]. Several sources have documented the analysis of chemical procedures carried out on alien plant explants. One such study conducted by Shooto [36] investigated the influence of hydrochloric acid and sodium hydroxide on the amounts of Cr and Pb heavy metal ions adsorbed by a biosorbent produced from rhizomes belonging to the *Acorus calamus* species. According to Shooto [36], the surface area of raw biomass can be increased through chemical treatment, and acid and base treatments result in an increase from 5.78 to 9.98 $cm^2\ g^{-1}$ and a near doubling, respectively. The study found that while both biosorbents had a higher adsorption capacity for Pb than for Cr, NaOH was the most effective modifier, followed by HCl. Specifically, upon treatment with an alkaline solution, the adsorption capacities of the biosorbent for Cr and Pb were found to increase by more than 10 $mg\ g^{-1}$ as compared to its untreated counterpart. In essence, chemical modifications lead to a favorable outcome in terms of reducing soluble organic compounds, which ultimately enhances pollutant removal efficiency.

9.6 The mechanism of biosorption

The term biosorption pertains to the non-active absorption of harmful substances by inanimate, lifeless, or biogenic materials [6]. Biosorption is a metabolic process that occurs outside of the cell membrane and involves various mechanisms for pollutant absorption, which may differ depending on the biomass applied. The physicochemical processes that underlie biosorption include absorption, adsorption, ion exchange, surface complexation, and precipitation. This spontaneous process does not rely on microorganisms' metabolisms to occur. Biosorbents are frequently utilized in biotechnology applications, as they enable the separation of organic or inorganic substances from solutions; this principle has also gained relevance with respect to environmental protection efforts.

The physicochemical characteristics of the metals and biosorbents and the process parameters influence the mechanisms of biosorption. Biosorbents derived from living organisms (microbes or plants) accumulate metals in capillaries and spaces of the cell walls and membranes when exposed to a concentration gradient [6]. The surfaces of biosorbents bear critical functional groups, viz. hydroxyls, sulfydryls, ketones, amines, carboxyls, and imidazoles, which help to facilitate the biosorption processes of specific metal ions [37]. The functional groups induce the detachment of metals from their typical binding sites, leading to a modification in the oxidation state of metallic ions and subsequently mitigating their harmful effects [38]. Nevertheless, there are various additional variables, such as the acidity level of the polluted environment, the arrangement of the biosorbents' structure, and thermal conditions that impact the performance of biosorbents. Zhang *et al* [39] stated that negatively charged biosorbents work best in sites contaminated with cationic metallic species. The optimum pH range is between 7.0 and 8.0. At lower pH values, competitive binding of hydrogen ions and metal ions ensues. At higher pH values, metals precipitate as hydroxides. Sites polluted with anionic metallic species such as chromium, arsenic, and molybdenum are best remediated at acidic pH values, i.e. 2.0–4.0 (figure 9.3 as adapted from [3]). High temperatures have also been discovered to drive biosorption by increasing the surface activity of biosorbents and the kinetic energy of the heavy metals to be absorbed [6].

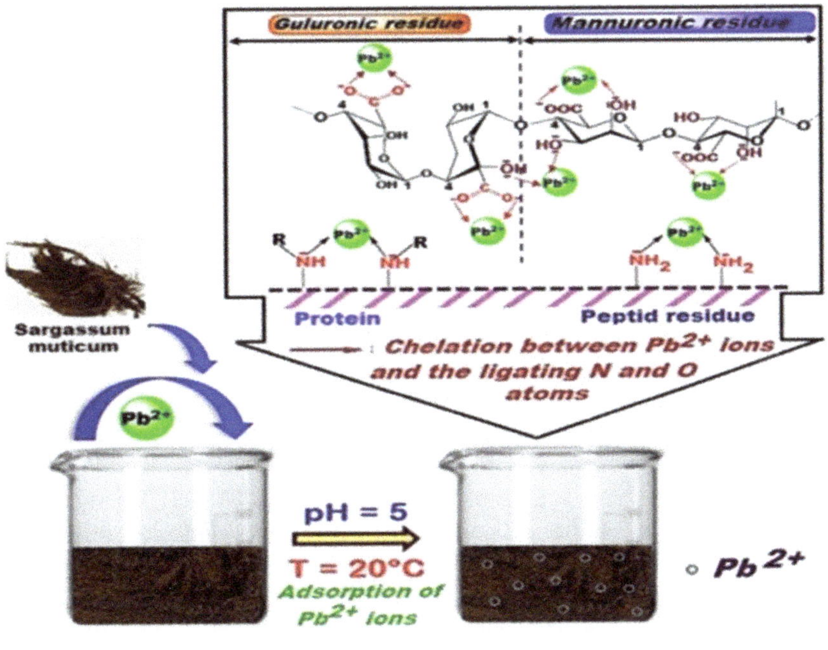

Figure 9.3. The adsorption mechanism of heavy metals by *Sargassum muticum*;lead was used in a case study. Adapted by permission from Springer Nature [3], Copyright (2021).

9.7 Conclusions

Invasive plants are generally known to be harmful to the environment. Nevertheless, these natural resources are versatile and can be used for various purposes such as food, medication, ornamental displays, and environmental remediation. In particular, they can address ecological concerns associated with the management of invasive species by serving as biosorbents. The chapter offered an overview of the different types of biosorbents classified according to their sources. Biosorbents can be extracted from invasive plants and subsequently chemically treated to introduce functional groups into them that can create biochar when modified. Such biosorbents have become useful materials for environmental healing purposes. Scientists remain keen to explore innovative applications of environmentally friendly techniques in order to use bioresources productively while also conserving nature. The use of invasive species for environmental reasons presents a promising solution with which to mitigate the detrimental effects of their proliferation on natural habitats. In addition, this approach addresses potential health hazards associated with ecosystem contamination. Various studies have documented cases in which non-native species efficiently acted as biosorbents in purifying contaminated areas. Notably, water hyacinths demonstrated exceptional capacity in absorbing heavy metals and storing energy from water sources, among numerous other examples described in the research literature.

Moreover, by utilizing biosorbents obtained from invasive plant species, researchers aim to improve their effectiveness in meeting environmental protection standards. A sustainable technique for a greener future with diverse combination options and the potential for application and research commercialization can be achieved through the ongoing development of biosorbents derived from such sources.

References

[1] Islam A, Teo S H, Taufq-Yap Y H, Ng C H, Vo D N, Ibrahim M L, Hasan M M, Khan A R, Alam S M and Awual M 2021 Step towards the sustainable toxic dyes and heavy metals removal and recycling from aqueous solution-a comprehensive review *Resour. Conserv. Recycl.* **175** 105849

[2] Zhang F, Wang X, Xionghui J and Ma L 2016 Efficient arsenate removal by magnetite-modified water hyacinth biochar *Environ. Pollut.* **216** 575–83

[3] Nguyen D T, Tran T V, Kumar P S, Din A T M, Jalil A A and Dai-viet N 2021 Invasive plants as biosorbents for environmental remediation: a review *Environ. Chem. Lett.* **20** 1421–51

[4] Tauqeer M, Khan M E, Ji R S, Bansal P and Mohammad A 2021 Metal and metal oxide nanomaterials for wastewater decontamination *Metal, Metal-Oxides and Metal-Organic Frameworks for Environmental Remediation* (New York: Springer)

[5] Sodhi K K, Kumar M, Agrawal P K and Singh D K 2019 Perspectives on arsenic toxicity, carcinogenicity and its systemic remediation strategies *Environ. Technol. Innov.* **16** 100462

[6] Singh P and Srivastava S 2022 A review on biosorbents *Int. J. Res. Anal. Rev.* **9** 370–82 http://ijrar.org/viewfull.php?&p_id=IJRAR22A2002

[7] Simberloff D *et al* 2013 Impacts of biological invasions: what's what and the way forward *Trends Ecol. Evol.* **28** 58–66
[8] Terry T J, Madsen M D, Gill R A, Anderson V J and St.-Clair S B 2021 Selective herbicide control: using furrows and carbon seed coatings to establish a native bunchgrass while reducing cheatgrass cover *Restor. Ecol.* **29** 13351
[9] Beesley L, Moreno-Jiménez E, Gomez-Eyles J L, Harris E, Robinson B and Sizmur T 2011 A review of biochars' potential role in the remediation, revegetation and restoration of contaminated soils *Environ. Pollut.* **159** 3269–82
[10] Feng Q, Wang B, Chen M, Wu P, Lee X and Xing Y 2021 Invasive plants as potential sustainable feedstocks for Biochar production and multiple applications: a review *Resour. Conserv. Recycl.* **164** 105204
[11] Yang L, Deng Y, Shu Z, Chen Q, Yang H and Tan X 2022 Application of invasive plants as biochar precursors in the field of environment and energy storage *Front. Environ. Sci.* **10** 902915
[12] Cui X *et al* 2022 Biochar from constructed wetland biomass waste: a review of its potential and challenges *Chemosphere* **287** 132259
[13] Buranov A U and Mazza G 2008 Lignin in straw of herbaceous crops *Ind. Crops Prod.* **28** 237–59
[14] Jahirul M, Rasul M, Chowdhury A and Ashwath N 2012 Biofuels production through biomass pyrolysis: a technological review *Energies* **5** 4952–5001
[15] Bridgwater A V, Czernik S and Piskorz J 2001 An overview of fast pyrolysis *Progress in Thermochemical Biomass Conversion* (New York, NY: Wiley)vol 2 977–97 pp
[16] Demirbas A H 2005 Yields and heating values of liquids and chars from spruce trunkbark pyrolysis *Energy Sources* **27** 1367–73
[17] Tippayawong N, Kinorn J and Thavornun S 2008 Yields and gaseous composition from slow pyrolysis of refuse-derived fuels *Energy Sources* **30** 1572–8
[18] Demirbas A and Arin G 2002 An overview of biomass pyrolysis *Energy Sources* **24** 471–82
[19] Garcia B, Alves O, Rijo B, Lourinho G and Nobre C 2022 Biochar: production, applications, and market prospects in Portugal *Environments* **9** 95–114
[20] Lian W, Yang L, Joseph S, Shi W, Bian R, Zheng J, Li L, Shan S and Pan G 2020 Utilization of biochar produced from invasive plant species to efficiently adsorb Cd(II) and Pb(II) *Bioresour. Technol.* **2020** 317–124011
[21] Dong L, Li S, Jin Y, Hu B and Sheng G 2021 Enhanced adsorption of Eu(III) from wastewater using *Solidago canadensis*-derived biochar Functionalized by Ca/Al-LDH and Hydroxyapatite *Appl. Surf. Sci.* **567** 150794
[22] Das D, Bordoloi U, Muigai H H and Kalita P 2020 A novel form stable PCM based bio composite material for solar thermal energy storage applications *J. Energy Storage* **30** 101403
[23] Zhuang H, Zhu H, Zhang J, Shan S, Fang C, Tang H and Xie Q 2020 Enhanced 2,4,6-trichlorophenol anaerobic degradation by Fe_3O_4 supported on water hyacinth biochar for triggering direct interspecies electron transfer and its use in coal gasification wastewater treatment *Bioresour. Technol.* **296** 122306
[24] Mo C, Zhang J and Zhang G 2020 Hierarchical porous carbon with three dimensional nanonetwork from water hyacinth leaves for energy storage *J. Energy Storage* **32** 101848

[25] Zhang L, Cheng H, Pan D, Wu Y, Ji R, Li W, Jiang X and Han J 2021 One-pot pyrolysis of a typical invasive plant into nitrogen-doped biochars for efficient sorption of phthalate esters from aqueous solution *Chemosphere* **280** 130712

[26] Zhao T, Chen R and Wang J 2020 A mild method for preparation of highly selective magnetic biochar microspheres. *Int. J. Mol. Sci.* **21** 3752–32466534

[27] Raj F R, Boopathi G, Kalpana D, Jaya N V and Pandurangan A 2022 Sustainable development through restoration of Prosopis juliflora species into activated carbon as electrode material for supercapacitors *Diam. Relat. Mater.* **121** 108767

[28] Liu D, Tang Y, Li J, Hao Z, Zhu J, Wei J, Liu C, Dong L, Jia B and Chen G 2021 Eupatorium adenophorum derived adsorbent by hydrothermal-assisted HNO_3 modification and application to Pb^{2+} adsorption. *J. Environ. Chem. Eng.* **9** 105972

[29] Rees F, Sterckeman T and Morel J L 2016 Root development of non-accumulating and hyperaccumulating plants in metal-contaminated soils amended with biochar *Chemosphere* **142** 48–55

[30] Cai J F, Zhang L, Zhang Y, Zhang M X, Li H L, Xia H J, Kong W and Yu F 2020 Remediation of cadmium-contaminated coastal saline-alkaline soil by *Spartina alterniflora* derived biochar *Ecotoxicol. Environ. Saf.* **205** 111172

[31] Huang X, Liu Y, Liu S, Li Z, Tan X, Ding Y, Zeng G, Xu Y, Zeng W and Zheng B 2016 Removal of metformin hydrochloride by *Alternanthera philoxeroides* biomass derived porous carbon materials treated with hydrogen peroxide *RSC Adv.* **6** 79275–84

[32] Du Y, Feng Y, Shu L, Ren Z J, Kong Q, Xu F and Wang Q 2018 A mesoporous biochar from bio-invasion alligator weed for adsorption of Rhodamine B from aqueous solution *Desalin. Water Treat.* **135** 341–50

[33] Ma Y, Oliveira R S, Freitas H and Zhang C 2016 Biochemical and molecular mechanisms of plant-microbe-metal interactions: relevance for phytoremediation *Front. Plant Sci.* **7**

[34] Anandrao A K and Reshma S P 2021 Adsorption of methylene blue by sulphonated biomass of *Ailanthus excelsa J. Emerg. Technol. Innov. Res.* **8** 446–58

[35] Khamparia S and Jaspal D 2016 Investigation of adsorption of Rhodamine B onto a natural adsorbent Argemone mexicana *J. Environ. Manage.* **183** 786–93

[36] Shooto N D 2020 Removal of toxic hexavalent chromium (Cr (VI)) and divalent lead (Pb (II)) ions from aqueous solution by modified rhizomes of *Acorus calamus*. *Surf. Interfaces* **20** 100624

[37] Chen X L, Li F, Xie X J, Li Z and Chen L 2019 Nanoscale zero-valent iron and chitosan functionalized *Eichhornia crassipes* biochar for efficient hexavalent chromium removal. *Int. J. Environ. Res. Public Health* **16** 3046

[38] Dong L, Li S, Jin Y, Hu B and Sheng G 2021 Enhanced adsorption of Eu(III) from wastewater using *Solidago canadensis*-derived biocharfFunctionalized by Ca/Al-LDH and Hydroxyapatite *Appl. Surf. Sci.* **567** 150794

[39] Zhang H, Xu F, Xue J, Chen S, Wang J and Yang Y 2020 Enhanced removal of heavy metal ions from aqueous solution using manganese dioxide-loaded biochar: behavior and mechanism *Sci. Rep.* **10** 6067

IOP Publishing

Adsorption Applications for Environmental Sustainability

Kingsley Eghonghon Ukhurebor, Uyiosa Osagie Aigbe and Robert Birundu Onyancha

Chapter 10

Microorganism-derived biosorbents in the sequestration of contaminants from the soil

Timothy Imanobe Oliomogbe, Joseph Onyeka Emegha and Kingsley Eghonghon Ukhurebor

The soil has the greatest impact on the critical zone of the Earth's surface. Ecosystems and people alike gain from the many benefits the soil offers. Both people and the entire environment are at risk from soil contamination. Biosorption is gaining popularity as a way of lessening this contamination. This approach is considered cost-effective and eco-friendly for eliminating impurities from water-based solutions. The goal of this chapter is to provide an overview of biosorption and its processes as well as the advantages and drawbacks of green biosorbents in addressing soil contamination.

10.1 Introduction

Industrialization and the mining of natural resources have caused extensive environmental degradation and contamination [1, 2]. While this process has taken place to advance human activities, it has also led to the release of heavy metals (HMs) and other substances into the entire ecosystem [1, 3]. In recent years, there has been a growing concern regarding the potential threat of pollutants infiltrating the food chain. As such, there is an increasing necessity to implement effective measures to clean up natural resources such as soil, water, and air [1–3]. This urgency has resulted in the development of innovative technologies that prioritize pollutant elimination rather than relying on traditional disposal methods. Examples of these innovative technologies include green or clean technologies that focus on reducing or eliminating the toxic solvents used in conventional manufacturing processes [1–4].

The critical zone of the Earth's surface is most significantly influenced by the soil [2]. Ecosystems and people alike gain from the many benefits the soil offers. Soil plays a

doi:10.1088/978-0-7503-5598-8ch10

crucial role in upholding the planet's biodiversity, enabling agricultural production and promoting healthy plant growth [2, 3]. It also provides suitable habitats for animals and humans alike while serving as an effective means of mitigating climate change by absorbing carbon and nitrogen compounds [2–4]. The proper management of toxic substances is crucial for preserving environmental quality [4]. This involves regulating their disposal, fate, and decontamination processes. Additionally, it serves a vital function in the academic realm by contributing to ongoing research efforts aimed at understanding the impact of these toxins on our environment [2, 5]. In addition to serving as a source of vital nutrients and carbon for the growth of microorganisms and plants, soil also functions as a reservoir for pollutants resulting from human activities and industrial processes. Agricultural operations themselves also introduce a multitude of toxins into the soil, largely through the use of pesticides, fertilizers, and sludge [5, 6]. There is also a chance that HMs can leak into subterranean water, which could contaminate it and harm people's health, particularly those who drink borehole water or groundwater [7].

As a result of these ongoing environmental concerns, scientists have created and researched a wide range of solutions. Microorganisms have developed strategies for surviving in environments contaminated with HMs, including bioaccumulation, biomineralization, biosorption, and biotransformation [8]. Among these, biosorption has continued to be cited as an exciting study subject that has expanded recently. In order to prevent or limit environmental contamination and the potential for its uptake in the food chain, it is crucial to eliminate or reduce HM pollution. The technique of bioremediation is employed to enhance the speciation or stability of metals, leading to a subsequent reduction in their bioavailability [9–11]. Due to its accessibility and effectiveness, biosorption has a number of advantages over other traditional approaches toward HM treatment [8].

Microorganism-derived biosorbents, which are both abundant and affordable, are commonly employed to remove metal contaminants [5–8]. An essential aspect of biosorption technology, which contributes to its profitability, is its superior performance compared to alternative methods of decontamination [12, 13]. To mitigate the detrimental impact of HMs on the environment, it is imperative to utilize economical methods for metal remediation. One such method is the use of biological techniques. According to empirical findings, the majority of living entities possess the ability to function as biosorbents and thereby exhibit an important passive mechanism for attracting HMs. However, this capacity does not extend to portable alkali metallic ions such as sodium and potassium ions [14].

Biosorption refers to a remediation approach with biological origins which entails the use of low-cost biomaterials for extracting metal species from solution [14]. Microorganisms have been demonstrated to show excellent sequestration properties and are being widely investigated and adopted. The microorganisms that fall under this category include yeast, algae, and plants as well as certain polysaccharide materials [13, 15].

This chapter presents a comprehensive survey of biosorption, encompassing topics such as the underlying mechanisms and key determinants that impact this process. In addition, we delve into various applications for biosorbents in environmental settings while exploring potential avenues for green innovations within this field.

10.2 Heavy metals and the environment

In recent times, an unrestricted flow of HMs into ecosystems has been caused by the growth of civilization [2, 3]. In aqueous and terrestrial environments, HMs can form ambiguous mineral and organometallic linkages [16]. HMs may flow through the environment with remarkable freedom. The ingestion of HMs by humans and animals, along with the absorption of these elements by plants, presents a considerable risk to all living organisms [16, 17]. Their harmful effects on various environmental components and tendency to accumulate in the food chain further exacerbate this perilous situation [17].

Soil pollution due to HMs is predominantly attributed to industrial and power engineering activities, air emissions, and disposal in landfills [3]. The use of pesticides as well as sewage sludge-based fertilizers has the potential to transfer metals into the soil [6]. Road networks contribute towards the contamination of soils through transportation activities. Soil features such as the pH level determine the ability of metals within the soil to dissolve, migrate, or be assimilated by different species [18]. In acidic soils, metals exist in a soluble state at lower pH values, which makes them more accessible to plants. The function of certain ecosystems is hampered by the constant disruption of chemical equilibrium caused by high metal concentrations in soils [18]. The growth of soil microorganisms is impeded by the presence of metals that have accumulated within the soil. The fundamental physiological processes of bacteria involved in the breakdown and transformation of organic materials are compromised [19, 20].

10.3 An overview of biosorption

The various fundamental mechanisms involved in biosorption are largely contingent on certain parameters that include the nature of the substance being absorbed, the chosen biosorbent, the prevailing environmental temperature, and whether or not there are live organisms that can engage in comprehensive metabolic processes [20, 21]. Quantifying biosorption is complex. The use of the prefix 'bio' in academic fields such as 'biotechnology,' 'bioengineering,' and 'bioprocessing' denotes the fundamental involvement of biological entities, including living organisms, components, or products that are derived from them [21]. The terms adsorption and absorption are commonly used interchangeably to refer to the process of sorption [20, 21]. When a substance transitions from one state to another, such as when a solid consumes liquids or water absorbs gases, it is considered to be absorbed. On the other hand, adsorption involves ions and molecules that adhere to another molecule's surface or a two-dimensional surface [21]. Biosorption is the process by which organic or inorganic substances are absorbed and retained by a living organism. These materials can be organic or inorganic, soluble or insoluble, and liquid or solid [19–21]. Electron transfer, ionic bonding, adsorption, and precipitation are a few of the physicochemical processes that take place in biosorption

[4–7]. The term 'biosorption' is commonly utilized by researchers to refer to all the biotic and abiotic methods employed for eliminating pollutants from solutions [20]. However, it should be noted that this generalization is not entirely accurate, particularly in cases that involve living cell systems [9, 12, 21]. Classic adsorption and absorption techniques are designed to remove organic and certain inorganic pollutants, including hazardous metals, from liquid and gas streams [4, 12, 17].

Over the years, several scholars have highlighted that a significant proportion of biosorption research has primarily concentrated on metals and their associated elements. They have established that biosorption refers to the process whereby biological material removes metal or metalloid species, compounds, or particles from an environment [22]. Considering the prevalence of research into biosorption, it is noteworthy that a significant proportion of this investigation has centered on microbial systems. Furthermore, certain alternative classifications also restrict their scope to microbial matter [23]. Studies have shown that biosorption is an effective, economical, and eco-friendly approach to removing HMs from polluted environments [4, 12]. Biosorption is a highly advantageous treatment technique when compared to conventional methods. It provides numerous benefits including cost-effectiveness, high efficacy in removing metal from dilute solutions, minimal generation of chemical and biological sludge, the absence of a need for additional fertilizer, regeneration potential for the biosorbent material, and even the possibility of recovering the metal [5, 20, 21]. These advantages make biosorption an attractive option with great academic potential [24]. Certain factors inhibit the performance and efficiency of the selected microorganism used as a biosorbent [24]. The effectiveness of metal transfer to the biosorbent is influenced by several chemical properties associated with the metal solution. In-depth analysis and understanding of these chemical properties are crucial for comprehending the mechanisms that underlie biosorption efficiency [25]. Biosorption, a physicochemical process, binds molecules with a negative charge (found in cell membranes) to metal ions (mostly present as cations) [24, 25]. To optimize this process, it is important to understand the methods of biosorption used to remove HMs. During sorption, numerous distinct mechanisms take place. Due to the intricate nature of biological materials, various mechanisms can occur simultaneously and operate at varying rates, leading to a multifaceted system [26]. The utilization of biosorbents, mainly microorganisms of biological origin, has gained significant popularity due to their effective elimination and retrieval of HMs [27, 28]. The process can be influenced by numerous factors such as the pH level, temperature variations, and the duration of contact during biosorption [29, 30]. Microorganisms have been widely explored as biosorbent materials due to their ability to selectively remove HMs across a diverse range of physicochemical conditions through the processes of adsorption and desorption. This approach presents an attractive alternative for addressing environmental contamination concerns, offering promising results for sustainable metal removal strategies [5, 12]. They also offer the advantage of having a large surface-to-volume ratio [12, 27].

Microorganisms such as *Bacillus cereus* have been employed in the biosorption process of metals such as cadmium [31] and copper [29]. Different forms of microbial biosorbents have been studied for their capacity to adsorb, such as bacteria [32], fungi [33], and marine and freshwater algae [25]. Studies have found that mixed microbial biosorbents in activated sludge can effectively store HMs [34]. Different types of biomasses are especially adept at this task.

When considering biosorbents for clean-up operations, mobility is a crucial consideration. The biosorptive characteristics of various microbial biomasses vary, and these discrepancies are also substantial within each group. Each biosorbent's capacity for biosorption, however, is impacted by its prehistory, pretreatment, and experimental settings [21–23]. To optimize biosorption, the organism must possess adaptability toward alterations in both the bioreactor structure and the physico-chemical environment [35]. Bacteria have emerged as a popular choice for biosorption owing to their abundant availability, minute size, ability to thrive in regulated environments, and remarkable resilience toward an extensive array of environmental factors [36, 37]. Their exceptional ability to absorb these metals is attributed to their large surface-to-volume ratios and numerous sites that can potentially undergo chemisorption, such as teichoic acid found on their cell walls [38].

10.4 Modification of biosorbents

In the context of biosorption, the cell wall plays a crucial role, as it serves as an indispensable component that governs the binding of metal ions with molecules [9, 13]. Any changes made to this vital element can have a profound impact on its ability to perform this function effectively. Therefore, understanding the cellular mechanisms and physiological properties that influence these alterations is essential in enhancing our knowledge of biosorption processes [13, 25]. Consequently, novel techniques have been introduced into existing biosorption-based technologies to improve their effectiveness by modifying cells through physical or chemical means [5, 39]. Physical treatments such as drying, freezing, defrosting, and lyophilization (freeze-drying) are examples of methods utilized to this end [39].

The capacity of *Chlorella* sp. cells to extract manganese was observed to be significantly enhanced when subjected to heating, as compared to the capacity of live cells [39]. Similarly, the ability of *Pseudomonas aeruginosa* biomass to bind zinc was found to increase upon exposure to high temperatures. This effect is presumably attributed to the disruption in cellular membrane integrity that generates additional binding sites for zinc within the cell structure [40]. Other chemical agents, such as acids or alkalis, can also be used to treat biomass. *Penicillium spinulosum* biomass that had undergone a detergent treatment exhibited more effective copper sorption, according to Ross and Townsely [41]. *Saccharomyces cerevisiae* and *Rhizopus arrhizus* showed comparable results in a study by Gadd *et al* [42]. Alkaline compounds increased metal removal by 5%–34% when added to cyanobacteria (*Phormidium laminosum*).

The adsorption of metals was, however, reduced following treatment with acidic chemicals. This was most likely brought on by increased proton–dissolved-metal competition [40]. All of these modifications were intended to increase biosorption effectiveness. While there are cases for the adoption and deployment of biosorption, there are two main concerns stalling its large-scale adoption. These include the complexity of collecting biomass from the water used for processing as well as the biomass's poor mechanical durability [40]. It is possible to comprehend biosorption better in a laboratory setting in which cells are floating freely. However, the cells should undergo immobilization before biosorption is applied on a larger scale.

The implementation of the biosorption process in reactors requires suitable flow rates, which result in an increase in pressure that may lead to the disintegration of suspended cells. Consequently, the use of microorganisms for sorption and desorption cycles is infeasible due to these complications. To overcome such difficulties, it is advisable to immobilize the cells using a technique that does not impede their ability to adsorb metal ions [40]. Using a wide range of methods known as sequestration or fixing, specific biological molecules and substances can have their movements entirely or partly restricted on a solid substrate or inside predefined structures. Biomasses can function very effectively in continuous-flow reactors because sequestration increases their elastic strength and resistance. It is possible to immobilize both living and dead bacterial cells [39]. Kuhn and Pfister [43] used *Zoogloea ramigera* cells immobilized by alginate to successfully remove cadmium (Cd), zinc (Zn), manganese (Mn), lead (Pb), copper (Cu), and strontium (Sr). Alginate beads exhibited sorptive characteristics; however, they were insufficient to bind metals to immobilized cells. Further research has shown that *Chlorella salina* cells immobilized by calcium alginate can remove higher quantities of cobalt, zinc, and manganese than non-immobilized cells [40].

10.5 The biosorption process

The ambiguous nature of the biosorption definition dictates that determining the mechanism(s) responsible for biosorption is typically difficult, except in basic laboratory systems [40]. Various processes may be at play within complex biological materials under favorable circumstances. Because biomass possesses diverse structural characteristics, multiple groups such as hydroxyls, carbonyls, phosphorous compounds, thiols and aminos can interact with metal species in several ways, depending on the effects of the physiochemical variables involved [21]. While the terms bioaccumulation and biosorption are often utilized interchangeably, their respective pollutant sequestration mechanisms exhibit notable distinctions. Biosorption refers to the utilization of physicochemical absorption pathways, such as electrostatic forces and ion/proton displacement; it is the process by which biological materials adsorb substances from solutions [44]. The efficacy of biosorption in removing a diverse array of HMs from water has been established through scientific demonstrations. This method is particularly effective against extremely harmful ions such as those of cadmium, chromium, lead, mercury, and arsenic [45, 46]. The cell walls of bacteria possess functional groups such as carboxyl, phosphonate, amine and hydroxyl groups which play a role in performing binding functions [47]. Hence, the heterogeneity of cell wall architectures is essential for biosorption to be productive. Gram-positive bacteria exhibit higher sorption ability owing to the presence of their thicker peptidoglycan layer, which enables them to bind more effectively to different compounds and materials [48, 49]. In recent studies, the utilization of engineered microorganisms that possess metal-binding proteins and peptides on their external surface has been explored as a means to enhance the efficacy and specificity of microbial sorbents [50, 51]. Biosorption is a sustainable technology for removing and recovering metals; it uses microorganisms (living or dead), agricultural products, and other industrial by-products as biosorbents [52, 53]. To maximize the effectiveness of biosorption, several variables have to be considered and studied. The binding of pollutants is determined by their chemical composition, the size of the biomass, the metal ions' interactions, and the ionic strength [15]. Biosorption is beneficial due to its simplicity, lack of a need for auxiliary nutrients, cost-effectiveness (due to reversibility), ease of desorption, and high adsorption rate. However, other factors must be taken into account, such as the potential toxicity of pollutants to bacterial cells when living cells are used in the process [54].

10.6 Biosorption mechanisms

Biosorption is a multifaceted procedure that involves the adhesion of a sorbate to a biosorbent. Numerous organic materials possess the ability to function as biosorbents through physical interactions such as electrostatic attraction or chemical processes such as ion exchange and chelation. These methods may involve the displacement of bound metal cations, reduction, proton displacement, complexation, and precipitation (see figure 10.1 as adapted from [55]). Biosorbents contain different chemical (functional) groups, including amine, phosphate, sulfonate, carbonyl, thioether, sulfhydryl, carboxyl, phosphodiester, phenolic, imine, imidazole, and amide groups. Biosorbents are effective in attracting and trapping metal ions due to their diverse properties. Extensive research has been conducted with the aim of examining the underlying processes of these mechanisms [55, 56]. The

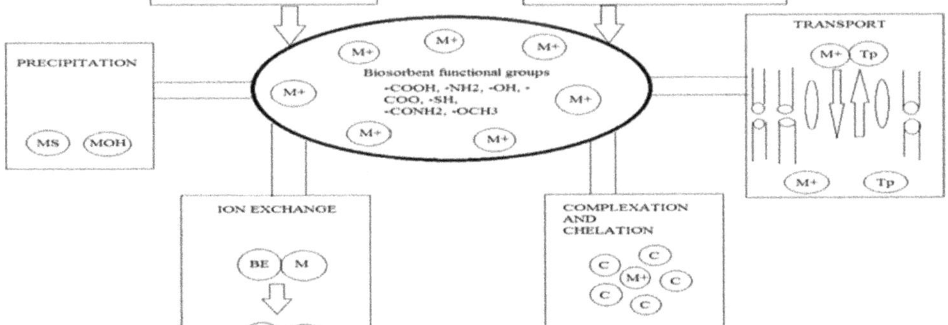

Figure 10.1. Different hypothesized mechanisms of biosorption. Reproduced from [55]. CC BY 3.0.

associated effects of the various factors influencing the formation of new metals due to biosorption are as follows [55]:

- The properties of metal ions, including their molecular weight, ionic radius, and oxidation states are factors that affect their chemical makeup, stereochemistry, and coordination.
- The characteristics of biosorbents include their composition and whether they are made from living or nonliving microorganisms.
- The category of the biological ligand binding location.
- Various elements are implicated in the course of action, including but not limited to the pH level, temperature, quantities of sorbate and sorbent employed, as well as the presence of other metal ions that may hinder progress.
- The presence of binding sites.

10.6.1 Complexation

Complexation refers to the process of creating a complex through the combination of multiple species. Mononuclear (monodentate) complexes occur when ligands bind with metal ions, positioning metal atoms at their centers. A complex with multiple metal ions at its core is called a polynuclear (multidentate) complex [55]. The charge on the metal atom can vary based on the number of binding ligands present, and it may be positive, negative, or neutral. It is more favorable to form a complex with a monodentate ligand than a multidentate because the latter has multiple ligands, which can result in binding with a number of species [56]. Covalent bonds are used by the metal ion to interact with the ligands.

According to a study [57], it was determined that carboxylic, hydroxyl, and amine groups played a role in enabling surface complexation during the process of antimony (III) biosorption. Other studies have reported a comparable biosorption mechanism for the elimination of Cd (II), Cr (VI), and Cd (II) by *Acidiphilium*, *Termitomyces clypeatus*, and alkali-modified sewage sludge, respectively [55].

10.6.2 Chelation

This term describes a procedure in which a chelating substance attaches to a metal ion at multiple points, forming a circular configuration called a chelate [55]. The majority of stable structures are created through the involvement of polydentate ligands and their ability to bond repeatedly. The stability of the structure is enhanced by an increase in ligand binding sites. Chelates exhibit greater stability than that of complexes due to their ability to bind with metal ions at more than one location. In one study, effluent was found to be a promising biosorbent for the elimination of Cd (II) that was biosorbed in addition to carboxylic acids found on the surface of the biosorbent [58]. In another study, soybean meal waste was found to contain carboxyl and hydroxyl groups on its surface helped to remove Cr (III) and Cu (II) through biosorption. The process involved was similar to that observed in other studies [59].

10.6.3 Coordination

A complex's metal atom undergoes a process of coordinate covalent bonding with adjacent nonmetallic atoms through the acceptance of a lone pair of electrons. This chemical interaction is crucial in establishing strong intermolecular connections and facilitating various chemical reactions within the compound [55]. The coordinating atom, also referred to as the donor, is a nonmetallic element. On the other hand, the metallic atom that receives an electron pair is recognized as the acceptor. Coordinate compounds are characterized by the presence of bonds with coordinating groups, which may include =NOH, –O–R, –OH, –NH, –S–, –N=, $-NH_2$, and =O [55].

10.6.4 Ion exchange

The process of ion exchange is a fundamental mechanism that plays an essential role in biosorption [23]. In this phenomenon, binary metal ions are exchanged with counterions present on the surface of the biosorbent. This complex process has significant implications and serves as a crucial area for research in various fields related to chemistry and biology. The most common application of ion exchange in biosorption is the removal of HM ions from wastewater [55]. The known purification methods operate using ion exchange as their mechanism. Ion exchange is a fundamental chemical process that can occur through two primary mechanisms—cation exchange and anion exchange [55]. Cation exchangers can be exemplified by carboxyl groups, whereas amino/imidazole groups are considered to be anion exchangers [60]. Previous research studies have reported an ion exchange mechanism for biosorption that involves the application of rice straw to eliminate cadmium through metal ion exchange as well as the removal of Cu (II), Zn (II), and Pb (II) by watermelon rind [58].

10.6.5 Precipitation

The functional groups on the microbial cell surface react with metal ions, leading to precipitation. In some instances, the metal precipitates of insoluble organic materials are formed. These precipitates can either remain intact or penetrate the cell itself. The majority of extracellular polymeric substances secreted by microbes contribute to the creation of organic deposits. Research has revealed that when *Mesorhizobium amorphae* is exposed to Cu (II), it leads to distortions, clumping, and harm on the cell surface, as seen in an SEM/EDX examination [61]. This same biosorption mechanism for metal ion removal has also been identified in similar studies that used green tomato husk, watermelon rind, and soybean meal for the elimination of metal ions [59, 62].

10.6.6 Reduction

This procedure involves the reactions between a metal and functional groups such as carboxyls, which result in reduction and ultimately crystal growth [55]. Gold and palladium are examples of metals that can be acquired through this method. Reduction occurs when metals bind to specific sites on biosorbents. The elimination of harmful hexavalent chromium may also be accomplished using this technique, since numerous organisms reduce Cr (VI) to Cr (III) via biosorption in an aqueous solution [63–65]. Different methods can be used to examine the biosorption mechanism. The Boehm method or potentiometric titration can determine the ion exchange properties and acidic/basic characteristics of functional groups on materials' surfaces [66]. Fourier transform infrared spectrometry (FTIR) is useful in identifying various types of functional groups as well as the bonds involved in biosorption [67]. The utilization of a scanning electron microscope (SEM) is an effective method for quantitatively assessing changes in both the structure and the morphology of a biosorbent before and after metal biosorption. In addition, information about an element's availability on the surface can be obtained through energy-dispersive x-ray (EDX) analysis. A quantitative spectroscopic technique called x-ray photoelectron spectroscopy (XPS) is also employed to analyze the chemical and physical phenomena occurring at the interface between a biosorbent and the metal ions it absorbs [68].

10.7 The pros and cons of biosorption

To comprehend biosorption effectively, it is necessary to evaluate its advantages and disadvantages.

10.7.1 Pros

- The advantages of biosorption are its low energy needs, its capacity to remove many pollutants at once, and its cost-effectiveness. Because it makes use of organic substances such as biological and agricultural wastes, biosorption is also regarded as a green technology. The fact that biosorption is simple to incorporate into existing wastewater treatment systems makes it a viable substitute for conventional techniques of pollutant removal.
- As nonliving beings, cells can indeed be managed under any conditions, in addition to those that stimulate cell division. In short, a variety of working conditions, including the pH level, concentration, and temperature of the metal, can be used. The process may not be executed in aseptic conditions. If the value and amount of metal recovered are substantial, it can be effectively desorbed and retrieved. If the biomass is substantial, metal-loaded biomass can be burned, avoiding the need for continued treatment [69].

10.7.2 Cons

- Little or no adoption of biosorption technology has been deployed at a large scale [70].
- Biosorption has been demonstrated to have high operating costs and to suffer from incomplete removal of contaminants [70].

10.8 Various factors that impact the biosorption processes

Various physical and chemical treatments can affect biosorption processes. Boiling, drying, autoclaving, and mechanical disruption are examples of physical treatments, while alkali treatment is a common chemical process [12]. In particular, this chemical treatment has been noted for its ability to enhance the biosorptive capacity of certain fungal systems. This improvement is attributed to the deacetylation process that chitin undergoes, which results in higher metal affinities through the formation of chitosan–glucan complexes [36]. For bioremediation to achieve success, a thorough understanding of the factors that affect biosorption is crucial. Various characteristics and parameters have been recognized as having an impact on biodegradation, which have been extensively discussed [9, 12–15, 27, 39, 55].

10.8.1 The effect of pH

The pH level is a significant biosorption variable. It significantly affects the solubility of metal ions and the capacity of biomass to bind sites. Metal biosorption is hindered by lower pH values [15, 55]. The quintessential pH range for metal uptake is 2.5–6. Above the latter value, the ability of biosorbents to extract metals is hindered [9]. When the pH is low (acidic), hydronium ions interact strongly with the active ligands of the biosorbent. The presence of binding sites creates a competitive environment in which protons and metal ions contend for occupancy [55]. At elevated pH values, the number of hydrogen ions is lower, which exposes the negative charge of functional groups' active sites and increases biosorption by attracting positively charged metal ions. When the pH is increased, there is a possibility for metals to precipitate and create hydroxides. This may hinder the process of biosorption [59]. Based on an investigation, *S. cerevisiae* biomass exhibited an enhanced uptake of Cr (VI) from wastewater as the pH level was raised from 1 to 4 [71]. In the case of pretreated *Aspergillus niger*'s ability to absorb Cr, its optimal pH level was determined to be 3 [72]. In addition, *Rhizopus cohnii* biomass had an elevated biosorption of cadmium as the pH rose from 2.0 until it reached a plateau between pH levels of 4.5 and 6.5 [55].

10.8.2 The effect of temperature

It is important to consider temperature as a factor in the biosorptive removal of HMs. The concept of temperature in relation to biosorption refers to the thermodynamics and kinetic energy associated with metal ions [73]. Depending on the circumstances, the temperature can impact biosorption positively or negatively. Variations in temperature affect the capacity of a biosorbent for adsorption. Elevated temperatures typically enhance the efficiency of pollutant elimination by biosorption, yet it should be acknowledged that such conditions may also compromise the integrity of the biosorbent material [38]. Hence, selecting an ideal temperature is imperative to enhance biosorption efficiency and augment the adherence of metal ions. To illustrate, a notable biosorption of up to 86% for cadmium ions was attained by utilizing *S. cerevisiae* at a temperature of 40°C [71].

There was also a sharp rise in the biosorption rates of Cr (VI) by *Streptococcus equisimilis* when the incubation temperature increased from 25°C to 40°C [55].

10.8.3 The effect of time

Bacterial biomass is capable of rapidly biosorbing HMs such as uranium, nickel, cobalt, copper, cadmium, chromium, lead, and zinc. Nonliving cells from this biomass are highly efficient at this process and it usually reaches equilibrium within minutes [74, 75]. To maximize the effectiveness of desorption and biomass regeneration, the contact time must be carefully considered.

The duration needed to attain the highest degree of biosorption is subject to variation and depends on a number of factors, such as the specific type of biosorbent utilized, the particular metal ion being targeted, and their overall amalgamation [76]. Initially, biosorption is expeditious (typically occurring within an hour), and almost 90% of the metal binding can be attributed to the unoccupied and accessible nature of all sites for the absorption of metal ions. However, over time, the rate of absorption lessens as the percentage saturation of metals in solution increases [77]. In the case of *A. niger* biomass, there is usually no further biosorption after 24 h [55].

10.8.4 The effect of the initial metal concentration

The potential to overcome the mass transfer resistance between the solid and liquid phases is influenced by the initial concentration of metal ions [55]. The biosorption capacity increases as the metal ion concentration increases up to a certain point where it reaches a saturation value. However, this also means that the biosorption efficiency decreases with increasing concentrations of the metal ion. At reduced ion concentrations, greater efficiency is achieved as a result of the comprehensive interplay between ions and their accessible binding sites, which serves to enhance efficacy rates. This phenomenon can be attributed to the complete interaction between these factors. As concentrations rise, an increase in the quantity of metal ions that remain unattached to binding sites occurs as a result of reaching maximum saturation [78]. The efficacy of *Hypnea valentiae*'s biosorption capacity was evaluated by subjecting it to varying initial Cd ion concentrations (25–500 mg L^{-1}) in a study. The most effective result obtained using simulated wastewater samples was an efficiency rate of 86.8%, which was obtained at the lowest concentration tested, namely 25 mg L^{-1} [79]. According to another study, an inverse relationship was found between copper ion concentration and cashew nut shell biosorption efficiency. Specifically, when the concentration of copper ions was raised from 10 to 50 mg L^{-1}, a decline in effectiveness was observed, as biosorption dropped from 86.03% to 76.17% [79–82].

10.8.5 The effect of agitation speed

Increasing the level of agitation results in a boost in biosorption capability for the biosorbent through a reduction in resistance to mass transfer. The increased agitation enhances the biosorbent's capacity for metal ion absorption [83, 84], though it can also damage the physical nature of the biosorbent. For the most homogeneous suspensions with a high capacity for biosorption, it is best to stick to a moderate speed. On the other hand, fast agitation can cause a vortex phenomenon, destroying the homogeneity of the suspension. Moreover, excessive turbulence has the potential to limit the duration of biosorbent and biosorbent interaction, leading to a decline in the biosorption rate [85]. It was reported in [85] that the ideal rotation rate for *A. niger* was 120 rotations per minute for the effective absorption of cadmium and zinc. Upon increasing the agitation speed from zero to 80 rotations per minute, there was a notable improvement in biosorption efficiency which increased from 32.4% to approximately double that value, i.e. around 65% [55].

10.9 Biosorbents for the removal of heavy metals from soil

Eliminating HMs via microbial processes is a multifaceted endeavor that entails considering various aspects, such as the chemical nature of the associated metal ions, microorganism cell membrane characteristics, cellular physiology, and additional physicochemical factors [86–89]. There are various kinds of microbial biomass which have been found to be effective in biosorption, such as bacteria [32], fungi [33], and marine and freshwater algae [25]. When selecting biomass for environmental clean-up operations, availability is a critical element. To maintain cost-effectiveness, biomass must originate from natural sources or residual materials [90]. Over the years, various biosorbents have been developed and utilized for HM removal. Various studies have demonstrated that these microbes can indeed eliminate hydrocarbon compounds, pesticides, and HMs [91].

Microorganisms can react to their biological surroundings and thrive under diverse conditions. Their main requirements are a source of carbon and energy [92]. A wide array of economical biosorbents utilized to remove metals from contaminated sources are commonly classified into distinct groups, including bacteria, fungi, algae, plants, industrial by-products, agricultural residues, and other polysaccharide materials [93]. These are described below.

10.9.1 Algae as biosorbents

Algae, which are aquatic organisms without proper roots and stems, vary in size from micro to macroalgae. They possess the ability to produce their food through photosynthesis. Interestingly, these plants can grow substantially even when limited nutrients are available. Due to their significant size, high absorption capacity, and nontoxic nature, algae are highly regarded as valuable biosorbent materials. This makes algae a promising option for applications such as bioremediation, especially for the removal of HMs from water sources.

Given their outstanding capacity for sorption and potentially unlimited availability, algae have attracted special interest from the developers of new biosorbent materials. They are subdivided into several distinct evolutionary pathways, each of which is populated by a different type of algae. The vast majority of biosorption studies have mainly focused on brown algae due to its unique components and qualities. In contrast to other microbial biosorbents, algae are autotrophic, consuming minimal amounts of nutrients and generating a great deal of biomass.

Algal substances exhibit a notable capacity for sorption and can be procured in significant amounts with ease [9, 94].

10.9.2 Fungi as biosorbents

In comparison to chemically bonded treatment processes, the application of fungi as biosorbents has been demonstrated to be a successful, economical, and environmentally friendly option. Given the variety of types of fungi, they can produce exoenzymes for the absorption of complex carbohydrates that were hydrolyzed earlier. This ability allows degradation on various levels, which results in the breakdown of, and the release of nutrients from pollutants such as pesticides or fertilizers. Moreover, one of the merits associated with fungal biomass is its ability to facilitate mass production owing to its ease of cultivation in fermenters. The filamentous configuration of this biomass further offers a convenient means for separation via filtration, adding to its advantages.

In contrast to yeasts, filamentous fungi exhibit greater resilience towards fluctuations in nutrient availability, aeration levels, pH, and temperature. Furthermore, their biomass contains relatively lesser amounts of nucleic acids [95, 96]. Studies have also demonstrated the potential metal-sequestering abilities of these fungi [96]. The impact of pretreatment applied to the biomass of *A. niger* has been examined with regards to its biosorption capacity for lead, cadmium, copper, and nickel. In relation to uranium uptake, deceased fungal biomass sourced from *Rizopus arrhizus* demonstrated the greatest efficiency at 180 mg U g^{-1} dry mass compared to *S. cerevisiae* yeasts, which managed approximately 157 mg U g^{-1} dry mass removal [97].

10.9.3 Bacteria as biosorbents

Bacteria, which are considered to be the most prevalent and versatile microorganisms on Earth, possess a diverse range of characteristics [18]. The surface structure of bacterial cells plays a pivotal role in biosorption processes [32]. The primary constituent of bacterial cell walls is peptidoglycan, which varies across different species and allows for their classification [12]. Gram-positive bacteria have thick peptidoglycans connected by amino acids, with teichoic acids that bind to cytoplasmic membrane lipids to form lipoteichoic acids responsible for robust membrane adhesion [56]. Gram-negative bacteria cells exhibit a reduced level of peptidoglycan within their thin cell wall structure. However, the inclusion of an outer membrane layer that is composed of phospholipids and lipopolysaccharides imparts a net negative charge to these microorganisms. This feature further supports the efficient metal ion binding capabilities of such bacteria [56, 98]. The presence of lipopolysaccharides, teichoic acids, and teichuronic acids accounts for the negative charge exhibited by gram-negative bacteria. In addition, certain extracellular polysaccharides are capable of binding with metals; however, these characteristics are not universal across all strains of gram-negative bacteria. Furthermore, it is possible to eliminate such properties from species that do possess them through either chemical washing or mechanical disruption techniques [96]. The functional groups found in the cell wall are responsible for attracting metal ions [18]. In general, this occurs when metal ions bind to these reactive groups on the bacterial cell wall, which is then followed by their uptake into the interior of the cell. Gram-positive bacteria have higher levels of glycoproteins compared to gram-negative bacteria, resulting in greater absorption of metals from their environment, while gram-negative bacteria exhibit less absorption due to the phospholipids present in their structures [18, 40].

10.10 Conclusions

Contaminants are generally difficult to remove from the environment. Biosorption can be used to clean up polluted environments. In the eradication of contaminants, the study of biosorption still shows highly promising results. It is a mechanism that presents an extensive range of possibilities and pairings, exhibiting significant adaptability in its use. Many research studies have examined the biosorption characteristics of various microbial biosorbents, and their outcomes provide strong evidence to support the use of this method for metal elimination. Industrial biosorption has not yet been fully exploited, and many consider it more of a hypothetical technology than an actual solution. However, with a greater understanding of the complexity involved in industrial processes, there is renewed interest in this field.

References

[1] Ukhurebor K E, Aigbe U O and Onyanche R B *et al* 2021 Developments, utilization and applications of nanobiosensors for environmental sustainability and safety *Bionanomaterials for Environmental and Agricultural Applications* (Bristol: IOP Publishing) 14–1–18

[2] Ukhurebor K E, Aigbe U O and Onyanche R B *et al* 2022 An overview of the emergence and challenges of land reclamation: issues and prospect *Appl. Environ. Soil. Sci.* **5889823** 1–14

[3] Edema-Sillo T and Emegha J O 2022 Effect of environmental degradation on the inhabitants of Obodo Community in Warri South local government area of delta state, Nigeria *J Appl. Sci. Environ. Manage.* **26** 1951–5

[4] Asha Latha P and Sandeep Reddy S 2013 Review on bioremediation—potential tool for removing environmental pollution *IJBACS* **3** 21–33 https://www.cibtech.org/J-CHEMICAL-SCIENCES/PUBLICATIONS/2013/Vol-3-No-3/JCS-03-011-ASHA-REVIEW-POLLUTION.pdf

[5] Sarkar B, Mukhopadhyay R, Ramanayaka S, Bolan N and Ok Y S 2021 The role of soils in the disposition, sequestration and decontamination of environmental contaminants *Philos. Trans. R. Soc. Lond. B: Biol. Sci.* **376** 20200177

[6] Hussein H, Ibrahim S F, Kandeel K and Moawad H 2004 Biosorption of heavy metals from waste water using *Pseudomonas* sp *Electron. J. Biotechn.* **7**

[7] Koller M and Saleh H M 2018 Introductory chapter: introducing heavy metals *Heavy Metals* (London: IntechOpen)

[8] Dhanarani S, Viswanathan E, Piruthiviraj P, Arivalagan P and Kaliannan T 2016 Comparative study on the biosorption of aluminum by free and immobilized cells of *Bacillus safensis* KTSMBNL 26 isolated from explosive contaminated soil *J. Taiwan. Inst. Chem. Eng* **69** 61–7

[9] Abbas S H, Ismail I M, Mostafa T M and Sulaymon A H 2014 Biosorption of heavy metals: a review *J. Chem. Sci. Technol.* **3** 74–102

[10] Akcil A, Erust C, Ozdemiroglu S, Fonti V and Beolchini F 2015 A review of approaches and techniques used in aquatic contaminated sediments: metal removal and stabilization by chemical and biotechnological processes *J. Clean. Prod.* **86** 24–36

[11] Ndeddy Aka R J and Babalola O O 2016 Effect of bacterial inoculation of strains of *Pseudomonas aeruginosa*, *Alcaligenes feacalis* and *Bacillus subtilis* on germination, growth and heavy metal (Cd, Cr, and Ni) uptake of Brassica juncea *Int. J. Phytoremed.* **18** 200–9

[12] Munees A and Mulgeta K 2013 Recent trends in microbial biosorption of heavy metals: a review *Indian J. Exp. Biol* **1** 19–26

[13] Abioye O P, Yusuf G B, Aransiola S A, Oyewole O A and Bala J D 2017 Biosorption of chromium and zinc by *Micrococcus virians* and *Staphylococcus aureus* isolated from soil *Int. Res. J. Environ. Sci.* **6** 1–7 http://www.isca.in/IJENS/Archive/v6/i5/1.ISCA-IRJEvS-2016-076.php

[14] Mustapha M U and Halimoon N 2015 Microorganisms and biosorption of heavy metals in the environment: a review paper *J. Microbe Biochemical Technol* **2015** 253–6

[15] Abdi O and Kazemi M 2015 A review study of biosorption of heavy metals and comparison between different biosorbents *J. Mater. Environ. Sci.* **6** 1386–99 http://www.jmaterenvironsci.com/Document/vol6/vol6_N5/164-JMES-1454-2015-Abdi.pdf

[16] Szymanski K 2009 Lead and chromium compounds in the natural environment and wastes *Rocz. Ochr. Środowiska* **11** 173 https://ros.edu.pl/images/roczniki/archive/pp_2009_011.pdf

[17] Filipiuk D, Fuks L and Majdan M 2006 Biosorption as a new method for sequestering heavy metals in industrial aqueous effluents *Przem. Chem.* **85** 417–22 (in Polish) https://sigma-not.pl/publikacja-20413-biosorpcja-jako-metoda-usuwania-i-odzysku-metali-ci%C4%99%C5%BCkich-z-wodnych-%C5%9Bciek%C3%B3w-przemys%C5%82owych-przemysl-chemiczny-2006-6.html

[18] Shamim S 2018 Biosorption of heavy metals *Biosorption* ed D Jan and V Branislav (London: IntechOpen)

[19] Okoh A I 2006 Biodegradation alternative in the cleanup of petroleum hydrocarbon pollutants *Biotechnol. Mol. Biol. Rev.* **1** 38–50 https://academicjournals.org/journal/BMBR/article-abstract/E94E08A40207

[20] Ahalya N, Ramachandra T V and Kanamadi R D 2004 Biosorption of heavy metals *Res. J. Chem. Environ.* **7** 71

[21] Geoffrey G M 2008 Biosorption: critical review of scientific rationale, environmental importance and significance for pollution treatment *J. Chem. Technol. Biotechnol.* **84** 13–28

[22] Lim M W, Von Lau E and Poh P E 2016 A comprehensive guide of remediation technologies for oil contaminated soil—present works and future directions *Mar. Pollut. Bull.* **109** 14–45

[23] Gadd G M 2000 Bio remedial potential of microbial mechanisms of metal mobilization and immobilization *Curr. Opin. Biotechnol.* **11** 271–9

[24] Dostalek P, Patzak M and Matejka P 2004 Influence of specific growth limitation on biosorption of heavy metals by *Saccharomyces cerevisiae Int. Biodeterior. Biodegr.* **54** 203–7

[25] Cossich E S, Tavares C R G and Ravagnani T M K 2002 Biosorption of chromium (III) by Sargassum sp. biomass *Electron. J. Biotechnol.* **5** 133–40

[26] Chojnacka K 2010 Biosorption and bioaccumulation—the prospects for practical applications *Environ. Int.* **36** 299

[27] Masindi V and Muedi K L 2018 Environmental contamination by heavy metals *Heavy Metals* (London: IntechOpen)
[28] Kumar R, Sharma A K, Singh P, Dhir B and Mehta D 2014 Potential of some fungal and bacterial species in bioremediation of heavy metals *J. Nucl. Phys. Mater. Sci. Radiat. Appl.* **1** 213–23
[29] Pugazhendhi A, Boovaragamoorthy G M, Ranganathan K, Naushad M and Kaliannan T 2018 New insight into effective biosorption of lead from aqueous solution using *Ralstonia solanacearum*: characterization and mechanism studies *J. Clean Prod.* **174** 1234–9
[30] Shanmugaprakash M, Venkatachalam S, Rajendran K and Pugazhendhi A 2018 Biosorptive removal of Zn(II) ions by Pongamiaoil cake (*Pongamia pinnata*) in batch and fixed-bed column studies using response surface methodology and artificial neural network *J Environ. Manag.* **227** 216–28
[31] Hajdu R, Pinheiro J P R, Galceran J and Slaveykova V I 2010 Modeling of Cd uptake and efflux kinetics in metal-resistant bacterium *Cupriavidus metallidurans Environ. Sci. Technol.* **44** 4597–602
[32] Ilhan S, Nurbas M, Bakhsh N, Kilicarslan S and Ozdag H 2004 Removal of chromium, lead and copper ions from industrial waste waters by *Staphylococcus saprophyticus Turkish Electron. J. Biotechnol.* **2** 51–7
[33] Sen M and GhoshDastidar M 2007 Biosorption of chromium (VI) by resting cells of *Aspergillus* sp *Iran J. Environ.Health. Sci. Eng.* **4** 9–12
[34] Hammaini A, Gonzalez F, Ballester A, Blazquez M L and Munoz J A 2006 Biosorption of heavy metals by activated sludge and their desorption characteristics *J Environ Manage.* **84** 419–26
[35] Fomina M and Gadd G M 2014 Biosorption: current perspectives on concept, definition and application *Bioresource. Technol.* **160** 3–14
[36] Ahemad M and Kibret M 2013 Recent trends in microbial biosorption of heavy metals: a review *Biochem. Mol. Biol.* **1** 19–26
[37] Srivastava S and AgrawalMondal M 2015 A review on progress of heavy metal removal using adsorbents of microbial and plant origin *Environ. Sci. Pollut. Res.* **22** 15386–415
[38] Mosa K A, Saadoun I, Kumar K, Helmy M and Dhankher O P 2016 Potential biotechnological strategies for the cleanup of heavy metals and metalloids *Front. Plant Sci.* **7** 1–14
[39] Wang J and Chen C 2009 Biosorbents for heavy metals removal and their future *Biotechnol. Adv.* **27** 195–226
[40] Singh V P and Stapleton R D 2002 *Biotransformations: Bioremediation Technology for Health And Environmental Protection* (Amsterdam: Elsevier)
[41] Modak J M and Natarajan K A 1995 Biosorption of metals using nonliving biomass: a review *SME* **12** 189–96
[42] Gadd G M, White C and De Rome L 1988 Heavy metal and radionuclide by fungi and yeasts ed P R Norris and D P Kelly *Biohydrometallurgy A* (London: Rowe, Chippenham, Wilts)
[43] Kuhn S P and Pfister R M 1989 Adsorption of mixed metals and cadmium by calcium-alginate immobilized *Zooglea ramigera Appl. Microbiol. Biotechnol.* **31** 613
[44] Volesky B 2003 *Sorption and Biosorption* (Montreal-St. Lambert: BV Sorbex, Inc.)
[45] Aryal M and Liakopoulou-Kyriakides M 2015 Bio removal of heavy metals by bacterial biomass *Environ. Monit. Assess.* **187** 4173

[46] Saba R Y, Ahmed M and Sabri A N 2019 Potential role of bacterial extracellular polymeric substances as biosorbent material for arsenic bioremediation *Bioremediat. J.* **23** 72–81
[47] Vijayaraghavan K and Yun Y S 2008 Bacterial biosorbents and biosorption *Biotechnol. Adv.* **26** 266–91
[48] Van Hullebusch E D, Zandvoort M H and Lens P N 2003 Metal immobilization by biofilms: mechanisms and analytical tools *Rev. Environ. Sci. Biotechnol.* **2** 9–33
[49] Ahluwalia S S and Goyal D 2007 Microbial and plant derived biomass for removal of heavy metals from wastewater *Bioresour.Technol.* **98** 2243–57
[50] Ayangbenro A S and Babalola O O 2017 A new strategy for heavy metal polluted environments: a review of microbial biosorbents *Int. J. Environ. Res. Public Health* **14** 94
[51] Ueda M 2016 Establishment of cell surface engineering and its development *Biosci. Biotechnol. Biochem.* **80** 1243–53
[52] Inoue K, Parajuli D, Ghimire K N, Biswas B K, Kawakita H, Oshima T and Ohto K 2017 Biosorbents for removing hazardous metals and metalloids *Materials* **10** 857
[53] Osman M S and Bandyopadhyay M 1999 Biosorption of lead (II) ions from wastewater by using a fungus *P. ostreatus J. Civ. Eng.* **27** 193–6
[54] Ojuederie O B and Babalola O O 2017 Microbial and plant-assisted bioremediation of heavy metal polluted environments: a review *Int. J. Environ. Res. Public Health* **14** 1504
[55] Kanamarlapudi S L R K, Chintalpudi V K and Muddada S 2018 Application of biosorption for removal of heavy metals from wastewater *Biosorption* (London: IntechOpen) 69–116
[56] Park D, Yun Y-S and Park J M 2010 The past, present, and future trends of biosorption *Biotechnol. Bioprocess Eng.* **15** 86–102
[57] Wu F, Sun F, Wu S, Yan Y and Xing B 2012 Removal of antimony (III) from aqueous solution by freshwater *Cyanobacteria microcystis Biomass. J. Chem. Eng.* **183** 172–9
[58] Ding Y, Jing D, Gong H, Zhou L and Yang X 2012 Biosorption of aquatic cadmium (II) by unmodified rice straw *Bioresour. Technol.* **114** 20–5
[59] Witek-Krowiak A and Reddy D H K 2013 Removal of microelemental Cr (III) and Cu (II) by using soybean meal waste–unusual isotherms and insights of binding mechanism *Bioresour. Technol.* **127** 350–7
[60] Chojnacka K, Chojnacki A and Gorecka H 2005 Biosorption of Cr^{3+}, Cd^{2+} and Cu^{2+} ions by blue–green algae *Spirulina* sp.: kinetics, equilibrium and the mechanism of the process *Chemosphere* **59** 75–84
[61] Mohamad O A, Hao X, Xie P, Hatab S, Lin Y and Wei G 2012 Biosorption of copper (II) from aqueous solution using non-living *Mesorhizobium amorphae* strain CCNWGS0123 *Microbes Environ.* **27** 234–41
[62] García-Mendieta A, Olguín M T and Solache-Ríos M 2012 Biosorption properties of green tomato husk (*Physalisphiladelphica lam*) for iron, manganese and iron–manganese from aqueous systems *Desalination* **284** 167–74
[63] Park D, Yun Y-S, Jo J H and Park J M 2005 Mechanism of hexavalent chromium removal by dead fungal biomass of *Aspergillus niger Water Res.* **39** 533–40
[64] Park D, Yun Y-S and Park J M 2005 Studies on hexavalent chromium biosorption by chemically treated biomass of *Ecklonia* sp *Chemosphere* **60** 1356–64
[65] Park D, Yun Y-S and Park J M 2005 Use of dead fungal biomass for the detoxification of hexavalent chromium: screening and kinetics *Process Biochem.* **40** 2559–65
[66] Michalak I and Chojnacka K 2010 Interactions of metal cations with anionic groups on the cell wall of the macroalga *Vaucheria* sp *Eng. Life Sci.* **10** 209–17

[67] Pistorius A, DeGrip W J and Egorova-Zachernyuk T A 2009 Monitoring of biomass composition from microbiological sources by means of FT-IR spectroscopy *Biotechnol. Bioeng.* **103** 123–9
[68] Michalak I, Chojnacka K and Witek-Krowiak A 2013 State of the art for the biosorption process—a review *Appl. Biochem. Biotechnol.* **170** 1389–416
[69] Abdi O and Kazemi M 2015 A review study of biosorption of heavy metals and comparison between different biosorbents *J. Mater. Environ. Sci.* **6** 1386–99 https://www.jmaterenvironsci.com/Document/vol6/vol6_N5/164-JMES-1454-2015-Abdi.pdf
[70] Feng D and Aldrich C 2004 Adsorption of heavy metals by biomaterials derived from the marine algae *Ecklonia maxima Hydrometallurgy* **73** 1–10
[71] Hlihor R, Diaconu M, Fertu D, Chelaru C, Sandu I and Tavares T 2013 Bioremediation of Cr (VI) polluted wastewaters by sorption on heat inactivated *Saccharomyces cerevisiae* biomass *Int. J. Environ. Res.* **7** 581–94 https://ijer.ut.ac.ir/article_638.html
[72] Mungasavalli D P, Viraraghavan T and Jin Y-C 2007 Biosorption of chromium from aqueous solutions by pretreated *Aspergillus niger*: batch and column studies *Colloids Surf. A: Physicochem. Eng. Asp.* **301** 214–23
[73] Park D, Yun Y S and Park J M 2010 The past, present, and future trends of biosorption *Biotechnol. Bioprocess Eng.* **15** 86–102
[74] Özar A and Özer D 2003 Comparative study of the biosorption of Pb (II), Ni (II) and Cr (II) ions onto *S. cerviciae*: determination of biosorption heats *J. Hazard Mater.* **100** 219–29
[75] Kapoor A and Viraraghavan T 1997 Heavy metal biosorption sites in *Aspergillus niger Bioresour. Technol.* **61** 221–7
[76] Abdel-Ghani N, Hegazy A and El-Chaghaby G 2009 *Typha domingensis* leaf powder for decontamination of aluminium, iron, zinc and lead: biosorption kinetics and equilibrium modeling *Int. J. Environ. Sci. Technol.* **6** 243–8
[77] Liu Y-G, Ting F, Zeng G-M, Xin L, Qing T, Fei Y, Ming Z, Xu W-H and Huang Y-E 2006 Removal of cadmium and zinc ions from aqueous solution by living *Aspergillus niger Trans. Nonferrous Met. Soc. China* **16** 681–6
[78] Naiya T K, Bhattacharya A K, Mandal S and Das S K 2009 The sorption of lead (II) ions on rice husk ash *JHM Lett.* **163** 1254–64
[79] Rathinam A, Maharshi B, Janardhanan S K, Jonnalagadda R R and Nair B U 2010 Biosorption of cadmium metal ion from simulated wastewaters using *Hypnea valentiae* biomass: a kinetic and thermodynamic study *Bioresour. Technol.* **101** 1466–70
[80] Senthil Kumar P, Ramalingam S, Sathyaselvabala V, Kirupha S D and Sivanesan S 2011 Removal of copper (II) ions from aqueous solution by adsorption using cashew nut shell *Desalination* **266** 63–71
[81] Kumar D and Gaur J 2011 Metal biosorption by two cyanobacterial mats in relation to ph, biomass concentration, pretreatment and reuse *Bioresour. Technol.* **2** 2529–35
[82] Abdel-Aty A M, Ammar N S, Ghafar H H A and Ali R K 2013 Biosorption of cadmium and lead from aqueous solution by fresh water alga *Anabaena sphaerica* biomass *J. Adv. Res.* **4** 367–74
[83] Venugopal V and Mohanty K 2011 Biosorptive uptake of Cr (VI) from aqueous solutions by *Parthenium hysterophorus* weed: equilibrium, kinetics and thermodynamic studies *J. Chem. Eng.* **17** 151–8

[84] Chong H, Chia P and Ahmad M 2013 The adsorption of heavy metal by Bornean oil palm shell and its potential application as constructed wetland media *Bioresour. Technol.* **130** 181–6
[85] Halttunen T, Salminen S and Tahvonen R 2007 Rapid removal of lead and cadmium from water by specific lactic acid bacteria *Int. J. Food Microbiol.* **114** 30–5
[86] Esposito A, Pagnanelli F, Lodi A, Solisio C and Vegliò F 2001 Biosorption of heavy metals by *Sphaerotilus natans*: an equilibrium study at different pH and biomass concentrations *Hydrometallurgy* **60** 129–41
[87] Pagnanelli F, PetrangeliPapini M, Trifoni M, Toro L and Veglio F 2000 Biosorption of metal ions on *Arthrobacter* sp.: biomass characterization and biosorption modeling *Environ. Sci. Technol.* **34** 2773–8
[88] Hassan S H A, Kim S J, Jung A-Y, Joo J H, Oh S E and Yang J E 2009 Biosorptive capacity of Cd(II) and Cu(II) by lyophilized cells of *Pseudomonas stutzeri J. Gen. Appl. Microbiol.* **55** 27–34
[89] Gabr R M, Hassan S H A and Shoreit A A M 2008 Biosorption of lead and nickel by living and nonliving cells of *Pseudomonas aeruginosa* ASU 6a *Int. Biodeterior. Biodegradat.* **62** 195–203
[90] Kao W-C, Huang C-C and Chang J-S 2008 Biosorption of nickel, chromium and zinc by MerP expressing recombinant *Escherichia coli J. Hazard Mater.* **158** 100–6
[91] Kumar A, Bisht B S, Joshi V D and Dhewa T 2011 Review on bioremediation of polluted environment: a management tool *Int. J. Environ. Sci.* **1** 1079–93 https://www.indianjournals.com/ijor.aspx?target=ijor:ijes&volume=1&issue=6&article=004
[92] Vidali M 2001 Bioremediation: an overview *Pure Appl. Chem.* **73** 1163–72
[93] Romera E, González F, Ballester A, Blázquez M and Munoz J 2007 Comparative study of biosorption of heavy metals using different types of algae *Bioresour. Technol.* **98** 3344–53
[94] Volesky B and Holan Z R 1995 Biosorption of heavy metal *Biotechnol. Prog.* **11** 235–50
[95] Leitão A L 2009 Potential of Penicillium species in the bioremediation field *Int. J. Environ. Res. Public Health* **6** 1393–417
[96] Li X, Li A, Long M and Tian X 2015 Equilibrium and kinetic studies of copper biosorption by dead *Ceriporia lacerata* biomass isolated from the litter of an invasive plant in China *J. Environ. Health Sci. Eng.* **13** 37
[97] Sag Y and Aktay Y 2002 Kinetic studies on sorption of Cr (VI) and Cu (II) ions by chitin, chitosan and *Rhizopus arrhizus Biochem. Eng. J.* **12** 143–53
[98] Kuyucak N and Volesky B 1988 Biosorbents for recovery of metals from industrial solutions *Biotechnol. Lett.* **2** 137–42

Chapter 11

Modeling the adsorption process of the biosorbent-based decontamination of aqueous solutions using kinetic and isotherm models

Philomina Nkeonye Okanigbuan, Uyiosa Osagie Aigbe and Kingsley Eghonghon Ukhurebor

Biosorption, which employs materials of natural origin, has shown great potential in the elimination of noxious contaminants from industrial effluent. In order to complete judicious designs for biosorption systems and determine the useful mechanism involved in these processes, it is important to perform mathematical modeling using various isotherm models (ISMs) and kinetic models (KIMs). This chapter gives an overall evaluation of the use of either the linear (LINR) or nonlinear (NLINR) forms of these models in assessing the sorption mechanisms involved in heavy metal (HM) and dye biosorption.

11.1 Introduction

Water is a supreme gift from nature owing to its key importance for the survival of various biota on the earth. As a result of its restricted geological availability, preserving its value is a worldwide demand that broadly increases with each passing day. Unfortunately, in the contemporary era, various catastrophes, disputes, and wars have affected our limited water sources. Furthermore, diverse water contaminants are associated with numerous human actions, being simultaneously released into water bodies and causing a serious decline in water quality [1, 2].

Water contamination is caused by an extensive variety of sectors, which are responsible for discharging numerous contaminants into the marine ecosystem, including HMs, phenols, dyes, pesticides, insecticides, and pharmaceuticals. HMs and dyes are considered to be extremely hazardous contaminants in water systems because they display carcinogenic, mutagenic, immunogenic, and teratogenic features. Hence, the World Health Organization (WHO) and ecological

doi:10.1088/978-0-7503-5598-8ch11

organizations have defined appropriate limits for contaminants in water reservoirs, and exceeding those limits may have considerable consequences. Effects such as neural toxicity, carcinogenicity, and multiplicative abilities can destroy both human and animal wellbeing. Furthermore, these contaminants result in eutrophication and the limitation of photosynthesis. Their lack of degradability and/or extremely accumulative nature are the cause of these problems [3]. Hence, it is imperative to carry out the proper remediation of wastewater containing various contaminants before it is discharged into aquatic bodies [4, 5].

As a result of this grave ecological issue, scientists are creating new technologies that will permit the elimination of unsafe compounds created by sources of ecological contamination. With various degrees of success, several procedures, such as coagulation, oxidation, chemical precipitation, ion exchange, membrane separation, reverse osmosis, and sorption have been offered to eradicate pollutants from industrial effluent. Nevertheless, many of these procedures have downsides, such as astronomical operational and maintenance costs, the creation of unsafe sludge, and a challenging treatment process. Hence, due to its outstanding ease of use and low vulnerability to unsafe pollutants such as dyes and HMs, sorption has become one of the most utilized confiscation processes [6].

The astronomical cost and incomplete reusability of activated carbon (AC) prevent its widespread use as an effective adsorbent. Owing to its low cost, the absence of a requirement for extra nutrients, ease of use, and absence of adverse ecological consequences, and given that it is an operational alternative to the traditional approaches used to treat contaminant-containing effluents, biosorption, i.e. the usage of biomaterials as biosorbents, has appeared as a capable technique for the remediation of HM/dye-polluted environments that has extraordinary efficacy even at minuscule concentrations (CTNs) of sorbent [7, 8].

The new biosorption technology for contaminant removal and recovery holds potential as an alternative type of theoretically economically attractive treatment and is especially suited to an extensive diversity of industrial contaminant-bearing point-source effluent discharges; its mode of operation is based on the sorption or complexation of liquefied contaminants owing to the chemical action of various biomasses [9]. Due in great part to the attraction between the sorbate and the biosorbent, the biosorption procedure positively encourages some forms of biomass to confiscate biological or inert pollutants via the ability to selectively bond with the ions that are present in water-based solutions. The usage of biosorption technology has the advantages of minimal operational expense, ecological friendliness, ease of implementation, the ability to reuse the biosorbent, instant action, extraordinary specificity, and lack of resultant noxious waste production. Examples of biosorbents include animals, microorganisms (such as bacteria, fungi, and algae), fruit skins, plant wastes, activated sludge, and biopolymers [3, 7].

The mechanisms of biosorption are categorized based on the metabolism dependence, which is the cell's biochemical breakdown (metabolism) dependence. The case in which the contaminant is eliminated from a solution is called metabolism-independent. The key biosorption mechanisms are sorption, chelation, complexation, microprecipitation, ion exchange, coordination, and diffusion via the cell walls and membranes. These depend on the species utilized, the origin of the biomass, the processing, and the chemistry of the solution [10, 11] This chapter discusses detailed studies that have assessed the efficacy and effectiveness of contaminant sorption using different species of biosorbent in diverse effluent backgrounds, with a key focus on the modeling of various biosorption processes (BPs) using different ISMs and KIMs.

11.2 The use of isotherm models and kinetic models for biosorption processes

ISMs and KIMs are systematically used to assess the sorption process (SP), its mechanism, and its implementation. ISMs offer data about the sorption procedure and biosorbent surface properties, and a conclusion is extracted from the correlation between the sorbate(s) and the biosorbent. To investigate exceptional SPs, various representation techniques are used together with theoretical computations which are reinforced with dominant and powerful theoretical models, as described in [1].

11.3 Isotherm models

Although most projected engineering treatments would use some sort of flow-through or nonstop procedure, mainstream biosorption research uses closed batch setups. Countless separate biosorption techniques have been used, making comparisons between findings generally difficult. The important process for establishing a biosorbent's capacity is thought to be a complete sorption ISM. In many of these studies, a straightforward defined system is employed, and biosorption is regularly assessed under a variety of distinct conditions, including altering the biomass's CTN, the pH, the contaminant's CTN, the presence of competing cations, etc. Although flow and other nonstop systems are more complex, breakthrough curves are frequently used for the results for columns to assess sorbent effectiveness [12].

The ISM is commonly applied to identify the link between the insoluble sorbate CTN in the fluid phase (liquid, to be specific) and the biosorbent in the solid phase at a constant (CT) temperature and at certain settings, such as particle size of the biosorbent, pH, ionic strength, and fixed mass. The results obtained for the sorbent capacity, surface features, ways of optimizing the sorption mechanism, and the practical modeling of the SP influenced by the sorption ISMs characterize the interactions between the biosorbent materials and the pollutants. The physicochemical considerations, in combination with the best thermodynamic theories, offer information about the sorption procedure, surface characteristics, and scale of sorbent affinity [2, 13–15].

ISMs are utilized to quantitatively quantify the biosorbent's sorption capacity (SC), affinity, sorption procedure, the sorbate's distribution on the biosorbent, and the mass solution. The biosorbent's effectiveness is evaluated in terms of the sorption rate, its mechanical strength, its capacity, and its renewal and recycling potentialities and prospects. The general principal factor in the sorption procedure is the SC, which is assessed using equilibrium experiments and sorption ISMs. The equilibrium sorption ISM is largely used to explain sorption procedures such as ion exchange, chemisorption, electrostatic interaction, surface complexation, cation–π interaction, anion–π interaction between the biosorbent and the sorbate, intraparticle diffusion, external diffusion, physisorption, van der Waals interactions (hydrophobic interactions, etc.), coverage type, and the interaction between the sorbate and the biosorbent. [2, 15, 16]

The well-recognized ISMs utilized for the BP are the Freundlich (FRH), Langmuir (LNR), Temkin (TMM), Redlich Peterson (RPM), Dubinin–Radushkevich (DRM), and Toth (TOHM) models, among which the LNR and FRH models are most often used to explain the sorption equilibrium between the biosorbent surface and the sorbate [13, 14, 18]. The LNR equation assumes that the optimum sorption relates to a monosaturated layer of sorbate molecules on the biosorbent surface, that the sorption energy remains CT, and that there is no transmigration of the sorbate on the biosorbent surface. It also assumes that all sorption sites are alike in terms of activity and that the intermolecular forces are reduced with increasing distance from the sorptive surface. The NLINR and linearized forms of the LNR are given by equations (11.1) and (11.2)

$$q_e = \frac{q_m K_L C_e}{K_L C_e + 1} \tag{11.1}$$

$$\frac{C_e}{q_e}\frac{1}{q_e}C_e + \frac{1}{q_m K_L} \tag{11.2}$$

K_L (l mg^{-1}), q_e (mg g^{-1}), and q_m (mg g^{-1}) represent the LNR CT, the equilibrium SC and the corresponding saturation SC. The K_L and q_e values are obtained from the slope and the intersection of the fitted line of the plot/graph of C_e/q_e against C_e. A characteristic of the LNR ISM is that it can specify a dimensionless factor known as the separation factor or balance factor, which is expressed in equation (11.3).

$$R_L = \frac{1}{(1 + (K_L \times C_i))} \tag{11.3}$$

R_L and C_i are the equilibrium factor (dimensionless) and the initial CTN in mg l^{-1}. The value of R_L indicates the expected form of sorption: $R_L > 1$ = unfavorable, $R_L = 1$ = linear sorption, $R_L = 0$ = irreversible and $0 < R_L < 1$ = favorable sorption [19–22].

On the other hand, the FRH model takes account of the sorbent surface's heterogeneity and the active spots as well as the exponential distribution of their energies. The NLINR and linearized forms of this model are shown as equations (11.4)–(11.5).

$$q_{e=} K_F C_e^{\frac{1}{n}} \tag{11.4}$$

$$\ln q_m = \frac{1}{n}\ln C_e + \ln K_F \tag{11.5}$$

Here, K_F (mg g^{-1}) and $1/n$ (the inverse of n) correspond to the FRH CT, which is an indicator of the SC and the SC extent. K_F and n are obtained from the intercept and gradient/slope of the graph of ln q_e against ln C_e. The value of $1/n$ should be in the range from zero to one; it depicts the degree of nonlinearity between the CTN of the solution and the sorption. If the $1/n$ value is equivalent to one, the sorption is LINR and elevated values of n describe a comparatively even surface. If the n value is low, it signifies the presence of increased numbers of active energy sites [23, 24].

The TMM comprises a parameter that reflects the sorption interaction between the sorbate and the species. It assumes that as the coverage of the surface improves, the sorption heat (thermal energy) of all particles in the layer decreases in a LINR manner instead of a logarithmic manner. The NLINR and linearized forms of this model are shown in equations (11.6) and (11.7).

$$q_m = \frac{RT}{b_T}\ln(K_T C_e) \tag{11.6}$$

$$q_m = \left(\frac{RT}{b_T}\right)\ln C_e + \left(\frac{RT}{b_T}\right)\ln K_F \tag{11.7}$$

Here, R (8.314 J·mol^{-1} ·K^{-1}), T (K), K_T (l·g^{-1}), and b_T correspond to the gas CT, the absolute temperature (in Kelvin), the equilibrium binding CT of the TMM, and the sorption heat (J·mol^{-1}) that is connected to the TMM CT. The CTs b_T and K_F are obtained from the slope and intercept of the plot of q_e against ln C_e [19, 25].

The DRM supposes a inconsistent surface or an irregular sorption potential. The DRM is a semiexperimental calculation wherein sorption arises on diverse surfaces by a pore-filling procedure whose energy distribution is Gaussian. This has been applied to quantitatively/numerically illustrate gas sorption on microporous biosorbents. The NLINR and the LINR formats of the DRM are given by equations (11.8)–(11.9).

$$q_e = q_s e^{(-K_{DR}\varepsilon^2)} \tag{11.8}$$

$$\ln q_e = \ln q_s - (K_{DR}\varepsilon^2) \tag{11.9}$$

and

$$\varepsilon = RT\ln\left(1 + \frac{1}{C_e}\right) \tag{11.10}$$

Here, K_{DR} (mol^2·kJ^{-2}) and q_s (mg g^{-1}) are the DRM CT and the theoretical DRM saturation capacity, respectively [19, 21, 22].

11.4 Kinetic models

KIMs represent the basic details of the rate of sorption data, stages of rate-limiting, and the diffusion procedure of the sorbate [13, 14] Sorption kinetics is deemed to be mandatory when selecting requirements for effluents and water remediation capabilities in order to complete scale-up procedures. Sorption kinetics approximates the rate at which contamination is removed during the SP; in addition, it gives practical data insights into the sorption reaction. It defines the solute uptake rate, which in turn affects the sorption reaction time. The sorption rate is a critical variable for the creation of batch sorption arrangements; to choose the perfect working settings for a large-scale batch procedure, data for the kinetics of solute uptake is needed [2].

Various KIMs are utilized to define the BP mechanism. Its kinetics is impacted by the biosorbent surface complexity, the CTN of the solute, and the flow. The main KIMs utilized for the sorbate–sorbent interaction are the pseudo first-order (PFO), pseudo second-order (PSO), Elovich (ELM), and intraparticle diffusion (IPD) models, among which the PFO and PSO models are the most utilized in almost every biosorption procedure [26]. The PFO model relies on the PFO equation and is mostly applied to the liquid–solid sorption technique. This model describes the sorption kinetics of biosorbent surface molecules using the following regular first-order differential equation, i.e. the NLINR and LINR equations shown in equations (11.11) and (11.12). On the other hand, the PSO model assumes that sorption takes place on two surface sites. Its NLINR and LINR model equations are given below (equations (11.13) and (11.14)) [27].

$$\frac{dq_t}{dt} = K(q_e - q_t) \tag{11.11}$$

$$\ln(q_e - q_t) = \ln q_e - \frac{K_1}{2.303}t \tag{11.12}$$

$$\frac{dq_t}{dt} = K_2(q_e - q_t)^2 \tag{11.13}$$

$$\frac{t}{q_t} = \frac{1}{K_2 q_e^2} + \frac{1}{q_e}t \tag{11.14}$$

Here, q_t (mg g^{-1}), K and K_1, and K_2 denote the SC at a contact time t, the PFO model rate CT, and the PSO model rate CT, respectively [6].

The linearization of the PSO model has led to more than five various types of PSO model. The use of the NLINR form of the function causes the error (ER) to spread over the whole curve. The PFO and PSO models are evaluated by plotting ln (q_e−q_t) and t/q_t against t. The PSO model may be impacted by the dosage, pH, particle size, and temperature; hence, this model evaluates the influence of the apparent rate factors. This model can also be applied to assess the preliminary uptake of the solute and the biosorbent's SC. When the CTN of the solute is low, the sorption mechanism is explained better by equation (11.14), and at elevated CTNs, the PFO model is favored because the value of ln (q_e−q_t) is enhanced, exponentially dominating the ER function at low initial CTNs, which is the opposite behavior to that observed for elevated initial CTNs. Although LINR model applications have advanced, they may be erroneous when used to create kinetic systems. These two models do not describe the diffusion of the solute into the biosorbent; hence, the diffusion mechanism should be studied before conclusions are reached [26].

Pollutant sorption from water-based solutions and the kinetic sorption of gases on solids are widely defined by the ELM equation [27]. The NLINR and LINR forms of this model are depicted by equations (11.15) and (11.16):

$$\frac{dq_t}{dt} = \alpha e^{-\beta t} \tag{11.15}$$

$$q_t = \frac{1}{\beta}(\alpha\beta) + \frac{1}{\beta}\ln(t), \tag{11.16}$$

where α (mgg^{-1} min^{-1})andβ (gmg^{-1}) relate to the initial rate, the level of surface coverage, and the activation energy for chemisorption [13].

The IPD model is generally employed to assess the rate-limiting phase of the sorption procedure. This involves the sorption of the solute in a solution via the transfer of the mass sorbate through film diffusion, pore diffusion, and surface diffusion. Surface and pore diffusion can possibly take place simultaneously, but film diffusion is an independent stage. The IPD model is implemented by evaluating Weber and Morris's (1963) model LINR equation (equation (11.17)):

$$q_t = K_{IPD}\sqrt{t} + C, \tag{11.17}$$

where K_{IPD} (mg g^{-1} min$^{1/2}$) and C are the rate CT and the boundary layer thickness, respectively; the C value defines the boundary layer impact, where elevated values denote larger impacts. A plot/graph of q_t against $t^{1/2}$ gives a LINR function; when the LINR plot/graph never passes via the source/origin and has several LINR parts, these parts relate to the various mechanisms responsible for sorption. The transmission of a solute from a solution to the sorbent is described by four key mechanisms, which are the mass transmission of the solute's constituent parts immediately after the sorbent is released into the solution (mass transfer), film diffusion (which comprises the slow transmission of the solute from the borderline layer to the surface of the biosorbent), the arrival of the solute at the biosorbent surface together with its movement to the biosorbent pores, and the swift sorptive sorbate attachment onto the active spots of the biosorbent pores. The first and last mechanisms are not considered during the kinetic engineering structural layout. If the arrangement is described by inadequate mixing, a trivial amount of solute, and a minimal CTN, IPD determines the degree of the limiting phase; if not, IPD controls the process. Pore diffusion is a slow process, which makes equation (11.17) follow a curved trend [26, 28].

Today, the tool most often used to define the best-fit sorption model is LINR regression, because it measures the sorbate's distribution, the sorption system, and the consistency of the hypothetical assumptions of the sorption ISMs and KIMs using suitable mathematical methods to predict the general sorption behavior under various operating settings. Hence, owing to the intrinsic bias caused by linearization, different ER functions are utilized to mitigate this gap. As a consequence of developments in computer technology, there has been vast use of NLINR isotherm and kinetic modeling [29].

The drawbacks of the linearization of NLINR equations include subtle variations in the structure of the ERs and violations of the ER variation and the normality theory of standard least squares. Wide-ranging, NLINR regression gives a more precise determination of model factors than the LINR regression technique. In recent years, the following have been used to determine the best fit between ISMs and KIMs: the average/mean relative ER, the sum of absolute ERs, hybrid ER functions, the sum/totality of the ERs squared, Marquardt's percent standard deviation, and the coefficient of determination (R^2). The utilization of NLINR procedures instead of LINR regression permits an effortless and accurate fitting of the experimental data with the hypothetical models [30]. NLINR regression is an iterative procedure, in contrast to LINR regression, in which model factor values are iterated based on a designated procedure to reduce the recognized ER function (in the case of R^2, to maximize it). Computing tools such as Matlab, OriginPro, Mathematica, and SPSS make it straightforward to carry out this regressive iteration [31].

In a study of the sequestration of malachite green dye by clove leaf activated with potassium hydroxide, the process of biosorption was found to be defined by the Dubinin-Radushkevich Model (DRM) and ELM models. The optimum SC (q_m) was determined to be 131.6 mg g^{-1} [32]. The PSO and LNR models were found to superlatively describe the sorption of Cr(VI) by *Pisum sativum* peels activated by triethylenetetramine. The determined optimum q_m was 312.0 mg g^{-1} [33]. A study of the use of a fungal-extract-based biosorbent for Cu(II) and Ni(II) biosorption found that this process could be described using the PSO and LNR models. The optimum q_m values were 7.18 and 8.50 mg g^{-1} for Cu(II) and Ni(II), respectively [34]. Another study explored the use of brewer's spent grain for the biosorption of various HMs, and it was found that the PSO and LNR models optimally described the process of biosorption [35]. Harvested *Chromochloris zofingiensis* (ChZ) microalgae were assessed for methylene blue (MB) biosorption; the results showed that the process was well described by the LNR and PSO models. The optimum q_m was estimated to be 135.4 mg g^{-1} [36]. The biosorption of brilliant green dye by a pistachio shell biosorbent was described by the PFO model and the Sips model (SIPM), and the optimum q_m value was determined to be 52.4 mg g^{-1} [37]. Table 11.1 gives a summary of the ideal ISMs and KIMs that have been used to describe the biosorption of some HMs and dyes by different biosorbents with their corresponding q_m values. Most of these processes were well-defined and represented using the LNR, FRH, and PSO models.

11.5 Conclusions

Models are systematically employed to assess the SP, its mechanism, and its implementation. In order to complete a judicious design of the BP system and assess the practical mechanism involved, mathematical modeling is important. This

Table 11.1. Sorption ISMs and KIMs used in the evaluation of biosorbents for the sequestration of various contaminants.

Biosorbent	Contaminant	ISM(s) and optimum q_m (mg g^{-1})	KIM	Reference
Modified sawdust biochar	Cu(II)	LNR/91.74, 112.36, and 133.33	PSO	[38]
Ozone, purified and sonicated sawdust biochars	MB	FRH and LNR/200, 526, and 769	PSO	[39]
Raphia hookeri fruit epicarp	Rhodamine B	FRH/666.67	PSO	[40]
Microparticles of wild plants	Lissamine Green B	FRH/3.61, 5.91, and 4.20	PSO	[41]
Poly(vinyl alcohol)–tricarboxi-cellulose	MB	LNR/806.5	—	[42]
Malt bagasse	Yellow Reafix BR2	LNR/68.8	PSO	[43]
Coconut husk	Crystal violet	FRH/0.7	PSO	[44]

chapter gave a general outline of the use of such models in either their LINR or NLINR forms in assessing the sorption mechanism that takes place during the biosorption of HMs or dyes. It was found from the studies reviewed that most BPs were defined using the LNR, FRHM, and PSO models in their LINR forms. To close the gap between experimental and predicted data in the future, there is an urgent need for the application of the NLINR forms of these models in different sorption systems.

References and further reading

[1] Aigbe U, Ukhurebor K, Onyancha R, Osibote O, Darmokoesoemo H and Kusuma H 2021 Fly ash-based adsorbent for adsorption of heavy metals and dyes from aqueous solution: a review *J. Mater. Res. Technol.* **14** 2751–74

[2] Aigbe U and Osibote O 2020 A review of hexavalent chromium removal from aqueous solutions by sorption technique using nanomaterials *J. Environ. Chem. Eng.* **8** 104503

[3] Elgarahy A, Elwakeel K, Mohammad S and Elshoubaky G 2021 A critical review of biosorption of dyes, heavy metals and metalloids from wastewater as an efficient and green process *Clean. Eng. Technol.* **4** 100209

[4] Aigbe U, Maluleke R, Lebepe T, Oluwafemi O and Osibote O 2023 Rhodamine 6G dye adsorption using magnetic nanoparticles synthesized with the support of Vernonia Amygdalina leaf extract (Bitter Leaf) *J. Inorg. Organomet. Polym. Mater.* pp. 1–20

[5] Agboola O *et al* 2021 A review on polymer nanocomposites and their effective applications in membranes and adsorbents for water treatment and gas separation *Membranes* **11** p. 139

[6] Ali K, Javaid M, Ali Z and Zaghum M 2021 Biomass-derived adsorbents for dye and heavy metal removal from wastewater *Adsorpt. Sci. Technol.* **2021** pp. 1–14

[7] Ordóñez J, Cortés S, Maluenda P and Soto I 2023 Biosorption of heavy metals with algae: critical review of its application in real effluents *Sustainability* **15** p. 5521

[8] Oyewole O, Adamu B, Oladoja E, Balogun A, Okunlola B and Odiniya E 2018 A review on heavy metals biosorption in the environment *Braz. J. Biol. Sci.* **5** pp. 225–36
[9] Abdi O and Kazemi M 2015 A review study of biosorption of heavy metals and comparison between different biosorbents *J. Mater. Environ. Sci.* **6** 1386–99 https://www.jmaterenvironsci.com/Document/vol6/vol6_N5/164-JMES-1454-2015-Abdi.pdf
[10] Mustapha M and Halimoon N 2015 Microorganisms and biosorption of heavy metals in the environment: a review paper *J. Microb. Biochem. Technol* **7** pp. 253–6
[11] Gouda S and Taha A 2023 Biosorption of heavy metals as a new alternative method for wastewater treatment: a review *Egypt. J. Aquat. Biol. Fish.* **27** pp. 135–53
[12] Gadd G 2009 Biosorption: critical review of scientific rationale, environmental importance and significance for pollution treatment *J. Chem. Technol. Biotechnol.: Inter. Res. Process, Environ. Clean Technol.* **84** pp. 13–28
[13] Aigbe U, Ukhurebor K, Onyancha R, Ama O, Okundaye B, Esiekpe E, Osibote O, Kusuma H and Osifo P 2021 Utility of bionanocomposites for wastewater treatment *Bionanomaterials for Environmental and Agricultural Applications* (Bristol: IOP Publishing)
[14] Aigbe U, Ukhurebor K, Onyancha R, Okundaye B and Osibote O 2022 Green nanomaterials for wastewater treatment analysis *Green Nanoarchitectonics* (Singapore: Jenny Stanford Publishing) 115–36 pp
[15] Yang J, Hou B, Wang J, Tian B, Bi J, Wang N, Li X and Huang X 2019 Nanomaterials for the removal of heavy metals from wastewater *Nanomaterials* **9** 424
[16] Ouni L, Ramazani A and Taghavi Fardood S 2019 An overview of carbon nanotubes role in heavy metals removal from wastewater *Front. Chem. Sci. Eng.* **13** pp. 274–95
[17] Djama C, Chebli D, Bouguettoucha A, Doudou I and Amrane A 2021 Statistical physics modelling of azo dyes biosorption onto modified powder of *Acorus calamus* in batch reactor *Biomass Convers. Biorefin.* **13** 1–16
[18] Bayuo J, Rwiza M, Sillanpää M and Mtei K 2023 Removal of heavy metals from binary and multicomponent adsorption systems using various adsorbents—a systematic review *RSC Adv.* **13** pp. 13052–93
[19] Abin-Bazaine A, Trujillo A and Olmos-Marquez M 2022 Adsorption isotherms: enlightenment of the phenomenon of adsorption *Wastewater Treatment* (London: IntechOpen)
[20] Jain C, Malik D and Yadav A 2016 Applicability of plant based biosorbents in the removal of heavy metals: a review *Environ. Process.* **3** pp. 495–523
[21] Jang E, Pack S, Kim I and Chung S 2020 A systematic study of hexavalent chromium adsorption and removal from aqueous environments using chemically functionalized amorphous and mesoporous silica nanoparticles *Sci. Rep.* **10** p. 555
[22] Basu A, Ali S, Hossain S and Asif M 2022 A review of the dynamic mathematical modeling of heavy metal removal with the biosorption process *Processes* **10** p. 1154
[23] Rahpeima S, Javanbakht V and Esmaili J 2018 Synthesis and characterization of activated carbon/maghemite/starch magnetic bionanocomposite and its application for permanganate removal from aqueous solution *J. Inorg. Organomet. Polym. Mater.* **28** pp. 195–211
[24] Aigbe U, Das R, Ho W, Srinivasu V and Maity A 2018 A novel method for removal of Cr (VI) using polypyrrole magnetic nanocomposite in the presence of unsteady magnetic fields *Sep. Purif. Technol.* **194** pp. 377–87
[25] Bayuo J, Rwiza M and Mtei K 2022 A comprehensive review on the decontamination of lead (II) from water and wastewater by low-cost biosorbents *RSC Adv.* **12** pp. 11233–54

[26] Kajjumba G, Emik S, Öngen A, Özcan H and Aydın S 2018 Modelling of adsorption kinetic processes—errors, theory and application *Advanced Sorption Process Applications* (London: IntechOpen) 1–19
[27] Osman A *et al* 2023 Methods to prepare biosorbents and magnetic sorbents for water treatment: a review *Environ. Chem. Lett.* **21** 2337–98
[28] Wu F, Tseng R and Juang R 2009 Initial behavior of intraparticle diffusion model used in the description of adsorption kinetics *Chem. Eng. J.* **153** pp. 1–8
[29] Ayawei N, Ebelegi A and Wankasi D 2017 Modelling and interpretation of adsorption isotherms *J. Chem.* **2017** 3039817
[30] Brouers F and Al-Musawi T 2015 On the optimal use of isotherm models for the characterization of biosorption of lead onto algae *J. Mol. Liq.* **212** pp. 46–51
[31] Tan K and Hameed B 2017 Insight into the adsorption kinetics models for the removal of contaminants from aqueous solutions *J. Taiwan Inst. Chem. Eng.* **74** pp. 25–48
[32] Sudarni D, Aigbe U, Ukhurebor K, Onyancha R, Kusuma H, Darmokoesoemo H, Osibote O, Balogun V and Widyaningrum B 2021 Malachite green removal by activated potassium hydroxide clove leaves agro-waste biosorbent: characterization, kinetics, isotherms and thermodynamics studies *Adsorpt. Sci. Technol.* **2021** pp. 1–15
[33] El-Nemr M, Aigbe U, Ukhurebor K, Onyancha R, El Nemr A, Ragab S, Osibote O and Hassaan M 2022 Adsorption of Cr^{6+} ion using activated Pisum sativum peels-triethylenetetramine *Environ. Sci. Pollut. Res.* **29** pp. 91036–60
[34] Yildirim A, Baran M and Acay H 2020 Kinetic and isotherm investigation into the removal of heavy metals using a fungal-extract-based bio-nanosorbent *Environ. Technol. Innov.* **20** p. 101076
[35] Wierzba S and Kłos A 2019 Heavy metal sorption in biosorbents–Using spent grain from the brewing industry *J. Clean. Prod.* **225** pp. 112–20
[36] Bayat M, Salehi E and Mahdieh M 2023 Chromochloris zofingiensis microalgae as a potential dye adsorbent: adsorption thermo-kinetic, isothermal, and process optimization *Algal Res.* **71** p. 103043
[37] Enache A, Cojocaru C, Samoila P, Ciornea V, Apolzan R, Predeanu G and Harabagiu V 2023 Adsorption of brilliant green dye onto a mercerized biosorbent: kinetic, thermodynamic, and molecular docking studies *Molecules* **28** p. 4129
[38] Eleryan A *et al* 2022 Copper (II) ion removal by chemically and physically modified sawdust biochar *Biomass Convers. Biorefin.* pp. 1–38
[39] Eldeeb T, Aigbe U, Ukhurebor K, Onyancha R, El-Nemr M, Hassaan M, Ragab S, Osibote O and El Nemr A 2022 Adsorption of methylene blue (MB) dye on ozone, purified and sonicated sawdust biochars *Biomass Convers. Biorefin.* 1–23
[40] Inyinbor A, Adekola F and Olatunji G 2016 Kinetics, isotherms and thermodynamic modeling of liquid phase adsorption of Rhodamine B dye onto *Raphia hookeri* fruit epicarp *Water Resour. Ind.* **15** pp. 14–27
[41] Al-Kadhi N 2019 The kinetic and thermodynamic study of the adsorption Lissamine Green B dye by micro-particle of wild plants from aqueous solutions *Egypt. J. Aquat. Res.* **45** pp. 231–8
[42] Tataru-Farmus R, Cimpoesu R, Nica I and Suteu D 2023 Biosorbent based on poly (vinyl alcohol)–tricarboxi-cellulose designed to retain organic dyes from aqueous media *Polymers* **15** p. 715

[43] da Silva B, Zanutto A and Pietrobelli J 2019 Biosorption of reactive yellow dye by malt bagasse *Adsorpt. Sci. Technol.* **37** pp. 236–59

[44] Imran M, Javed T, Areej I and Haider M 2022 Sequestration of crystal violet dye from wastewater using low-cost coconut husk as a potential adsorbent *Water Sci. Technol.* **85** pp. 2295–317

Chapter 12

Green biosorbents for the degradation of petroleum contaminants

Joseph Onyeka Emegha, Timothy Imanobe Oliomogbe, Joel Okpoghono, Adeoye Victor Babalola, Chioma Adaku Ejelonu, Dennis Eyetan Elete and Kingsley Eghonghon Ukhurebor

Recently, petroleum hydrocarbons (PHs) have become a major threat to human health and ecosystems, demanding immediate and efficient solutions in order to keep them under control. Even though several approaches such as physical or chemical techniques are used to clean up petroleum-contaminated sites, microorganisms are becoming highly valued for their economical and sustainable decontamination methods. This chapter presents an overview of current research into the bioremediation of PHs by green technology or green-derived biosorbents (GDBs), such as bacteria, fungi, and microalgae that are renowned for their ability to remove hydrocarbon contaminants.

12.1 Introduction

In recent years, environmentalists and scientists have become increasingly aware of the difficulties posed by environmental pollutants [1–4]. PHs and crude oil spills are among the primary sources of hydrocarbon contaminants in the environment today. The continuous degradation of the ecosystem by petroleum contaminants is of serious global concern, as it causes great danger to humans, aquatic life, and the environment [1, 4]. As an important source of energy, petroleum enters the environment via human activities such as accidental spills, transportation, storage, refining, production, exploitation, and industrial discharge [5–7]. Petroleum hydrocarbons include several chemical compounds that can be divided into four categories: aliphatic, aromatic, resinous, and asphaltic base hydrocarbon complexes. These compounds are difficult to control due to their dynamic nature [5, 8]. Each of these chemical compounds possesses unique physicochemical properties that are susceptible to degradation [8].

doi:10.1088/978-0-7503-5598-8ch12

Polycyclic aromatic hydrocarbons (PAHs), which are comprised of specific organic contaminants, have gained recognition for their mutagenic, carcinogenic, and teratogenic characteristics [1, 8]. These hydrocarbon complexes are toxic to both terrestrial and aquatic organisms and require several years or even decades for recovery [6, 8]. Many PAHs, such as phenanthrene (PHE) and pyrene (PYR) (a polycyclic aromatic hydrocarbon composed of three or four rings) are volatile particles or gaseous molecules that can be deposited in soil and water bodies. In addition, molecules of naphthalene (a polycyclic aromatic hydrocarbon composed of two rings) are ever-present environmental pollutants that cause diseases such as cataracts, hemolytic anemia, damage to the liver, as well as neurological damage [8]. Consequently, increasing exposure to these polycyclic aromatic hydrocarbons has caused great environmental concern worldwide. Pollution from PHs has shaped many aspects of our lives, including tourism, the economy, and leisure activities. When PHs are spilled into the aquatic environment, the oil spreads and sometimes forms oil–water emulsions [6]. Due to both natural and man-made effects (winds, water currents, etc.) the spilled oil can reach the shoreline, where it results in both economic and environmental damage [8]. This type of pollution, which affects both the aquatic ecosystem and human health, has attracted considerable interest from the scientific community as well as the sociopolitical class.

Researchers are voicing their concern because PH contamination of soils is becoming a major issue. Environmentalists and scientists are directing more attention towards finding solutions to this dilemma. It has been noted that oil-contaminated soil or PH-contaminated soil significantly impacts the ecosystem, which underscores the importance of effectively cleaning up oil-contaminated sites. Today, a good number of the remediation approaches for PH-contaminated soil and water bodies require physical or mechanical systems, as these can be simple and quick solutions to contaminant removal [9]. One of these treatment approaches is the soil washing technique (SWT) [6]. As a physical approach, the proposed SWT involves the washing of soil surfaces to separate contaminants from them. This is sometimes achieved by dissolving contaminants in a wash solution using a reagent [6, 10]. Recently, SWT has emerged as a prominent technology that can be applied either on-site or off-site. The process involves the use of aqueous or water-based surfactants to draw out pollutants into the mass liquid phase without requiring any chemical modifications [11]. Although this approach has advantages, there are also numerous disadvantages associated with the approach. To illustrate, the application of high-pressure washing to eliminate PHs can eradicate micro-organisms, while the application of harsh chemicals such as sorbents and dispersants may disturb the environment. In addition, this technique is time-consuming for the technicians, tedious, and only results in the quasi removal of PH contaminants [5, 6]. Therefore, it is difficult to achieve efficient cleanup using these conventional approaches in a short period of time [6, 12]. Hence, investigations of other eco-friendly cleanup techniques or systems to remove PH contaminants are necessary.

Today, biodegradation and green technology are likely to be eco-friendly and proficient alternatives for the comprehensive decontamination or remediation of environments polluted by PHs. They are also anticipated to be efficient and economical ways of preventing damage caused to oil-spill environments when compared with other mechanical or conventional methods [1, 13]. Green technology and GDBs have attracted tremendous interest from scientists and researchers because they are widely available, nontoxic, low cost, and eco-friendly. Studies have shown that various GDBs or microorganisms as well as fungi, algae, and bacteria can convert petroleum hydrocarbons into less toxic materials that are energy and carbon sources [13, 14]. These indigenous microorganisms (algae, bacteria, and fungi) have the capability to degrade PHs and PAHs that may be present on water and soil surfaces [13]. Therefore, removing microorganisms that cause massive damage from their natural habitats is an effective and viable option. The process of degrading and removing PAHs and crude oil pollutants varies with the microbial strain as well as the different properties of the pollutants. Some green microorganisms can exploit cyclic or branched alkanes or straight and aromatic hydrocarbons as their main source of carbon [15]. The success of this system is rooted in the capability of microorganisms to break down most hydrocarbons.

Multiple studies have investigated the use of microorganisms to degrade petroleum products, observing an array of degraders as well as how the characteristics of the surroundings (such as nutrients, contaminant concentration, pH, temperature, etc.) can influence the process. This chapter examines the use of GDBs or green biosorbents for the complete elimination or removal of PH-based pollutants. It discusses how different biosorbents such as algae, fungi, and bacteria can be used to remove harmful substances such as PHs and PAHs using GDBs. In addition, the chapter suggests ideas for further research in this area.

12.2 An overview of green biosorbents

The complex concept of biosorption has developed over recent decades [16, 17]. Its definition is multifaceted due to the various mechanisms, environmental conditions, and metabolic processes involved [16–18]. The term itself has been applied to many substances including metals, radionuclides, and organics. For specialist communication to be successful, it is necessary for accurate terminology to be adopted [16, 17]. Sorption is a physicochemical process in which substances attach to each other [16]. Biosorption, despite its 'bio' prefix, remains a purely physical and chemical method of using biological material to eliminate substances from solution [19]. However, attaching a material to a biological substance can be more complex than it appears. A key distinction is made here between bioabsorption and bioadsorption [17, 19]. The term sorption is employed interchangeably to describe both absorption and adsorption. Absorption occurs when one state merges with another of a different type (for example, when solids absorb liquids or water absorbs gases) to form a 3D matrix. Adsorption, however, refers to ions and molecules that physically bond to the surface of another molecule—essentially forming a 2D layer [16, 17, 20]. According to various researchers, biosorption can be classified as a form of adsorption wherein the sorbent is organic rather than non-biological [17]. The terms adsorption and sorption are often used interchangeably, but there is a difference between the two processes. Adsorption refers to an interaction in which molecules of gas or liquid stick (adsorb) onto a solid material (e.g. an atom, molecule, or molecular ion) by accumulating at the interface between a sorbent and another material when they come into contact with each other [21]. In addition, when adsorption occurs and progresses to the extent of generating a fresh three-dimensional surface entity, it may be classified as surface precipitation [16]. There are many different systems that fit somewhere into this continuum from adsorption to precipitation [19].

Essentially, biosorption is a physical process of absorption and adsorption that involves ion exchange, surface complexation, and precipitation [16, 17, 20]. The biotechnology industry has employed a technique known as biosorption to separate both inorganic and organic substances from solutions through the use of surface adsorption, chemisorption, and complexation [16, 22]. The functional groups present on the surface of biosorbents facilitate the physicochemical interactions that lead to the absorption of organic compounds. Since organic pollutants vary in their composition, they can easily pass through cell membranes, thus allowing them to enter the organic cellular structure [22]. Figure 12.1 illustrates the core principle of

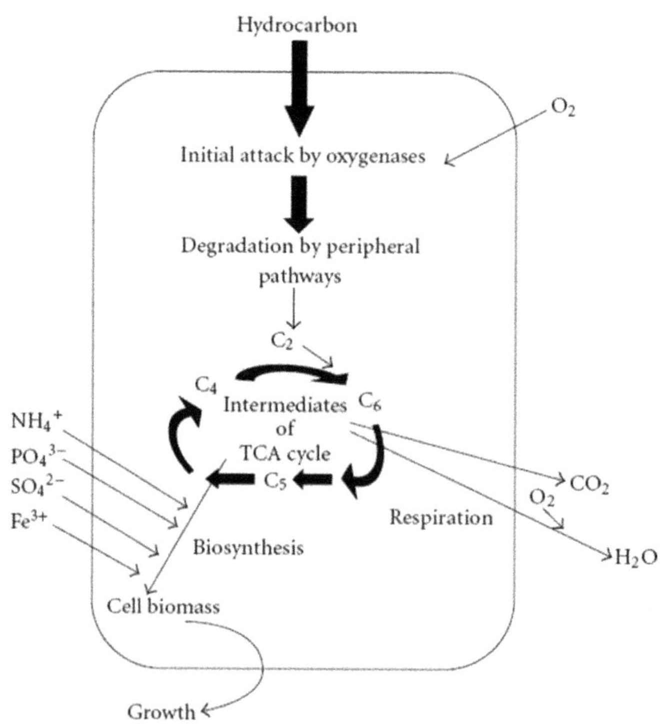

Figure 12.1. The aerobic degradation of PHs by green biosorbents. Reproduced from [22]. CC BY 3.0.

aerobic hydrocarbon degradation, as adapted from [22]. Usually, this principle of the aerobic degradation of hydrocarbons also applies to organic pollutants [22]. There is no doubt that aerobic conditions (i.e. those in which air and water are well oxygenated) degrade organic pollutants the quickest. Since oxygenases and peroxidases are used under aerobic conditions to catalyze an oxidative process, the degradation of these pollutants is much faster and more efficient under aerobic conditions [23]. In conclusion, biomass is generated through a series of chemical transformations that begin with the conversion of pollutants into intermediate metabolites. Metabolites such as acetyl coenzyme A (CoA) and pyruvate are produced by earlier steps in the metabolism and function as precursors. These precursors are then used in gluconeogenesis to generate the sugars required for development.

Moreover, the degradation of PHs is mediated by the enzymatic breakdown of xenobiotics as well as their attachment to microbial cells and biosurfactant production. In figure 12.2 [22], oxygenases are shown to attack xenobiotics first [23]. The most important mechanisms for the recovery of organic matter are the attachment of microorganisms to substrates and the production of biosurfactants [24]. Although the mechanism that enables cells to adhere to oil droplets is not fully understood, researchers have extensively studied how biosurfactants are produced [22]. This process has been shown to be effective in the removal of many petroleum contaminants that have been documented in the research literature [17, 22–24]. It is also important to mention that the advantages of the biosorption method over conventional methods in crude oil decontamination are based on a number of factors, including its high adsorption rate, simplicity of operation, and reversibility (the ability to go back from the sorbed state)—the two most important requirements for an

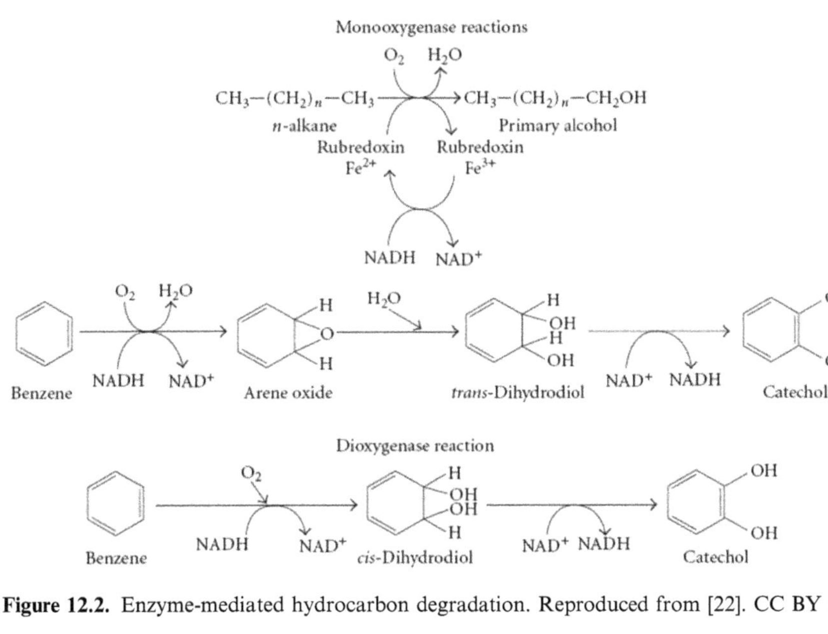

Figure 12.2. Enzyme-mediated hydrocarbon degradation. Reproduced from [22]. CC BY 3.0.

industrial desalination method. High efficiency and cost-effectiveness are also desirable characteristics [17, 25].

12.3 The characteristics of green biosorbents

Green microorganisms exhibit unique characteristics due to their different sources, which include animals and plants. Studies have shown that both living and nonliving organisms, such as primitive forms of algae, fungi, and bacteria, can be used for the biosorption process [5, 26]. It should be noted that an ideal biosorbent should possess excellent qualities such as regeneration, availability, large-scale usability, nontoxicity, and high binding capacity [27]. The adsorbent nature of any biosorbent is dependent on its structural characteristics, such as functional groups, pore distribution, and specific surface area [28]. Green biosorbents must be understood in light of their physical and chemical properties, along with their surface chemistry, since these features are affected by seasonal and environmental influences [26–28]. This means that these features are subject to a variety of seasonal and environmental variables. Several specialized techniques are available for this purpose, depending on the properties of the biosorbent. For example, the following techniques can be used; Fourier transform infrared (FTIR) spectroscopy, fluorescence spectrophotometry, x-ray photoelectron spectroscopy (XPS), scanning electron microscopy (SEM), nitrogen sorption, and so on. These characterization techniques are employed together or separately in determining the morphologies and chemical compositions of biosorbents [6, 29]. Most of the GDB species are composed of many functional groups, including amide, carboxyl, hydroxide, phosphate, and amino groups, which suggests the potential of biosorbents [17, 20, 26]. Since the cellular structures in the various GDBs are mainly responsible for binding, the realization of the biosorption process varies based on the diversity of the cell structures in the GDBs. Recent investigations have shown that gram-positive bacteria have great sorption ability due to their thicker and stronger peptidoglycan cell walls [30, 31].

12.4 Algae as biosorbents

As a large group of multicellular and unicellular organisms, algae are found in large quantities in both terrestrial and aquatic environments [5, 6, 29, 32]. However, the utilization of these algae as a good biosorbent material is due to their excellent characteristics; for example, their large surface-area-to-volume ratio, high sorption capacity, easy availability, high potential for regeneration, and eco-friendliness [6, 32, 33]. There are three major classes of algae in the world, namely macroalgae (marine or brown algae), microalgae (freshwater or green algae), and red algae [6]. Since they are autotrophic, they can exist as large micro and macro-organisms with less nutrition [34]. The various groups of algae each have unique cell wall compositions. However, the brown algae above all have shown to have good biosorption potential owing to the existence of well-structured cell walls [35]. These cell walls are complex cellulose and hemicellulose networks of biopolymers consisting of proteoglycans, polysaccharides, and other similar molecular components [5, 35]. Owing to the complex nature of these biopolymers, there is a possibility that several interactions can simultaneously occur between hazardous pollutants and macroalgae [33, 35–37].

Numerous studies have found that algae contain a variety of different functional groups, including hydroxyl, sulfate, amino, and carboxyl groups, which can be classified by their complex compositions (proteins, polysaccharides, and lipids), based on their structural characteristics [5, 6, 18–21, 34]. These complex components are associated with the adsorbent natures of the various functional groups [5, 34]. Therefore, several authors have reported that the efficiency of microbial algae biosorbents ranges from 15.3% to 84.6%, which is higher than those of other biomasses [6, 32].

12.4.1 Algae-based biosorbents for petroleum pollutant removal

12.4.1.1 Algae-based biosorbents for crude oil removal

Some previous studies have indicated that algae-derived adsorbents can be used to remove crude oil from water sources. The results of [38] show that dead biomass derived from *Scenedesmus abundans* and *Spirulina* sp. can be used to biosorb diesel and lubricating oil. Different biomasses were tested for sorption systems involving various types of biomasses and oil. Different biomasses had varying sorption rates, but *Spirulina* sp. had the ability to remove diesel within a short period of time. As a result, the study focused on examining how much oil (ranging from 0.1% to 2%) and biomass (also at 0.1%) were absorbed in batch systems, which determined their sorption rates and extent. Furthermore, the Freundlich and Langmuir models were also used to explain the sorption of diesel on algae, but not the sorption of lubricant oil. As per a model with three parameters, which was applicable for all isotherms, the highest rates at which diesel and lubrication oils could be absorbed by biosorption were roughly 12 g and 14 g, respectively, per gram of diesel or lubrication oil, contingent on the specific isotherm.

Research [39] has shown that the green macroalgae *Encephalitozoon intestinalis* can effectively remove crude oil from water through their biosorbent properties. Based on their findings, researchers discovered that sea water is better at absorbing crude oil and spent oil than fresh water. They also found that used oil with a higher viscosity was absorbed better than lower-viscosity crude oil under similar conditions. Moreover, it was found that the adsorption data were consistent with both the Langmuir isotherm and pseudo second-order kinetic models.

Several ways to remove harmful substances through biosorption have been suggested [35, 36]. Other processes, such as surface precipitation and adsorption, have been identified as well [6, 32]. Biosorption is a subprocess of bioremediation, and the type of biomass used can determine its mechanism [40]. Nonliving cells do not require metabolic energy for this process [41]. Functional groups (such as amino groups, carboxyls, sulfates, phosphates, and imidazoles) and substances found on algal cell walls (such as alginic acid and proteins) can pull in harmful pollutants [42]. In nature, most algal cell wall surfaces have a negative charge, which can attract cations [43].

12.4.1.2 The removal of polycyclic aromatic hydrocarbons using algae-derived biosorbents

The removal of PAHs through the use of algae-derived biosorbents has been studied, and the findings suggest that it is species specific and toxicant dependent. A study [44] tested four types of microalgal biomass for their effectiveness in removing PAHs such as *fluoranthene* and *pyrene* as well as their combinations. It was found that *Selenastrum capricornutum* had the highest efficiency for PAH removal compared to the other three options, while *Chlorella vulgaris* had the lowest efficiency rate (48%). A further study of biosorption [45] found that higher shaking speeds and temperatures increased the sorption capacity of *Sargassum hemiphyllum* for phenanthrene, while [46] asserted that hydrophobic substances bound to cell wall contents, such as cellulose and lipids, affected the process itself.

To gain greater insight into the interaction between organic pollutants and biomass, several investigations have been performed to examine the contributions made by different fractions derived from algae, such as proteins and fatty acids released by broken membranes—mainly under light—for example, [47]. Zhang *et al*'s study [48], on the other hand, revealed that after lipids were isolated from green algae, there was a 25% decrease in their adsorption capacity for phenanthrene, benzopyrene, and naphthalene. The results reported in [49] suggested that porous dead cells have a high capacity for adsorbing PAHs because metabolic protection is absent in living cells. It was also found that dead *S. capricornutum* is remarkably proficient at binding pollutants such as *phenanthrene, fluoranthene*, and *pyrene* via an adsorptive process—although [41] also noted that organic pollutants could bind to live cells due to the surface binding sites available in both live and dead algae. A summary of all these investigations can be seen in table 12.1.

12.5 Bacteria and fungi as biosorbents

Bacteria are considered to be environmentally friendly biosorbents for a number of reasons. First, bacteria lack nuclei and their cell walls are composed of macromolecules called peptidoglycans [6]. In general, bacteria can be classified according to the configuration of their cell walls. Bacteria are divided into two primary

Table 12.1. Algae biosorbents for petroleum contaminant removal.

S/N	Oil-degrading microorganisms	Hydrocarbon contaminants	Reference(s)
1.	*Scenedesmus abundans, Spirulina* sp.	Lubricating oil and diesel	[38]
2.	*Encephalitozoon intestinalis*	Spent oil and petroleum	[39]
3.	*Chorella vulgaris*; *Selenastrum capricornutum*	Pyrene; fluoranthene and pyrene	[44, 49]
4.	*Sargassum hemiphyllum*	PAH (phenanthrene)	[45]
5.	*Ulva prolifera*	Benzopyrene, naphthalene, and phenanthrene	[48]

categories: gram-positive and gram-negative. The cell walls of the former possess a thicker peptidoglycan, while those of the latter have a thinner structure [50]. Cell walls contain various acids (e.g. teichoic acid) that are connected to lipids in the cytoplasmic membrane. These lipids are known to create strong bonds inside a membrane [50, 51]. It is believed that the overall negative charge in the biosorption process is caused by the presence of phosphodiester bonds within the acidic monomers. Conversely, gram-negative bacteria have a thin cell wall made up of macromolecules called peptidoglycans. Even though bacteria develop various machineries and structures for decontamination to maintain their existence, some bacteria produce capsule-like coatings on their cell walls that contain charged polysaccharides that help to decontaminate the water supply [50]. Since bacteria have a high surface-to-volume ratio, they are excellent biosorbents for decontaminating polluted soils and waters.

Research regarding bioremediation has largely focused on bacteria; however, fungi are well-regarded for their economic and eco-friendly biosorption capabilities, which result from their characteristics such as easy growth, high biomass yield, and their ability to be chemically and genetically modified [52]. They also feature distinct cell wall properties such as chitin, lipids, polyphosphates, and proteins that make them excellent binders [53]. Their cell walls are composed of polysaccharides and glycoproteins, which contain a variety of binding groups such as hydroxyls, amines, carboxyls, and phosphates. They can easily be industrialized to biosorb oil from large volumes of polluted water by fermentation processes. This means that the biomass—created from inexpensive growth media or fermentation by-products—is less susceptible to changes in nutrient level and other factors such as temperature fluctuations, pH balance, and aeration [54]. Furthermore, the utilization of fungi has been gaining attention due to their production of various enzymes that can decompose a variety of hazardous chemicals [55–57]. Fungi are often seen to have a typical apical growth, which allows them to reach deeper into polluted sediments than other microorganisms can [58], and their metabolic capabilities enable them to assimilate hydrocarbons in challenging environments that have low levels of nutrients and low pH values [59].

12.5.1 Bacteria and fungi-based biosorbents for petroleum pollutant removal

12.5.1.1 Bacteria and fungi-based biosorbents for crude oil removal

The utilization of deceased microbial cells (such as those of fungi and bacteria) as GDBs has been researched due to their frequent occurrence in aquatic environments [60]. In one study [61], it was seen that the adsorption of crude oil on the surface of dead bacterial cells decreased over a certain period of time, likely since some compounds in the oil were adsorbed, such as aromatics and asphaltenes. It was also observed that some components of crude oil were quickly absorbed by dead cells but then later released as some polar organic pollutants underwent reversible adsorption.

Table 12.2. A list of microorganisms used as fungal and bacterial biosorbents.

Oil-degrading microorganisms	Reference
Pseudomonas sp.	
Bacillus sp.	[61]
Ochrobactrum S	
Aspergillus sydowii	
Penicillium citrinum	[67]
Acremonium selerotigenum	
Aspergillus flavu	
Penicillium chrysogenum	
Penicillium polonicum	[68]
Penicillium mononematosum	
Penicillium cyclopium	
Marine fungus (NIOCC #312)	[72]
Phialocephala sp.	
Penicillium sp.	[73]
Cladsporium sp.	

In addition, a variety of genera including *Cladosporium*, *Cunninghamella*, *Fusarium*, *Penicillium*, *Aspergillus*, and *Mucor* have been identified as having the capacity to break down aliphatic and aromatic hydrocarbons [62]. Furthermore, species including *Phanerochaete chrysosporium*, *Agaricus bisporus*, *Trametes versicolor*, and *Pleurotus ostreatus* have been found to be effective for decontamination purposes [63]. However, limited research has been performed on fungal bioremediation potential in aquatic environments, but such studies have suggested that fungal species from various marine ecosystems are capable of degrading petroleum hydrocarbons [64–66] (table 12.2). For example, *Penicillium citrinum* has been identified by [67] as being able to reduce the volume of crude oil by 77% and the volume of n-alkane by 95%. Similarly, in [12, 68], fungal species from *Penicillium*, *Aspergillus*, and *Cladosporium* genera were shown to have the capacity to break down crude oil while generating biosurfactants.

12.5.1.2 The removal of polycyclic aromatic hydrocarbons using bacterial and fungal biosorbents

Research has been undertaken to assess the impact of transportation on the ability of various hazardous organic pollutants to adhere to both living and nonliving microbial samples [6, 69]. An investigation [70] indicated that biomass obtained from deceased fungal agents can effectively remove hazardous pollutants. Further studies of the effectiveness of fungal biomass in eliminating organic pollutants, such as phenolics, have shown that nonliving biomass yields better results [71]. It has been noticed that biosorbents derived from fungi and bacteria can be used to acquire polycyclic aromatic hydrocarbons. A study [61] examined the effect of four different microbial substances, namely *Pseudomonas* sp., *Bacillus* sp., *Ochrobactrum* sp., and *Pseudomonas* sp., on the biosorption of petroleum hydrocarbons (table 12.2). A consortium of microbes that had been heated was observed to have a consistent tendency to adsorb PAHs. Pyrene showed higher hydrophobicity than those of phenanthrene and benzopyrene, resulting in more effective adsorption. The adhesion of naphthalene, phenanthrene, pyrene, and crude oil to microbial biomass was observed in the following order: naphthalene, phenanthrene, pyrene, and crude oil. These bacteria were also seen to take up the substances quickly. A study [72] used NIOCC #312, a marine fungus, as a biosorbent for the removal of PAHs such as phenanthrene from a contaminated aquatic environment. The results showed that phenanthrene had the capacity to quickly adhere to the fungal cell surface, suggesting that heat-treated fungal biomass has a possible application in removing PAHs from sites suffering from pollution.

Extensive research and analysis have also shown that certain isolated strains, such as *Phialocephala* sp., *Penicillium* sp., *Cladosporium* sp. *1, 2, 3, Penicillium brevicompactum*, and *Aureobasidium* sp. can break down hexadecane and 1-hexadecene into simpler carbon sources which they can assimilate for growth purposes [73] (table 12.2). The results of the gene expression analysis indicated that the genes responsible for transmembrane transporters were upregulated. This finding implies that the tested hydrocarbons were incorporated into the cell, which suggests their potential use in bioremediation processes aimed at reducing hydrocarbon pollution [1]. Accordingly, *Mucorracemosus* (CBMAI 847) and *Aspergillussclerotiorum* (CBMAI 849) demonstrated an impressive ability to metabolize pyrene and benzo [a]pyrene in the range between 50% and 90% [74]. Also, three marine basidiomycetes have been identified as having the capability to break down aromatic compounds effectively [75]. In particular, these microorganisms demonstrated an impressive ability to degrade nearly 100% of pyrene over a 48-hour incubation period. The investigation of the intermediate compounds produced during the breakdown of pyrene has provided an insight into the fundamental processes responsible for hydrocarbon decomposition. The data indicate that aromatic compounds may potentially be broken down using mycoremediation through the use of enzymes such as the cytochrome P450 system and epoxide hydrolases.

12.6 Major factors which can impact the process of biosorption

The outcome of biosorption can be influenced by several factors, including the type and nature of the biomass, temperature, ionic strength, acidity or alkalinity, and the presence of free-floating cells [1, 5, 6]. The use of chemical treatments, such as alkalis, has often been found to increase biosorbents' capacity. This is particularly noticeable in certain microbial systems due to the deacetylation of chitin, which leads to the creation of more effective chitosan–glucan complexes with stronger affinity [27]. In order for bioremediation to be successful, comprehensive knowledge of the factors that influence GDBs is essential. Numerous characteristics and parameters have been identified that can impact the biodegradation of PHs, most of which have been extensively considered [1, 6, 76].

12.6.1 Temperature

The temperature of the environment can affect the biodegradation of PHs by altering the chemical components of the pollutants and the diversity of microbial flora. In addition, higher temperatures can result in increased surface activity and kinetic activity, allowing pollutants to be removed more efficiently [77]. However, this can also lead to the breakdown of the physical structure of the biosorbent [78].

Algae have also been found to absorb oil more quickly and easily at higher temperatures due to the reduced thickness of the oil [6]. Moreover, elevated temperatures may also make it easier for adsorbate molecules to move [6]. Despite this, increasing the temperature of the biomass adsorbent could cause a significant increase in the porosity and the number of pores.

12.6.2 The moisture content of the soil

Another important parameter that affects the rate at which petroleum compounds can be biodegraded is soil moisture, and it is also one of the most important parameters to consider. A smaller amount of water in soil pores indicates that a smaller number of microbes are present, which results in a slower rate of biodegradation [79]. Microorganisms can be found residing in the water between soil pores and are involved in the process of biodegradation. Due to the impact of water scarcity on microbial growth and metabolic regulation, the ability of terrestrial ecosystems to break down hydrocarbons may be limited. In contrast, soils have a water potential ranging from 0.0 to 0.99, while the water activity in aquatic environments is stable at around 0.98 [80]. Soil with a water retention capacity in the range of 30%–90% and at a temperature of 20 °C or higher provides ideal conditions for the biodegradation of oil sludge [81]. The results of this research reveal that a 5% (wt/wt) application of oil sludge hydrocarbons at 100 000 litres per hectare of soil is an effective balance between rapid biodegradation and efficient land utilization. A total of all hydrocarbon classes investigated have been shown to have the highest biodegradation rate. The results of a study suggested that most temperate zone disposal sites require two 100 000 l applications of hydrocarbons per growing season or four 50 000 l applications of hydrocarbons per growing season in order to dispose of waste [76]. It was proposed that the absence of inhibition of degradation was due to a reduction in the soil's water-holding capacity caused by the addition of hydrocarbons.

12.6.3 Acidity or alkalinity

It has been shown that biosorption in many systems is pH dependent [6]. When microalgal biomass is present in the soil, pH variations may affect the movement of oil from the soil to the aqueous phase. The acidity of the solution, together with the presence of sorbate ions and the chemical components of adsorbents, can affect multiple elements, such as the activity of functional groups in biosorbents. Variations in the pH level of a biosorbent can affect its activity and ability to sorb. Additionally, hydrogen ions and the sorbate may compete against each other, depending on alterations in the pH scale [82].

Studies have shown that highly alkaline conditions (a pH above 10.0) lead to a decrease in biosorption efficiency [12]. It has been observed that the zeta potentials (or electrokinetic potentials) of oil and algae differ, depending on the pH level; this is an important factor in eliminating oil [12]. The sorption of oil by algae has been attributed to various causes, such as the absorption and distribution of oil in organic material, as well as the adsorption of oil onto the algae's surface due to particular interactions [6].

12.6.4 The chemistry of hydrocarbons

The biodegradability of hydrocarbons depends on their molecular structure and the mass of the hydrocarbon molecule. A variety of microorganisms are capable of degrading and assimilating aliphatic hydrocarbons, and the most rapid degradation is seen for n-alkanes that have lengths between C10 and C24 [76]. Many microorganisms are sensitive to short-chain alkanes of less than C10, and, as these are highly volatile, they tend to dissipate quickly. Generally, long-chain alkanes are fairly resilient to breakdown; however, when they have branches, the rate of decomposition is decreased. Aromatics are commonly oxidized, but only a few types of bacteria can metabolize them.

The decomposition of PAHs is a particularly time-consuming process. Then again, alicyclic compounds can be broken down through co-metabolism. The degree to which hydrocarbons are degraded is influenced by various elements, including their bioavailability (which is largely dependent on their physical form), concentration, solubility, hydrophobicity, and volatilization [76]. It is commonly thought that hydrocarbons vary in their susceptibility to microbial activity and that they can be arranged in order of decreasing resistance from n-alkanes to cyclic alkanes [83]. According to research, it has been shown that saturated aromatics are the most biodegradable materials, followed by light aromatics. It has been found that aromatics and polar compounds of high molecular weight have extremely low degradation rates [76]. However, [84] observed that naphthalene was degraded more than hexadecane, which does not conform to the general pattern.

12.6.5 Oxygen availability

The majority of studies have indicated that the biodegradation of hydrocarbons is an aerobic process, yet anaerobic decomposition has also been documented. This implies that, instead of oxygen being the electron acceptor, molecular oxygen is consumed [76, 85]. In the absence of oxygen, metal ions such as nitrates, iron, bicarbonates, nitrogen dioxide, and so on have been demonstrated to be viable substitutes for electron acceptors. In many cases, oxygen is the main limit on aerobic biodegradation. In contrast to anaerobic conditions, aerobic environments are much more favorable for the breakdown of petroleum hydrocarbons. The addition of oxygen therefore greatly contributes to the remediation of the site. In the catabolism of organic compounds, such as aromatic, aliphatic, and cyclic hydrocarbons, the first step involves the oxidation of substrates by oxygenases which require molecular oxygen to function [76]. It has been noted that the availability of oxygen is the rate-limiting variable in the biodegradation of PHs. The concentration of oxygen in water and soils is influenced by several elements, including the amount of oxygen used by microbes, the type of soil, the presence of substances that can be used for respiration, and whether the soil is muddy, all of which can lead to a reduction in oxygen levels [76].

It has been demonstrated that the degradation of PHs by microorganisms happens at a very low rate, and it has generally been considered to have a minor ecological significance [76]. Despite this, much evidence has been found to suggest that microbes can break down oxidized compounds such as benzoates and halobenzoates under anaerobic conditions. Furthermore, research has indicated that microbes are capable of degrading chlorophenols and polychlorinated biphenyls [85]. Recent studies have revealed that the microbial communities in soil and sludge are able to break down both non-substituted and alkyl-substituted compounds, for instance, benzene, xylene, toluene, acenaphthene, 1,3-dimethylbenzene, and naphthalene, even when there is no oxygen present [76].

12.6.6 The degree of salinity

The presence of salt in water allows microorganisms to degrade hydrocarbons to a considerable extent. An analysis of Hudson River sediment revealed that the mineralization rates and salinity levels of naphthalene were impacted by the surrounding salinity levels [76]. According to data from a research study by Ward and Brock [86], the rate of PH metabolism in salt-enriched ponds decreased (by 3.3%–28.4%) with increasing salinity, a phenomenon the authors attributed to decreasing microbial metabolism in the pond.

12.6.7 The ionic strength

A correlation between sorption and ionic strength was reported by Chung *et al* [45]. However, the results of [38] showed that ionic strength variations were greatly impacted by the kinds of organisms involved. In their study, the researchers found, for instance, that *Spirulina* sp. is capable of absorbing both diesel and lubrication oil. As a result, the concentration of $NaNO_3$ in the background electrolyte was greatest at 0.01 M, diminishing as the concentration was raised, and then increasing considerably at 0.5 M before dropping off again. Also, *S. abundans* steadily absorbed less and less lubricating oil and diesel; however, when the concentration of $NaNO_3$ was raised to 1 M, the absorption levels remained unchanged.

According to [45], brown seaweed was capable of absorbing phenanthrene at maximum temperature and low ionic strength. This study found that the expected trend of oil and diesel uptake based on ionic strength was strongly influenced by the kind of organism.

12.6.8 Bioavailability

The effective biodegradation of pollutants relies heavily on bioavailability, which is the amount of a substance that microorganisms can access through physical and chemical means [76]. The utilization of external nonionic surfactants, such as those present in oil-spill dispersants, significantly impacts the rate at which alkane degradation occurs [87]. The use of surfactants in circumstances involving oil contamination can result in three potential effects on the bacterial degradation of oil components, namely stimulatory, inhibitory, or neutral [88]. The application of chemical dispersants to control spilled oil within marine ecosystems has caused ecological harm, whereas biosurfactants derived from microorganisms are comparatively less toxic and partially biodegradable [76]. Various types of surfactants, including ionic and nonionic varieties available in the commercial market as well as biosurfactants and bacteria capable of producing such substances, have been subjected to investigations of their capacity to augment bioavailability. This assessment was conducted by Van Dyke *et al* [89]. Meanwhile, an experiment involving a land farm revealed that biodegradation rates can be considerably enhanced through the physical disruption of soil aggregates via sonication. Furthermore, [90] found evidence indicating that the constituents of soil have notable effects on the bioavailability levels exhibited by the contaminants therein.

12.7 The benefits of the biosorption process

Using biosorption to eliminate pollutants from the environment has numerous benefits compared to conventional removal techniques. Its versatility, which allows it to be operated under a variety of physicochemical conditions including variations in temperature, pH, and other ions, along with its capacity to rapidly treat large volumes of wastewater, make it an attractive option. Additionally, biosorption utilizes affordable and renewable biomaterials and does not require expensive reagents, which can cause storage issues or difficult disposal. Furthermore, this process exhibits high selectivity for certain contaminants. It necessitates minimal capital expenditure and has comparatively low overhead costs. Finally, it efficiently recovers bound hydrocarbons from biomass while minimizing hazardous waste production [5, 6, 37, 76].

12.8 Conclusions

In this chapter, we investigated how petroleum pollutants adhere to various biomass-derived adsorbents. It turned out that oxygen availability, temperature, bioavailability, pH, ionic strength, and many other factors have significant influences on the adsorption process. Prior research has mainly focused on the biosorption of single compounds rather than a mix of hydrocarbons. Knowledge about the removal of other hydrocarbons and the effects of more complex environmental conditions is yet to be explored. Further investigation is necessary to gain a better understanding of the components in biomass-derived biosorbents and their efficacy in adsorption or removal. Biomass-derived biosorbents offer a low-cost solution for the removal of pollutants. It is expected that more selective adsorbents will be found in future studies, which will also help to identify their underlying biosorption mechanisms.

References

[1] Dell' Anno F, Rastelli E, Sansone C, Brunet C, Ianora A and Dell' Anno A 2021 Bacteria, fungi and microalgae for the bioremediation of marine sediments contaminated by petroleum hydrocarbons in the omics era *Microorganisms* **9** 1695

[2] Ukhurebor K E, Aigbe U O and Onyanche R B *et al* 2021 Developments, utilization and applications of nanobiosensors for environmental sustainability and safety *Bionanomaterials for Environmental and Agricultural Applications* (Bristol: IOP Publishing)

[3] Edema-Sillo T and Emegha J O 2022 Effect of Environmental Degradation on the Inhabitants of Obodo Community in Warri South Local Government Area of Delta State, Nigeria *J. Appl. Sci. Environ. Manage.* **26** 1951–5

[4] Ukhurebor K E, Aigbe U O and Onyanche R B *et al* 2022 An overview of the emergence and challenges of land reclamation: issues and prospect *Appl. Environ. Soil. Sci.* **5889823** 1–14

[5] Shen T, Pi Y, Bao M, Xu N, Li Y and Lu J 2015 Biodegradation of different petroleum hydrocarbons by free and immobilized microbial consortia *Environ. Sci.: Process. Impacts* **17** 2022–33

[6] Kanwal M, Ullah H, Gulzar A, Sadiq T and Ullah M *et al* 2022 Biodegradation of petroleum hydrocarbons and the factors effecting rate of biodegradation *Am. J. Biomed Sci. Res.* **16** AJBSR.MS.ID.002182

[7] Vahabisani A 2020 *Exploring the effects of microalgal biomass on the oil behavior in a sand-water system* Concordia University, Montréal https://spectrum.library.concordia.ca/id/eprint/987883/

[8] Al-Hawash A B, Zhang X and Ma F 2018 Removal and biodegradation of different petroleum hydrocarbons using the filamentous fungus *Aspergillus* sp. RFC-1 *Microbiology* **8** 1–14

[9] Broman D, Ganning B and Lindblad C 1983 Effects of high pressure, hot water shore cleaning after oil spills on shore ecosystems in the Northern Baltic proper *Mar. Environ. Res.* **10** 173–87

[10] Wasko M and Bratt R 1991 Properties of a biosurfactant produced by the fuel contaminant *Ochrobactrum anthropii Int. Biodeterior* **27** 265–73

[11] Urum K, Pekdemir T and Çopur M 2004 Surfactants treatment of crude oil contaminated soils *J. Colloid Interface Sci.* **276** 456–64

[12] Lim M W, Von Lau E and Poh P E 2016 A comprehensive guide of remediation technologies for oil contaminated soil—present works and future directions *Mar. Pollut. Bull.* **109** 14–45

[13] Zhang T and Zhang H 2022 Microbial consortia are needed to degrade soil pollutants *Microorganisms.* **10** 261 PMID: 35208716; PMCID: PMC8874626.

[14] Johnsen A R, Wick L Y and Harms H 2005 Principles of microbial PAH-degradation in soil *Environ. Pollut.* **133** 71–84

[15] Luan T, Keith S, Zhong Y, Zhou H, Lan C and Tam N F 2006 Study of metabolites from the degradation of polycyclic aromatic hydrocarbons (PAHs) by bacterial consortium enriched from mangrove sediments *Chemosphere* **65** 2289–96

[16] Fomina M and Gadd G M 2013 Biosorption: current perspectives on concept, definition and application *Bioresour. Technol.* **160** 3–14

[17] Michalak I, Chojnacka K and Witek-Krowiak A 2013 State of the art for the biosorption process—a review *Appl. Biochem. Biotechnol.* **170** 1389–416
[18] Volesky B 2007 Biosorption and me *Water Res.* **41** 4017–29
[19] Gadd G M 2009 Biosorption: critical review of scientific rationale, environmental importance and significance for pollution treatment. *J. Chem. Technol. Biotechnol.* **84** 13–28
[20] Derco J and Vrana B 2018 *Biosorption* (London: IntechOpen) 78961
[21] Borda M J and Sparks D L 2008 Kinetics and mechanisms of sorption–desorption in soils: a multiscale assessment *Biophysico-Chemical Processes of Heavy Metals and Metalloids in Soil Environments* ed A Violante, P M Huang and G M Gadd (Hoboken, NJ: Wiley) 97–124
[22] Das N and Chandran P 2011 Microbial degradation of petroleum hydrocarbon contaminants: an overview *Biotechnol. Res. Int.* **2011** 941810
[23] Fritsche W and Hofrichter M 2000 Aerobic degradation by microorganisms *Biotechnology: Environmental Processes II* ed H-J Rehm and G Reed (Weinheim: Wiley-VCH) 144–67
[24] Hommel R K 1990 Formation and phylogenetic role of biosurfactants *J. Appl. Microbiol.* **89** 158–19
[25] Al-Masri M S, Amin Y, Al-Akel B and Al-Naama T 2010 Biosorption of cadmium, lead, and uranium by powder of poplar leaves and branches *Appl. Biochem. Biotechnol.* **160** 976–87
[26] Silva A, Delerue-Matos C, Figueiredo S A and Freitas O M 2019 The use of algae and fungi for removal of pharmaceuticals by bioremediation and biosorption processes: a review *Water* **11** 1555
[27] Wang J and Chen C 2006 Biosorption of heavy metals by *Saccharomyces cerevisiae*: a review *Biotechnol. Adv.* **24** 427–51
[28] Ramrakhiani L, Majumder R and Khowala S 2011 Removal of hexavalent chromium by heat inactivated fungal biomass of *Termitomyces clypeatus*: surface characterization and mechanism of biosorption *Chem. Eng. J.* **171** 1060–8
[29] Arief V O, Trilestari K, Sunarso J, Indraswati N and Ismadji S 2008 Recent progress on biosorption of heavy metals from liquids using low cost biosorbents: characterization, biosorption parameters and mechanism studies *Clean (Weinh)* **36** 937–62
[30] Morikawa M, Hirata Y and Imanaka T 2000 A study on the structure–function relationship of lipopeptide biosurfactants *Biochim. Biophys. Acta* **1488** 211–8 PMID: 11082531
[31] Okoh A I 2006 Biodegradation alternative in the cleanup of petroleum hydrocarbon pollutants *Biotechnol. Mol. Biol. Rev.* **1** 38–50 https://academicjournals.org/journal/BMBR/article-abstract/E94E08A40207
[32] Vahabisani A and An C 2021 Use of biomass-derived adsorbents for the removal of petroleum pollutants from water: a mini-review *Environ. Syst. Res.* **10** 1–10
[33] Chojnacka K 2009 *Biosorption and Bioaccumulation in Practice* (London: Nova Science Publishers)
[34] Shamim S 2018 Biosorption of heavy metals *Biosorption* ed D Jan and V Branislav (London: IntechOpen)
[35] Bilal M, Rasheed T, Sosa-Hernández J E, Raza A, Nabeel F and Iqbal H 2018 Biosorption: an interplay between marine algae and potentially toxic elements—a review *Mar Drugs* **16** 65
[36] Chojnacka K 2010 Biosorption and bioaccumulation—the prospects for practical applications *Environ. Int.* **36** 299–307

[37] Chen B, Wang Y and Hu D 2010 Biosorption and biodegradation of polycyclic aromatic hydrocarbons in aqueous solutions by a consortium of white-rot fungi *J. Hazard. Mater.* **179** 845–51
[38] Mishra P K and Mukherji S 2012 Biosorption of diesel and lubricating oil on algal biomass *3 Biotech.* **2** 301–10
[39] Boleydei H, Mirghaffari N and Farhadian O 2018 Comparative study on adsorption of crude oil and spent engine oil from seawater and freshwater using algal biomass *Environ. Sci. Pollut. Res.* **25** 21024–35
[40] Flouty R and Estephane G 2012 Bioaccumulation and biosorption of copper and lead by a unicellular algae *Chlamydomonas reinhardtii* in single and binary metal systems: a comparative study *J. Environ. Manage.* **111** 106–14
[41] Avery S V, Codd G A and Gadd G M 1998 Microalgal removal of organic and inorganic metal species from aqueous solution ed Y S Wong and N F Y Tam *Wastewater Treatment with Algae. Biotechnology Intelligence Unit.* (Berlin: Springer)
[42] Crist R H, Oberholser K, Shank N and Nguyen M 1981 Nature of bonding between metallic ions and algal cell walls *Environ. Sci. Technol.* **15** 1212–7
[43] Marbelia L, Mulier M, Vandamme D, Muylaert K, Szymczyk A and Vankelecom I F 2016 Polyacrylonitrile membranes for microalgae filtration: influence of porosity, surface charge and microalgae species on membrane fouling *Algal Res.* **19** 128–37
[44] Lei A P, Hu Z L, Wong Y S and Tam N F Y 2007 Removal of fluoranthene and pyrene by different microalgal species *Bioresour. Technol.* **98** 273–80
[45] Chung M, Tsui M T, Cheung K, Tam N F and Wong M H 2007 Removal of aqueous phenanthrene by brown seaweed sargassumhemiphyllum: sorption-kinetic and equilibrium studies *Sep. Purif. Technol.* **54** 355–62
[46] Swackhamer D L and Skoglund R 1994 Bioaccumulation of PCBs by algae: kinetics versus equilibrium *Environ. Toxicol. Chem.* **12** 831–8
[47] Widrig D L, Gray K A and McAuliffe K S 1996 Removal of algal-derived organic material by preozonation and coagulation: monitoring changes in organic quality by pyrolysis-GC-MS *Water Res.* **30** 2621–32
[48] Zhang C, Lu J and Wu J 2019 Adsorptive removal of polycyclic aromatic hydrocarbons by detritus of green tide algae deposited in coastal sediment *Sci. Total Environ.* **670** 320–7
[49] Luo L, Wang P, Lin L, Luan T, Ke L and Tam N F Y 2014 Removal and transformation of high molecular weight polycyclic aromatic hydrocarbons in water by live and dead microalgae. *Process Biochem.* **49** 1723–32
[50] Kanamarlapudi S L R K, Chintalpudi V K and Muddada S 2018 Application of biosorption for removal of heavy metals from wastewater. *Biosorption* (London: IntechOpen) 69–116
[51] Paknikar K, Pethkar A and Puranik P 2002 Bioremediation of metalliferous wastes and products using inactivated microbial biomass. *Indian J. Biotechnol.* **3** 426–43 http://nopr.niscpr.res.in/handle/123456789/11335
[52] Mulligan C, Yong R and Gibbs B 2001 Remediation alternative treatment option for heavy metal bearing wastewaters: a review *Bioresour. Technol.* **53** 195–206
[53] Javaid A, Bajwa R, Shafique U and Anwar J 2011 Removal of heavy metals by adsorption on *Pleurotus ostreatus Biomass Bioenergy* **35** 1675–82
[54] Leitão A L 2009 Potential of penicilliumspecies in the bioremediation field *Int. J. Environ. Res. Public Health.* **6** 1393–417

[55] Daccò C, Girometta C, Asemoloye M D, Carpani G, Picco A M and Tosi S 2020 Key fungal degradation patterns, enzymes and their applications for the removal of aliphatic hydrocarbons in polluted soils: a review *Int. Biodeterior. Biodegrad.* **147** 104866
[56] Lei A P, Wong Y S and Tam N F Y 2002 Removal of pyrene by different microalgal species *Water Sci. Technol.* **46** 195–201
[57] Deshmukh R, Khardenavis A A and Purohit H J 2016 Diverse metabolic capacities of fungi for bioremediation *Indian J. Microbiol.* **56** 247–64
[58] Hyde K D, Xu J, Rapior S, Jeewon R, Lumyong S, Niego A G T, Abeywickrama P D, Aluthmuhandiram J V S, Brahamanage R S and Brooks S *et al* 2019 The amazing potential of fungi: 50 ways we can exploit fungi industrially *Fungal Divers* **97** 1–136
[59] Horel A and Schiewer S 2020 Microbial degradation of different hydrocarbon fuels with mycoremediation of volatiles *Microorganisms.* **8** 163
[60] Cheng Z, Su Y, Ye S and Chen D *et al* 2020 Novel biosorbents synthesized from fungal and bacterial biomass and their applications in the adsorption of volatile organic compounds *Bioresour. Technol.* **300** 122705
[61] Xu N, Bao M, Sun P and Li Y 2013 Study on bioadsorption and biodegradation of petroleum hydrocarbons by a microbial consortium *Bioresour. Technol.* **149** 22–30
[62] Amend A, Burgaud G, Cunliffe M, Edgcomb V P, Ettinger C L, Gutiérrez M H, Heitman J, Hom E F Y, Ianiri G and Jones A C *et al* 2019 Fungi in the marine environment: open questions and unsolved problems *mBio* **10** 1–15
[63] Adenipekun C 2012 Uses of mushrooms in bioremediation: a review *Biotechnol. Mol. Biol. Rev.* **7** 62–8 https://academicjournals.org/journal/BMBR/article-abstract/2A84D4E12478
[64] Al-Nasrawi H 2012 Biodegradation of Crude Oil by Fungi Isolated from Gulf of Mexico *J. Bioremed. Biodegrad.* **3** 3
[65] Bik H M, Halanych K M, Sharma J and Thomas W K 2012 Dramatic shifts in benthic microbial eukaryote communities following the deepwater horizon oil spill *PLoS One* **7** e38550
[66] Sanyal O, Shinde V L, Meena R M, Damare S and Shenoy B D 2016 The ITS-based phylogeny of fungi associated with tarballs *Mar. Pollut. Bull.* **113** 277–81
[67] Barnes N M, Khodse V B, Lotlikar N P, Meena R M and Damare S 2018 Bioremediation potential of hydrocarbon-utilizing fungi from select marine niches of India *3 Biotech.* **8** 21
[68] Maamar A, Lucchesi M E, Debaets S, Van Long N N, Quemener M, Coton E, Bouderbala M, Burgaud G and Matallah-Boutiba A 2020 Highlighting the crude oil bioremediation potential of marine fungi isolated from the Port of Oran (Algeria) *Diversity* **12** 196
[69] Kim N, Seo J H, Yun Y S and Park D 2020 New insight into continuous recirculation-process for treating arsenate using bacterial biosorbent *Bioresour. Technol.* **316** 123961
[70] Kumar R, Bishnoi N R and Bishnoi K 2008 Biosorption of chromium (VI) from aqueous solution and electroplating wastewater using fungal biomass. *J. Chem. Eng.* **135** 202–8
[71] Rao J and Viraraghavan T 2002 Biosorption of phenol from an aqueous solution by *Aspergillus niger* biomass *Bioresour. Technol.* **85** 165–71
[72] Raghukumar C, Shailaja M, Parameswaran P and Singh S 2006 Removal of polycyclic aromatic hydrocarbons from aqueous media by the marine fungus NIOCC 312: involvement of lignin-degrading enzymes and exopolysaccharides *Indian J. Mar. Sci.* **35** 373–9 http://nopr.niscpr.res.in/handle/123456789/4169

[73] Velez P, Gasca-Pineda J and Riquelme M 2020 Cultivable fungi from deep-sea oil reserves in the Gulf of Mexico: genetic signatures in response to hydrocarbons *Mar. Environ. Res.* **153** 104816
[74] Passarini M R Z, Rodrigues M V N, da Silva M and Sette L D 2011 Marine-derived filamentous fungi and their potential application for polycyclic aromatic hydrocarbon bioremediation *Mar. Pollut. Bull.* **62** 364–70
[75] Vieira G A L, Magrini M J, Bonugli-Santos R C, Rodrigues M V N and Sette L D 2018 Polycyclic aromatic hydrocarbons degradation by marine-derived basidiomycetes: optimization of the degradation process *Braz. J. Microbiol.* **49** 749–56
[76] Chandra S, Sharma R, Singh K and Sharma A 2013 Application of bioremediation technology in the environment contaminated with petroleum hydrocarbon *Ann. Microbiol.* **63** 417–31
[77] An C and Huang G 2012 Stepwise adsorption of phenanthrene at the fly ash–water interface as affected by solution chemistry: experimental and modeling studies *Environ. Sci. Technol.* **46** 12742–50
[78] Pathak B, Gupta S and Verma R 2018 Biosorption and biodegradation of polycyclic aromatic hydrocarbons (PAHs) by microalgae *Green Adsorbents for Pollutant Removal* (Berlin: Springer) 215–47
[79] Dibble J T and Bartha R 1976 The effect of iron on the biodegradation of petroleum in seawater. *Appl. Environ. Microbiol.* **31** 544–50
[80] Bossert I and Bartha R 1984 The fate of petroleum in soil ecosystems ed R M Atlas *Petroleum Microbiology* (New York: Macmillan) pp 434–76
[81] Atlas R M 1981 Microbial degradation of petroleum hydrocarbons: an environmental perspective *Microbiol. Rev.* **45** 180–209
[82] Puranik P, Modak J and Paknikar K 1999 A comparative study of the mass transfer kinetics of metal biosorption by microbial biomass. *Hydrometallurgy* **52** 189–97
[83] Perry J J 1984 Microbial metabolism of cyclic alkanes ed R M Atlas *Petroleum Microbiology.* (New York: Macmillan) pp 61–98
[84] Cooney J J, Silver S A and Beck E A 1985 Factors influencing hydrocarbon degradation in three freshwater lakes *Microb. Ecol.* **11** 127–37
[85] Chen M, Hong C S, Bush B and Rhee G Y 1988 Anaerobic biodegradation of polychlorinated biphenyls by bacteria from Hudson River sediments *Ecotoxicol. Environ. Saf.* **16** 95–105
[86] Ward D M and Brock T D 1978 Hydrocarbon biodegradation in hypersaline environments *Appl. Environ. Microbiol.* **35** 353–9
[87] Rahman K S M, Rahman T J, Kourkoutas Y, Petsas I, Marchant R and Banat I M 2003 Enhanced bioremediation of n-alkane in petroleum sludge using bacterial consortium amended with rhamnolipid and micronutrients *Bioresour. Technol.* **90** 150–68
[88] Liu Z, Jacobson A M and Luthy R G 1995 Biodegradation of naphthalene in aqueous nonionic surfactant system. *Appl. Environ. Microbiol* **61** 145–51
[89] Van Dyke M I, Lee H and Trevors J T 1991 Applications of microbial surfactants *Biotechnol. Adv* **9** 241–52
[90] Weissenfel W D, Klewer H J and Langhoff J 1992 Adsorption of polycyclic aromatic hydrocarbons (PAH) by soil particles: influence on biodegradability and biotoxicity. *Appl. Microbiol. Biotechnol.* **36** 689–96

Chapter 13

The use of the response surface methodology for the adsorption of heavy metals and dyes by biosorbents

Juliana O Iji, Ayodeji O Oyedeji, Salami A Kazeem, Maryann I Uzochukwu, Cynthia U Odili, Abdullahi S Argungu, Fatima B Badiru, Stefan Csaki and David O Obada

Globally, there is an increasing search for alternative options to reduce the challenges of the discharge of untreated wastewater from various industrial activities such as tanneries. Large amounts of wastewater are continuously produced through various human activities, from which water can still be recovered and reused. This chapter examines the different pollutants in tannery wastewater and the possible ways of treating these contaminants and reviews the application of the response surface methodology approach for experiments in wastewater treatment. It also presents different techniques and raw materials used in producing porous ceramics, which are viable and cost-effective materials for wastewater treatment.

13.1 Introduction

Wastewater (WW) is often defined as used water that has been affected by commercial, domestic, and industrial activities. WW includes tannery, textile, and refinery effluents, storm runoff, and toxic road runoff. Recent developments in the tannery industry are one reason why the industry is a major source of industrial pollution. For instance, based on statistics, over 300 million cowhide pieces enter the global leather market yearly, which often results in the generation of about 300 million tons of WW [1]. In Kano, Nigeria, 39 tanneries have released their liquid and solid wastes into the environment [2]. Therefore, the discharge of tannery WW into water bodies during leather production has become a pressing issue that restricts the growth of the tannery industry. However, in some developed countries, strict guidelines have been imposed on the release of tannery effluents. For example, the USA and several countries in Europe have created standard procedures for

doi:10.1088/978-0-7503-5598-8ch13

tannery WW treatment. These countries defined the required standards in terms of the chemical oxygen demand (COD), pH value, biochemical oxygen demand (BOD), sulfides, oil, and chromium (Cr) in the WW bore released into the water body. Further, few countries have stated nitrogen limits for ammonia, which is also a toxic product [3].

Tannery WW is made by the tanning processes used in the production of leather. In some industries, tannery WW is collected separately and then mixed with WW from other sources. The total collected WW is subjected to treatment. The concentrations of BOD, COD, sulfides, suspended solids, and ammonia in tannery WW are generally very high [4]. The tanning process is complex and uses multiple substances and many chemicals. Tannery WW is highly alkaline and saline and contains suspended solids and organic matter. It has high oxygen consumption and a deep color because of the organic matter present.

Based on the treatment degree, tannery WW treatments can be classified as primary, secondary, or tertiary WW treatments. Primary treatment involves general chemical reactions and physical interaction that is completed in the pretreatment stage. The process involves the effective removal of suspended solids in tannery WW through the reduction of the organic load in the treatment plant. The treatment process includes screening and grille processes. In addition, these processes also remove thick materials such as scurf and undesirable substances such as floating organics, sediment, and coagulation products [1]. Secondary treatment for tannery WW is a biological process that involves aerobic and anaerobic secondary treatments. The aerobic treatment includes biofilm, natural biological, and activated sludge methods. Tertiary treatment is a deep and advanced treatment process. A single secondary treatment often fails to remove phosphorus and nitrogen completely. Hence, a combined treatment process is required. The focus of tertiary treatment is to confiscate pollutants that secondary treatment cannot remove completely.

In environmental engineering, optimizing the performance of WW treatment systems is crucial. However, the traditional method of optimization through multiple variables can be inflexible, unreliable, and time-consuming. This is the reason that an alternative, more effective, and adaptable method is preferred. Response surface methodology (RSM) is one such technique that is widely used for system performance optimization. This mathematical and statistical tool can help analyze the effects of various independent factors on a treatment process in order to obtain maximum output. RSM has been efficaciously applied to optimize WW treatment procedures for various industries, including the textile, tannery, paint, olive oil, and palm oil industries. It has also been used for landfill leachate treatment. This chapter aims to provide an assessment of the usability, effectiveness, and limitations of RSM for modeling and optimizing process parameters in WW treatment. By gathering information about the interactions among parameters that can optimize the treatment of WW, RSM can optimize parameters during the process of treating WW.

The process of filtration or separation involves passing a liquid or a gas (i.e. a fluid) through a porous substance to collect particulate matter from suspended solids as well as suspended substances, such as fine particles, microbes, polymer molecules, mist, and aerosols. Such porous substances are known as 'filters,' and ceramic materials have been extensively used to produce these filters. This chapter highlights a few of the different techniques and raw materials used in producing porous ceramics.

Hence, this chapter presents the different pollutants found in tannery WW and the use of various techniques to enhance the effective removal of tannery WW contaminants and heavy metals (HMs). RSM is discussed as a widely used method for system performance optimization, particularly in WW treatment. This chapter also discusses the use of porous ceramics for water filtration applications. This includes their processing techniques, the materials used to fabricate porous ceramics, and industrial applications of WW treatment.

13.2 Tannery wastewater characteristics

Leather production processes are classified into four parts: beam house, tanning, post-tanning, and the finishing stage. Chemicals (Cr salts, sulfates, biocides, sulfides, etc) are used in large quantities during the tanning processes, hence the large quantity of WW generated is characterized by serious physicochemical conditions, i.e. dark brownish WW with a high concentration of Cr, total dissolved solids (TDS), COD, BOD, total nitrogen (TN), total phosphorus (TP), HMs, strong odor, and high pH [5]. The treatment of tannery WW for domestic consumption imposes a series of restrictions into the treatment train which vary according to the supplier's requirements, the source's type, and vulnerability. Coagulation, flocculation, filtration, and oxidation systems are all part of the treatment train, while chlorine is the most common oxidative disinfectant [6]. International recommendations and national standards are used to assess the safety of treated water, and the World Health Organization (WHO) Guidelines for drinking-water quality are the most important of all [7]. The various contaminants in WW are presented in the following subsections. The preceding subsection presented some of the pollutants in tannery WW.

13.2.1 Total nitrogen

The eutrophication of water is caused by excessive nitrogen. During the tanning process, the TN is introduced from two sources. The methods used to treat the TN include chemical nitrogen removal (i.e. ion exchange, catalytic oxidation, electrolysis, and ozone oxidation), biological nitrogen removal (i.e. the anaerobic-anoxic-oxic (A_2/O), anaerobic oxic (A/O), and sequencing batch reactor (SBR) methods), and physicochemical nitrogen removal (i.e. air and steam stripping and chemical precipitation). Flocculent precipitation has also been effectively used in nitrogen removal when biological flocculation was used in the treatment of the TN in tannery WW [8].

13.2.2 Total phosphorus

The total phosphorus level describes the eutrophication of water. Its value is a significant parameter used to define water that is contaminated or polluted. Phosphorus is not present in the raw fur and raw hides during fur processing and tanning. However, phosphorus-containing compounds such as lecithin fatigue might be used during processing. An appropriate amount of phosphorus is sometimes added to balance the nitrogen in the WW [3]. Generally, there are three kinds of phosphorus found in tannery WW, namely: (i) polyphosphate, (ii) organic phosphorus, and (iii) orthophosphate. Membrane technology (MT) and chemical precipitation are the main techniques used in treating phosphorus-containing WW.

13.2.3 Chloride ions

The newly established and existing standards for chloride ion concentration in tannery effluents are set at around 3000 mg l^{-1}, while 4000 mg l^{-1} is permitted for fur processing. Regardless of improvements in technology, it is still difficult to achieve chloride-free tanneries. From a cleaner production point of view, several technologies are used to reduce chloride, from processing fresh hide directly, using salt-free pickling, or the cyclic utilization of WW. However, these methods might not guarantee the complete removal of chlorides. Hence, conventional technologies such as reverse osmosis (RO), ion exchange, and electroosmosis are commonly used [3].

13.2.4 Chromium salt

About 8% of tannery WW consists of chrome tanning WW. A high concentration of Cr, namely 3000–6000 mg l^{-1} [1], occurs when Cr combines with protein during cell metamorphosis. Following treatment, its flocculability is completely destroyed in the sludge, which harms the treatment process. A combination of biochemical processes and physical chemistry (ion exchange, electrolysis, and adsorption) is generally used to treat Cr-containing WW. After the physical process is complete, the treated WW is further treated in a biochemical treatment system [9].

13.2.5 Sulfur-containing compounds

Sulfates and sulfides are the main sulfur-containing compounds in WW produced by the leather industry. The sulfides are in the form of hydrogen sulfides, which often have a bad smell and toxicity when compared with cyanates. A high sulfate concentration might also interfere with anaerobic treatments. Methods such as flotation, flocculation, sedimentation, and adsorption have been used for sulfide treatment [10].

13.2.6 Heavy metals

The effluent from leather production contains HMs (i.e. manganese (Mn), iron (Fe), zinc (Zn), copper (Cu), and cobalt (Co)). Elements with atomic weights ranging between 63.50 and 200.60 and densities of more than 5.0 g m^{-3} are called HMs [11]. HMs play significant roles in biochemical/organic processes in the human body, and excess intake has negative consequences for humans. Other HMs, such as cadmium (Cd), Cr, mercury (Hg), arsenic (As), and lead (Pb), are dangerous even at low concentrations (parts per billion), since the accumulation of these elements over time harms humans [12]. Although the human body can tolerate certain HM levels without experiencing any serious health issues, excess exposure leads to physiological system failure and eventually death. Table 13.1 presents the sources and effects of HMs on humans as adapted from [7, 13].

Table 13.1. Sources of HMs and their health effects on humans.

HM	Sources of the HM	The effects of the HM on humans	Maximum contaminant level (MCL)
Pb	Storage batteries, piping, tank exterior coatings, plating, gasoline, Pb processing industries, insecticides, and medicinal flavorings	Damages the kidney and liver and reduces hemoglobin function. This leads to infertility and abnormalities in pregnancy and central nervous system disorders which might lead to death.	US EPA: 0.015 mg l^{-1} WHO standard: 0.01 mg l^{-1}
Cd	Electroplating, paint pigments, phosphate fertilizers, plastics, smelting, silver Cd batteries, mining, and alloy industries.	Irritation in the respiratory tract. Prolonged exposure cause anemia and a yellowish strain that appears between the teeth.	US EPA: 0.005 mg l^{-1} WHO standard: 0.003 mg l^{-1}
Cu	An environmental contaminant; its causes are mining, printed circuits, and fiber production. Agricultural sources: fertilizers, animal waste, and fungicidal sprays.	It causes jaundice and enlarges livers. Increases the occurrence of lung cancer	US EPA: 1.3 mg l^{-1} WHO standard: 2.0 mg l^{-1}
Zn	Zn salts are used for wood preservatives, photographic papers, ceramics, textile catalysts, batteries, and steel production.	Excessive exposure to Zn causes gastrointestinal distress and diarrhea.	US EPA: 5 mg l^{-1} WHO standard: 5.0 mg l^{-1}
Nickel (Ni)	Sources are washing electroplating baths used in the storage battery industry.	It causes cancer of the nose, bone, and lungs. Acute Ni (II) poisoning causes dizziness, headache, shortness of breath, and dry cough.	US EPA: 0.1 mg l^{-1} WHO standard: 0.05 mg l^{-1}

(*Continued*)

Table 13.1. (*Continued*)

HM	Sources of the HM	The effects of the HM on humans	Maximum contaminant level (MCL)
As	Sources include industrial deposits, leaching from pesticides, natural deposits, coal power plants, wood preservatives, and semiconductor manufacture.	Serious skin, lung, and kidney challenges.	US EPA: 0.010 mg l^{-1} (or ppm) WHO standard: 0.010 mg l^{1}

WHO: World Health Organization, US EPA: US Environmental Protection Agency

13.3 RSM techniques for wastewater treatment

The effectiveness of a treatment process relies on different processing factors, and optimizing these processes results in reduced costs and reduced consumption of resources such as materials and energy. One-factor-at-a-time is the traditional approach for optimizing processes. This studies the effect of one variable while the other variables are kept constant. However, this technique does not reveal the interactions among the variables, and the number of experimental runs required is too large, making this process costly and time-consuming [14]. To overcome these challenges, the design of experiment (DOE) method is used since it requires fewer experimental runs and the interaction between factors can be represented graphically. The DOE methods used with the RSM includes full factorial design (FFD), Box–Behnken design (BBD), and central composite design (CCD) [15]. The RSM is a collection of mathematical and statistical methods for the DOE framework whose aim is to examine the effects of the different factors using a few experimental runs. The RSM is primarily used to improve the yields obtained by chemical and processing companies [16]. However, the RSM is now used widely in other fields, such as aerospace, agricultural settings, the life sciences, automotive applications, and many others [14]. Recently, the RSM has become an important tool for the analysis and optimization of different WW treatment procedures, such as adsorption, disinfection, coagulation–flocculation, electrochemical processes, and advanced oxidation processes. The utilization of different computer software (i.e. Minitab, Design Expert) makes it simpler to apply the RSM.

13.3.1 Procedures for RSM

The mathematical model used for the RSM is obtained using the data from the DOE, the relationships between the process factors (which are the independent variables), and the response (which is composed of the dependent variables). The model thus obtained is then employed for the examination of the effect of the factors, the interaction between the factors and the variables, and the optimization of the process. The results or findings are presented as three-dimensional (3D) and two-dimensional (2D) graphs/plots. The RSM necessitates the utilization of optimization methods, statistical/mathematical experimental design, and linear regression modeling. There are several steps in the application of the RSM, and these include the identification of the factors and the selection of their ranges, experimental design selection, experimental procedures used to obtain the response parameters, results analysis, the generation of a regression model for the experimental results, determination of the model's competence, graphical representation of the model, and obtaining the optimum values [14].

13.3.1.1 Independent variable selection

Several factors or independent variables affect the response of a given system. However, it is not possible to include all factors in the design due to the high cost of materials production. Hence, the factors are first screened to determine which factors make a significant contribution, and this is done by carrying out experiments using the 'one-factor-at-a-time technique' or 'two-level factorial design.' Using this approach, the proper factor range can be selected, since this affects the overall precision of the developed model. The prior information obtained by the study also helps in selecting narrower ranges for the independent variables.

13.3.1.2 Experimental design

Numerous designs exist that can be used to conduct any study. These designs differ from one another in the ranges of factors used and the number of experimental runs. The main widely used DOEs include: (i) the CCD, (ii) the FFD, and (iii) the BBD. The development of each design model is done through a mathematical matrix. The CCD consists of a two-level (2^k) factorial point combined with a central point (C_0), '$2k$' axial points, and experimental points at a distance α from the central point. The experimental runs are calculated using equation 13.1, and the quadratic equation is estimated from '$2k$' and '2^k,' and the central point is used to estimate the error.

$$N = 2k + 2^k + C_0 \tag{13.1}$$

The BBD consists of a two-level factorial design together with a design block that is not completely fitted to a second order coded as (+1, 0, and −1). The BBD is an important technique and serves as an effective alternative to the CCD when the factors (k) equal three. The experimental runs of the BBD can be calculated using equation (13.2).

$$N = 2k(k - 1) + C_0 \tag{13.2}$$

In the FFD, three levels represent the minimum (−1), the mean (0), and the maximum (+1). The experimental run is defined by n^k, where n is the number of levels and k is the number of factors. After choosing any of the designs for the experiment, the next step is to carry out the adsorption or filtration test; the final removal efficiency is then recorded and the mathematical model is generated [14].

13.3.1.3 Optimization of the response surface

Response surface analysis is conducted to obtain the optimum conditions and minimum or maximum responses. In response surface techniques, optimization involves coefficient estimation and response prediction using a mathematical model, as shown in equation 13.3

$$Y = f(X_1, X_2, \ldots, X_n) \pm e, \tag{13.3}$$

where Y is the response, f is the function of the response, X_i is an independent variable (i= 1,2,..., n), and e is experimental error.

The function f may take a very complicated form that results in a higher-order polynomial (i.e. a quadratic model, as shown in equation (13.4)), and this depends on the relationship that exists between the independent variables, the response, and the interaction between factors. When the responses are considered, the optimum conditions can be graphically attained by overlaying contour plots of the RS regression models; all the input parameters then simultaneously meet the required conditions.

$$Y_m = b_o + \sum_{(i=1)}^{k} b_i X_i + \sum_{(i=1)}^{k} b_{ii} X_i^2 + \sum_{i}^{(i>j)} \sum_{j} b_{ij} X_i X_j \tag{13.4}$$

where Y_m is the response variable, X_i and X_j are independent variables, and b_o, b_i, b_{ii}, and b_{ij} are offset terms.

The contour plot shows the feasible area of the responses in the factor axis (X_1 and X_2) and the regions that best fit the optimization criteria. However, it becomes more difficult to obtain the simultaneous response conditions when more than three factors are considered; in this case, the multicriteria methodology can be used. In the multicriteria methodology, the desirability function is used and each one of the responses is converted into a desirability function within the range from zero to one. When a response is within the acceptable region, one is assigned as the desirability function, while the response outside the region is assigned to zero [69].

13.3.2 Applications of RSM in wastewater treatment

Recently, several research studies have described the utilization of the RSM for optimizing and analyzing different procedures used in WW treatment. Some of these studies are presented in the following subsections.

13.3.2.1 Coagulation–flocculation

Flocculation and coagulation are used to remove contaminants and HMs. Coagulation is a chemical process used in the removal of solid particles from WW. Coagulants such aluminum and ferric chloride are introduced into the WW to destabilize the forces keeping the solid particles, oily materials, or colloids together. In the case of flocculation, the process involved is the agglomeration of destabilized particles into microflows. The removal of coagulants is affected by the pH, the concentration of the co-coagulants, the coagulant dose, and the initial turbidity.

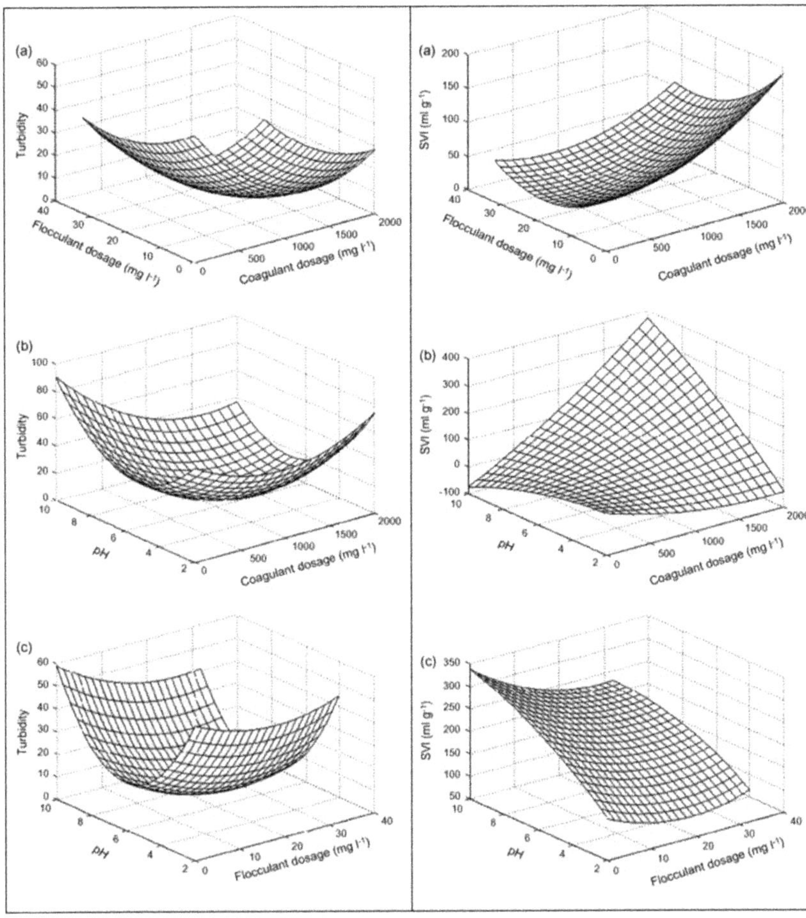

Figure 13.1. Response surface plots of turbidity and sludge volume index against the three input factors.

Using RSM for process optimization is advantageous, since obtaining the optimal conditions is easier, as reported by several studies. Wang *et al* [17] used a coagulation–flocculation process to treat WW, and the processes were optimized using the RSM. The independent variables were the coagulant dosage, the flocculate dosage, and the pH, while the responses were the turbidity (in nephelometric turbidity units, NTU) and the sludge volume index (ml g^{-1}). The resulting contour plot for the responses is shown in figure 13.1 as adapted from [17]. The plot shows the effects of two factors on a response while the third factor is kept constant. Moghaddam *et al* [18] applied the RSM to optimize the initial concentration, coagulant dosage, and initial pH in the removal of dye from an aqueous solution, in which they used the RSM technique to determine which of the factors affected the effectiveness of the removal coagulation/flocculation procedure. The result shows that the three factors significantly affect the efficiency of the process. For instance, an increase in the coagulant dosage increased the removal efficiency, and the maximum removal rate was attained using the maximum dose of the coagulant. Liu *et al* [19] also optimized the coagulation/flocculation performance using the RSM; the response parameters were traditional indexes such as turbidity removal, color, and the COD. Mensah-Akutteh *et al* [20] used the RSM to optimize coagulation–flocculation; the parameters studied included the alkalinity, pH, and alum dose. The optimal condition for the responses was presented and the mathematical model obtained from the study was suitable for predicting responses that were not included in the experiment. Iber *et al* [21] utilized the Box–Behnken approach in the RSM for the coagulation–flocculation of aquaculture WW treatment to determine the optimal operational values of the pH, chitosan dosage, and settling time before establishing and authenticating quadratic models.

13.3.2.2 Adsorption

Adsorption is a widely known technique for the treatment of HM effluents due to its low cost of production and ease of use. In terms of design and operation, the adsorption process is adaptable, and it is highly effective in the treatment of WW in a variety of situations. Because adsorption is a reversible technique, adsorbents can be reused using a suitable desorption process. Examples of materials used for adsorption include activated carbon (AC), carbon nanotubes (CNTs), low-cost adsorbents, and bioadsorbents. HMs are mainly removed from WW using AC adsorbents due to their huge mesopore and micropore contents and their high adsorbent surface areas. Several researchers have investigated the applicability of AC absorbents for HM removal [22, 23]. CNTs have great potential for removing HM ions such as Cu [24], and Ni [25] from WW. Although AC is widely used, it is very expensive. Therefore, the search for accessible and low-cost adsorbents has become a major research priority for HM removal. Several studies have been published on the topic of low-cost adsorbents over time, and adsorbents have been made from natural substances, waste products from industrial activities, and agricultural wastes. The use of low-cost biosorbents and their high efficacy in confiscating HM ions are due to the advantages of biosorption. Biosorption systems are particularly well suited to dilute HM effluents. The factors that play a significant role in the adsorption procedure are the adsorbent/adsorbate ratio, the ionic strength, the pH, and the nature of the adsorbent surface [26]. The 'RSM has been used in the optimization of the process parameters' according to Yang *et al* [27]. Their research considered the effect of various parameters, such as the coexistence time, the pH, the initial concentrations of fluoride and humic acid, and the adsorption time [27]. The results of the RSM showed which of the parameters had the largest influence on the removal of fluoride [27]. Nguyen *et al* [28] employed the RSM approach to optimize the adsorption capacity of biochar derived from spent coffee grounds; they considered parameters including the ions present, the norfloxacin level, and the pH. Mourabet *et al* [29] examined the confiscation of fluoride from an aqueous solution utilizing the BBD in the RSM. The independent variables considered had three levels, and these included the pH, adsorbent dose, temperature, and initial concentration. The optimal values of an initial concentration of 49.06 mg l^{-1}, a pH of 5.36, and an adsorbent dose of 0.15 g yielded the optimum removal efficiency of the fluoride, which was 90.83%. Iqal *et al* [30] investigated Cd removal from WW and an aqueous solution. The impact of factors such

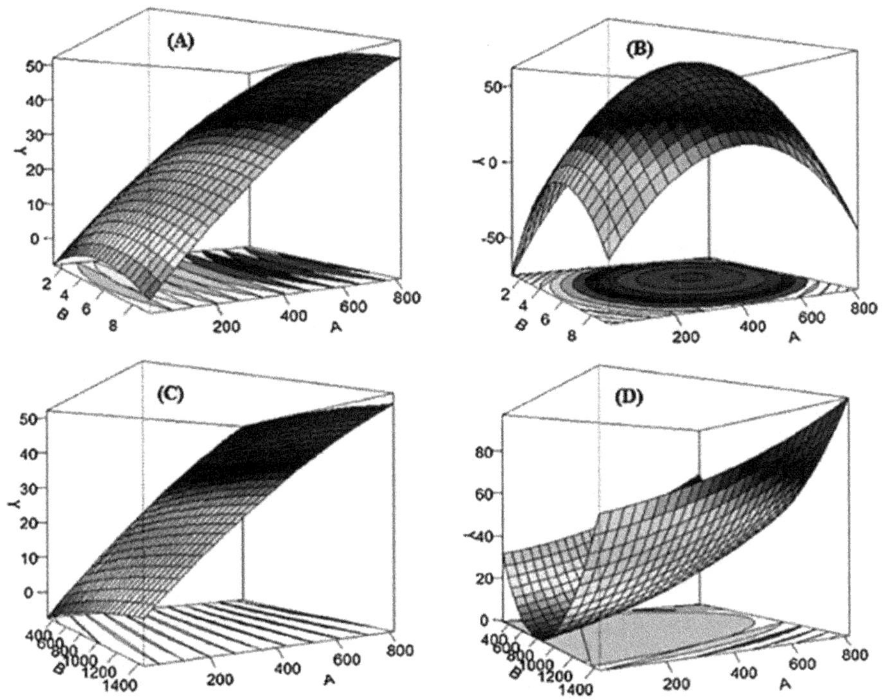

Figure 13.2. (A) The effect of the adsorbent dose and initial ion concentration on the removal of Cd using type I material; (B) the effect of the adsorbent dose and initial ion concentration on the removal of Cd by type II material; (C) the effect of the contact time and pH on the removal of Cd by type I material; (D) the effect of the contact time and pH on the removal of Cd by type II material. (Captions for figures 13.2(A) and (B), A: initial concentration, B: Absorbent dose. Captions for figures 13.2(C) and (D), A: pH, B: contact time. Y is the Cd adsorption rate (mg g^{-1}).)

as the metal ion concentration, adsorbent dosage, pH, and contact time were modeled using the RSM. The effect of each independent variable in terms of its effect on the removal efficiency was studied using contour plots. For example, the effect of 0.1–2.1 (g) of adsorbent dose in combination with the contact time, pH, and initial concentration on the adsorbent is illustrated in figure 13.2 as adapted from [30].

13.3.2.3 Membrane technology

Due to their high effectiveness and ease of use, membrane filtering methods based on various membrane types have the potential to remove HMs. Membrane technologies such as RO, ultrafiltration (UF), electrodialysis, and nanofiltration (NF) are used for HM removal. UF membranes are mainly used to remove colloidal and dissolved particles and operate at low transmembrane pressures. UF membranes have pores with larger diameters, and dissolved HM ions in the form of hydrated ions might easily pass through them. RO is a water treatment (purification) technique that employs a semipermeable membrane to filter out larger particles

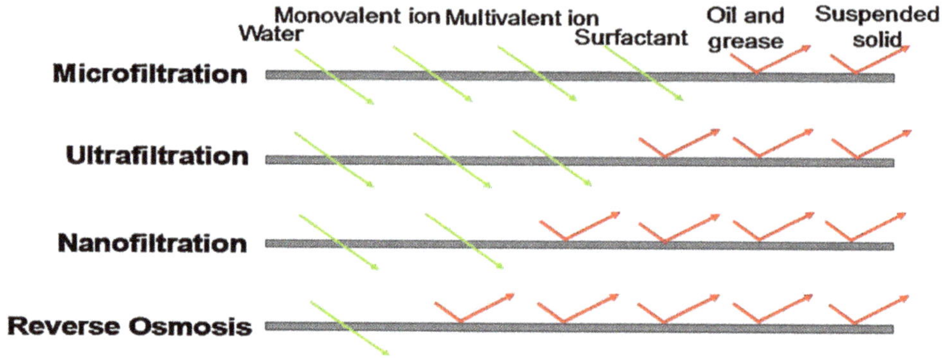

Figure 13.3. The rejection capabilities of membranes.

and pollutants such as salt, soil, and chlorine from drinking (domestic) water. NF is an intermediate between RO and UF. Ni [31], As [32], Cu [33], and Cr [34] are HM ions that can be removed from WW via NF. The NF process has the advantages of reliability, low energy consumption, and ease of operation as well as excellent pollution removal efficiency. Figure 13.3 depicts the rejection capabilities of various membranes as adapted from [35].

Recently, Waqas *et al* [36] optimized biological and filtration efficiency using the RSM; the sludge retention time, disk rotation, and hydraulic retention time were the input parameters considered. The WW filtration was conducted using MT. This research study demonstrated the effectiveness of the RSM in analyzing different filtration parameters, i.e. the COD, turbidity, ammonia removal rate, and MT. Wang *et al* [37] optimized the fabrication of layer-by-layer membranes. In their study, the relationship between the factors was determined by the BBD in the RSM. Maher *et al* [38] studied 'the elimination of HM from drinking water' using a NF MT; their procedure was optimized utilizing the RSM. The optimization was performed to determine the HM rejection rate of the membrane. The optimum nickel rejection rate was 93%, and 86% was achieved for Pb ions. Goyal *et al* [39] optimized 'the process parameters of an emulsion ionic liquid membrane via RSM and experimental work which was designed using BBD.' The factors considered included the treatment ratio, internal phase concentration, pH, internal-to-organic phase ratio, and the homogenization speed, while the response was the Cr removal rate. The interaction between the homogenization speed and the internal phase concentration is presented in the response surface plot in figure 13.4 as adapted from [39]. The plot shows that an optimum Cr removal rate of 92% can be obtained using homogenization speeds and internal phase concentrations from 0.5 to −0.5.

The future of tannery WW treatment should focus on multifunction treatments, intensification, and recycling. The treatment of tannery WW can be classified into the following stages: pretreatment, major treatment, and advanced treatment. The first stage (pretreatment) is designed to remove sand and large matter from the WW, while the main treatment stage should possess an exceptional capability to detoxify

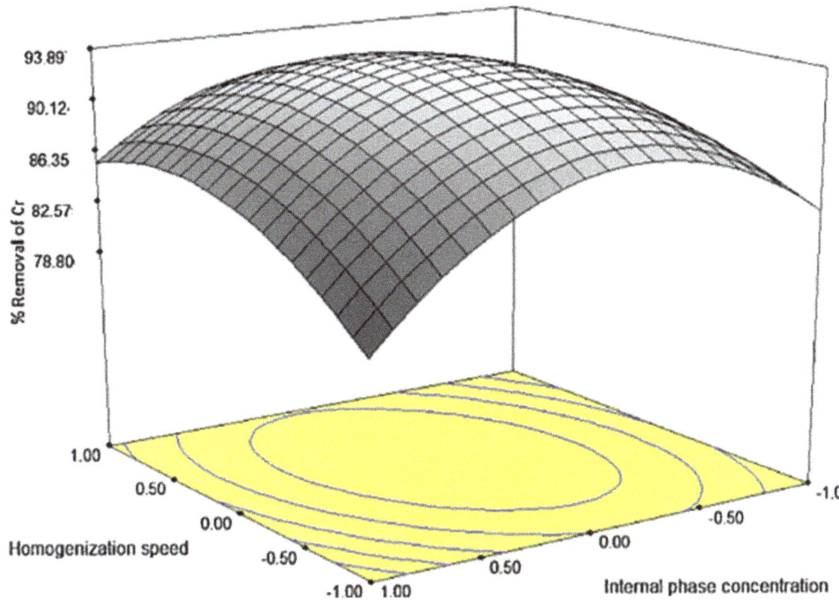

Figure 13.4. A response surface plot showing the dependence of the Cr removal rate on the internal phase concentration and the homogenization speed.

or break down the pollutants. The advanced treatment stage is used for sterilization, disinfection, and deep mineralization. Hence, a single treatment process might not be effective in purifying WW. For instance, electrocoagulation can be used simultaneously with electrooxidation to precipitate inorganic pollutants and degrade organic pollutants [40].

13.4 The application of porous ceramics in tannery WW remediation

Ceramic membranes are utilized in tannery WW treatment owing to their effectiveness, low cost, and good mechanical strength [41]. Ceramic membranes are divided into three types based on pore size (figure 13.5 as adapted from [42]): microporous ($d < 2.0$ nm), macroporous ($d > 50.0$ nm), and mesoporous (2.0 < nm $d < 50.0$ nm) [42]. Porous ceramics are classified as closed cell or open cell (or reticulated) depending on their basic pore structure, and this property is critical in determining the functions of diverse materials [43]. The pore structure parameters such as pore shape, size, distribution, and connection all play a role in porous ceramic materials' overall functionality. An open-cell shape with interconnected pores is perfect for use in fluid (liquid or gas) transfer applications due to its liquid permeability.

13.4.1 Processing techniques for porous ceramics

13.4.1.1 Partial sintering

Partial sintering is one of the commonest and broadly utilized techniques for fabricating porous ceramics. This technique requires the use of low temperatures

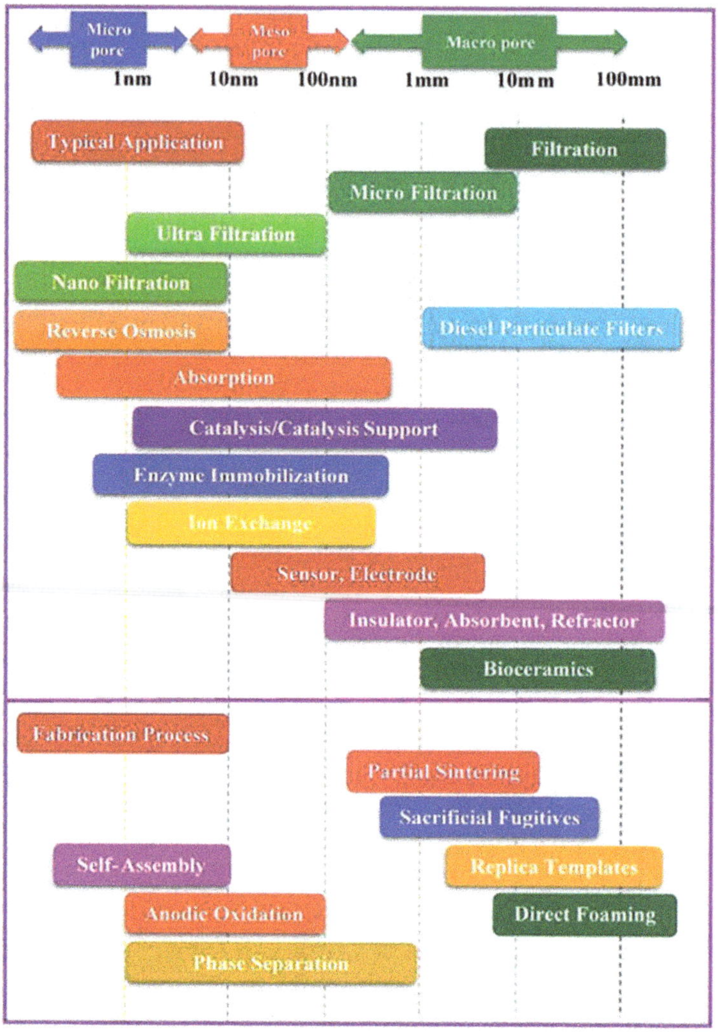

Figure 13.5. Representative fabrication processes of macroporous ceramics.

and reduced processing times on the compacted green body in order to generate its dense counterpart [44–46]. The size and volume of the starting powders as well as the sintering settings determine the microstructure of porous ceramics. The density of the green body and the usage of additives have a big impact on the resulting material's pore size and porosity [42–46]. Meanwhile, the technique's exceptional merit is thought to be due to the interconnected open porosity, extremely narrow size distribution, and ease of dimensional control of the produced structures. Porous ceramics have a low effective thermal conductivity due to these microstructural characteristics when compared to unsintered powder compacts. As a result, their conductivity drops rapidly, although it may still be customized across a wide range. As a result, partial sintering is regarded as a powerful approach for controlling the final product's thermal properties [47]. Advanced sintering processes, such as spark plasma sintering (SPS) and partial hot pressing, have been developed to improve grain bonding, which in turn improves the mechanical strength, thus satisfying the ever-increasing demands.

Partial sintering occurs when the compacted powder is bonded together by surface diffusion, full densification, and recrystallization; the use of low sintering temperatures results in structures with pores produced by the coarse powder network [48]. Partial sintering techniques result in a porous structure with three interconnected pore stages: (i) pores between agglomerates, whose sizes are less than 10.0 µm; (ii) pores inside agglomerates, whose sizes lie between 100.0 and 1000.0 nm; and (iii) pores within particles, whose sizes are 100.0 nm [49]. Overall, the porosity, pore size, and shielding effects of aligned fibrous grains all influence porous ceramics' mechanical strength [50].

13.4.1.2 Replica templates

Porous ceramics with interesting properties such as high surface areas and linked pores can be manufactured through the replica method [31]. In this method, a polymeric foam is impregnated into the ceramic slurry to form a template, which is then followed by heat treatment. The organic polymeric constituents disintegrate during the heating cycle, but the ceramic remains [52]. The porous ceramics fabricated using this process have the same structure and shape as the template used in their creation [53]. The replica process allows for precise control of the final morphological and microstructural properties. Solid polymeric sponges (i.e. polyurethane) and natural sponges (i.e. cellulose, marine sponge, and wood) can be utilized as templates to produce porous ceramics with open porosities of more than 70%. Furthermore, the choice of slurry and medium template has a substantial impact on the physical and mechanical characteristics of the finished porous ceramics [54].

13.4.1.3 Sacrificial templates

The sacrificial template approach differs from the replica method in that the latter copies the original template configuration, whilst the former creates a negative reproduction of the original pore features. As a result, the sacrificial template approach can create pores that are either open cell or closed cell. The sacrificial template procedure is used to produce porous ceramics with suitable mechanical strength and porosity by homogeneously inserting a sufficient amount of sacrificial porogen, which is then eliminated via thermal breakdown. The sacrificial porogen burns off, evaporates, or etches due to the thermal treatment. The raw materials used include a powder mixture (combination) of two constituents, a two-phase suspension, a preceramic polymer, or a ceramic slurry [55].

At high temperatures, the parent materials must not react with the sacrificial constituents. The commonly used sacrificial porogens include synthetic organics [56] and natural sources [57]. In comparison to their liquid equivalents, the first three sources are classified as non-liquid templates. The morphology of porous ceramics can be altered by using the right templates. The percentage, kind, and dimensional features of the sacrificial template play a critical role in establishing the shape, size, and structure of the pores in this process, eventually dictating the qualities of the resultant materials [58, 59]

13.4.1.4 Direct foaming

Direct foaming techniques involve the use of parent foamed materials (i.e. a ceramic suspension containing surfactant) that incorporate air bubbles obtained via chemical blowing or mechanical stirring; the foaming process is then followed by a sintering process and stabilization. This technique produces porous ceramics with closed and open pores and porosities close to 97%, which can be obtained through fast, low-cost, and very simple production stages [60, 61]. Meanwhile, the ability to mass produce these ceramics is an additional benefit. The stabilization of the air bubbles included in the suspension is the most essential issue in direct foaming, especially when the slurry has a large bubble content; under these conditions, the wet foam is prone to cracking and might eventually collapse [62].

13.4.1.5 Additive manufacturing

Porous ceramics have precisely designed structures with important characteristics such as pore size, porosity, and pore interconnectivity to meet the demands of sophisticated applications [63]. However, after decades of work, the strategies outlined above are unlikely to be suitable for the task. Additive manufacturing (AM), which is an evolving technology, enables the production of precisely customized porous scaffolds with complicated structures utilizing a layer-by-layer procedure based on a 'computer-aided design (CAD)' without the use of special tooling. The mechanical strength of a scaffold can be attained by designing materials with appropriate porosities and pore sizes [64]. To make porous ceramics perform better, fused deposition has been used. In one case, the initial material filament was created from silicon. A composite with a compressive strength of 689±795 MPa was created using pressureless reactive metal infiltration with aluminum.

13.4.2 Materials used to fabricate porous ceramics

Affordable and traditional materials such as alumina or zirconia have been investigated as suitable materials with which to produce filters for utilization in water filtration. Naturally derived materials such as zeolites, clays, quartz, apatite, and industrial wastes are among the low-cost resources used. Materials made from unprocessed alumina and silica are commonly employed as affordable precursors in the production of inorganic filtration membranes but achieving high performance levels at low cost is still a major issue [65].

13.4.2.1 Kaolin

There are many different types of clay, each with its own mechanical, chemical, and physical features. Kaolinite, illite, and montmorillonite are the most common clay minerals [66]. Clays are abundantly available around the world and require very little processing before they can be used to prepare membranes. As a result, major efforts have been undertaken to prepare less expensive ceramic membranes using various clay types, due to the mechanical strength and pore architectures that can be attained after heat processing. Kaolin is the most common clay type; its mineral form is kaolinite. As a result, kaolin is an important component of emerging low-cost MT and is particularly ideal for membrane production. Kaolin ($Al_2Si_2O_5(OH)_4$) is widely studied as the basic parent material used to produce filters in inorganic membranes, such as those used in microfiltration [67], support [68], and ultrafiltration layers [69]. Kaolin has a particle size that is small enough for membrane manufacture [65].

13.4.2.2 Zeolite minerals

Hydrated aluminosilicates $[(SiO_2)(AlO_2)_x]M_yH_2O$ make up the majority of naturally occurring zeolite minerals. The three-dimensional framework structure of these natural materials is characterized by nanoscale porosity [70]. Zeolites are found in a variety of rocks, although they are most prevalent in volcanic elastic sediments, which contain the largest and purest deposits. They are used in a variety of industries, including water treatment, construction, catalysis, agriculture, and medical applications [71]. Grinding, shaping, and sintering are all procedures used in the fabrication of natural zeolite membranes to obtain durable bulk materials with the required aluminosilicate phases. Roque-Malherbe *et al* [72] exploited 'natural zeolite' to make permeable backing layers for membranes.

13.4.2.3 Hydroxyapatite

Hydroxyapatite (HAp) occurs naturally and has applications in the pharmaceutical, chemical, environmental, and biological industries. In the context of water treatment, hydroxyapatite can be used for HM removal from WW using a cation exchange method in which Ca^{2+} replaces the metal ions. HAp is not only efficient in adsorbing metal pollutants and dyes, but it can also adsorb anionic and cationic dyes effectively [73]. In addition, HAp can be produced from waste materials or obtained through naturally deposited low-cost forms [74, 84–99]. To date, studies have demonstrated that naturally obtained minerals are cheaper than synthetic materials in the production of hydroxyapatite-based microfiltration membranes [75, 76].

Apart from dyes and metals, there are emerging pollutants with structures that are highly stable and whose effects are completely unknown. The common emerging pollutants are pharmaceuticals, cosmetics, pesticides, and herbicides. Various secondary and primary treatment processes are less efficient in removing these pollutants. Adsorption using HAp has been shown to have a significant impact in removing pollutants. For instance, HAp sourced from codfish bones and treated with titanium salt has been used to degrade pollutants. The results of a study showed that HAp had a superior adsorption capacity of more than 90% [77].

13.4.2.4 Quartz sand

Quartz sand occurs naturally in sedimentary rock; its constituent is crystalline silica (SiO_2). Quartz is one of the most plentiful and extensively dispersed minerals on the planet [78], since chemical and mechanical weathering has no impact on it. Sands made up almost entirely of quartz grains have occasionally been deposited by

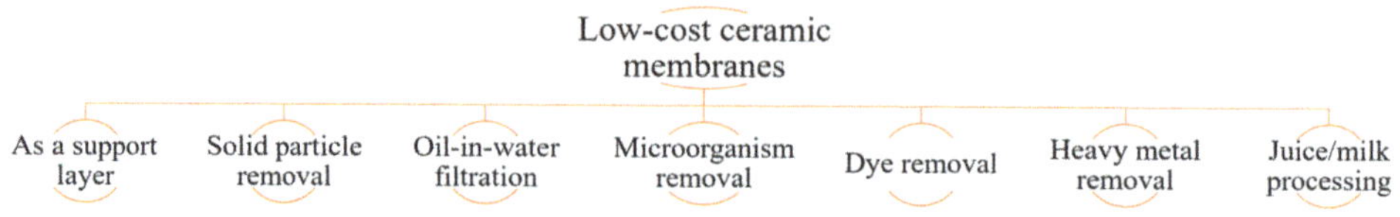

Figure 13.6. Applications of low-cost membranes.

geological processes. These deposits have been discovered to be high-quality silica sand sources. In their natural state, silica sands are known for their chemical uniformity and great purity. Several investigations into the production of filtration membranes produced from quartz sand have been carried out [79].

13.4.2.5 Bauxite

This is a sedimentary rock mostly composed of aluminum minerals, with an estimated global reserve of 55–75 billion tons [80]. The predominant high-temperature residue of bauxite is alumina, along with certain additional Fe and silicon oxides. Bauxite is classified as an alumina precursor that contains some contaminants (in the form of impurities), which are similarly possible contenders for membrane manufacture.

13.5 Industrial applications of membranes for water treatment

MT has numerous applications owing to its cost-effectiveness and simplicity compared to other filtration technologies such as absorption, distillation, and adsorption [81]. The development of innovative technologies to purify water before it is discharged or used is necessitated by water pollution caused by industrial growth, as well as limited access to fresh water. MT is unquestionably one of the most successful methods for treating WW and desalinating saltwater. Low-cost membranes produced from various natural sources and wastes have been widely used for water filtration applications. According to several research studies, low-cost membranes are mainly used for support layers, oil droplet separators, dye removal for the textile industry, microfiltration, and ultrafiltration membranes, as shown in figure 13.6 (adapted from [65]).

Most research into the fabrication of less expensive membranes has focused on the production of a support layer and then covered the support layer with another active layer for additional ultrafiltration, microfiltration, or even desalination applications [82]. Low-cost membranes are used to separate water–oil mixtures, which is one of the most explored water treatment challenges.

13.6 Issues that need to be resolved and possible solutions

The presented treatment techniques have been found to have the potential to improve tannery effluent quality to a great extent. However, there are some challenges caused by the treatment plants, including high salinity and Cr pollution. The real problem arises from the by-products of the treatment, which require advanced segregation techniques. These by-products include adsorbents,

Cr-containing sludge, deposits, etc. which are usually poorly managed, yet they are extremely hazardous. Given the cost of treatment, the Cr content is usually moved to landfill treatment, which causes a significant transfer of pollution.

The reuse or recycling of the by-products is more desirable for various industrial applications than conventional landfill. Strategies such as the use of novel Cr tanning agent carriers, chrome-free organic tanning agents, and Cr exhaustion are examples of ways in which Cr segregation can be eliminated from Cr pollution in the WW treatment processes. High salinity is another serious problem caused by tannery WW treatment, since it undermines the whole treatment process. Several efforts have been made to reduce this effect using biological treatment, membrane filtration, and electrochemical treatment using promising techniques. However, the treated tannery WW still has high salinity and requires post-treatments that include RO, electrodialysis, and electro-deionization [83].

13.7 Conclusions

This chapter evaluated the different pollutants found in tannery WW and the use of different purification techniques in the removal of HMs and industrial effluents. The pollutants in tannery WW include TN, total phosphorus, chloride ions, and sulfur-containing composites; excessive concentrations of these pollutants have significant effects on humans and the environment. This chapter summarized the various methods used for the removal of tannery WW contaminants. Flocculation and coagulation are effective techniques for removing particulates and impurities. Flotation is also used to remove particles. Ion exchange is used in HM removal due to its high treatment capacity, quick kinetics, and high removal efficiency. Adsorption and membrane filtration are the most practical and cost-effective methods for treating HMs in tannery effluent. The merit of these techniques include reliability, low consumption, ease of operation, and excellent pollution removal efficiency. This chapter also presented the application of RSM techniques for WW treatment. The RSM helps in the determination of the optimum conditions needed to maximize the removal efficiency and the interaction between the factors, which can be seen in the graphical illustrations and the models. As seen from the reviewed papers, large amounts of information can be obtained from very few experimental runs, and this results in low production costs and time savings. The response surface techniques aid in optimizing the important variables and in finding the appropriate combination of process factors (e.g. coagulant dose, pH, and initial concentration). This chapter also provided a detailed explanation of various techniques used in the synthesis of porous ceramics, the raw materials used, and the industrial applications of MT. It highlighted the potential for the utilization of porous ceramics in the treatment of tannery WW.

References and further reading

[1] Zhao C and Chen W 2019 A review for tannery wastewater treatment: some thoughts under stricter discharge requirements *Environ. Sci. Pollut. Res* **26** 26102–111

[2] Akan J C, Abdulrahman F I, Ayodele J T and Ogugbuaja V O 2009 Impact of tannery and textile effluent on the chemical characteristics of Challawa River, Kano State, Nigeria *Aust. J. Basic Appl. Sci.* **3** 1933–47
[3] Chen Y, Wang N, Ola O, Xia Y and Zhu Y 2021 Porous ceramics: light in weight but heavy in energy and environment technologies *Mater. Sci. Eng.: R: Rep.* **143** 100589
[4] Gutterres M, Benvenuti J and Fontocra J T 2015 Characterisation of raw wastewater from tanneries *J. Soc. Leather Technol. Chem.* **99** 280–7
[5] Cooman K, Gajardo M, Nieto J, Bornhardt C and Vidal G 2003 Tannery wastewater characterization and toxicity effects on *Daphnia* spp *Environ. Toxico.: An Inter.* J **18** 45–51
[6] Fawell J and Nieuwenhuijsen M J 2003 Contaminants in drinking waterenvironmental pollution and health *Br. Med. Bull.* **68** 199–208
[7] World Health Organization, WHO, and World Health Organisation Staff 2004 *Guidelines for Drinking-Water Quality* (Geneva: World Health Organization)1 https://www.who.int/publications/i/item/9789241549950
[8] Zhang Y, Yang Q, Gao H, Zhao Y, Tang X, Zhao C and Fang C 2021 Application of a modified biological flocculant in total nitrogen treatment of leather wastewater *Water Sci. Technol.* **83** 2901–10
[9] Li H 2014 Study on treatment of tannery wastewater and chromium sludge for Cr (III) recovery *Doctoral Dissertation* (Harbin: Harbin Institute of Technology)
[10] Sitko R, Janik P, Zawisza B, Talik E, Margui E and Queralt I 2015 Green approach for ultratrace determination of divalent metal ions and arsenic species using total-reflection X-ray fluorescence spectrometry and mercapto-modified graphene oxide nanosheets as a novel adsorbent *Anal. Chem.* **87** 3535–42
[11] Srivastava N K and Majumder C B 2008 Novel biofiltration methods for the treatment of heavy metals from industrial wastewater *J. Hazard. Mater.* **151** 1–8
[12] Żak S 2012 Treatment of the processing wastewaters containing heavy metals with the method based on flotation *Ecol. Chem. Eng.* S **19** 433–8
[13] Ismail M G, Weng C N, Rahman H A and Zakaria N A 2013 Freundlich isotherm equilibrium equations in determining effectiveness a low cost absorbent to heavy metal removal in wastewater (leachate) at Teluk Kitang Landfill, Pengkalan Chepa, Kelantan, Malaysia *J Geogr Earth Sci.* **1** 1–8
[14] Nair A T, Makwana A R and Ahammed M M 2014 The use of response surface methodology for modelling and analysis of water and wastewater treatment processes: a review *Water Sci. Technol.* **69** 464–78
[15] Montgomery D C 2017 *Design and Analysis of Experiments* (New York: Wiley)
[16] Box G E and Wilson K B 1951 On the experimental designs for exploring response surfaces *Ann. Math. Stat* **13** 1–45
[17] Wang J P, Chen Y Z, Ge X W and Yu H Q 2007 Optimization of coagulation–flocculation process for a paper-recycling wastewater treatment using response surface methodology *Colloids Surf.,* A **302** 204–10
[18] Moghaddam S S, Moghaddam M A and Arami M 2010 Coagulation/flocculation process for dye removal using sludge from water treatment plant: optimization through response surface methodology *J. Hazard. Mater.* **175** 651–7
[19] Liu X, Li X M, Yang Q, Yue X, Shen T T, Zheng W, Luo K, Sun Y H and Zeng G M 2012 Landfill leachate pretreatment by coagulation–flocculation process using iron-based coagulants: optimization by response surface methodology *Chem. Eng. J.* **200** 39–51

[20] Mensah-Akutteh H, Buamah R, Wiafe S and Nyarko K B 2022 Optimizing coagulation–flocculation processes with aluminium coagulation using response surface methods *Appl. Water Sci.* **12** 188
[21] Iber B T, Torsabo D, Chik C E N C E, Wahab F, Sheikh Abdullah S R, Abu Hassan H and Kasan N A 2023 Response surface methodology (RSM) approach to optimization of coagulation-flocculation of aquaculture wastewater treatment using chitosan from carapace of giant freshwater prawn *Macrobrachium rosenbergii Polymers* **15** p 1058
[22] Kang K C, Kim S S, Choi J W and Kwon S H 2008 Sorption of Cu^{2+} and Cd^{2+} onto acid- and base-pretreated granular activated carbon and activated carbon fiber samples *J. Ind. Eng. Chem.* **14** 131–5
[23] Jusoh A, Shiung L S, Ali N and Noor M J M M 2007 A simulation study of the removal efficiency of granular activated carbon on cadmium and lead *Desalination* **206** 9–16
[24] Li Y, Liu F, Xia B, Du Q, Zhang P, Wang D, Wang Z and Xia Y 2010 Removal of copper from aqueous solution by carbon nanotube/calcium alginate composites *J. Hazard. Mater.* **177** 876–80
[25] Pillay K, Cukrowska E M and Coville N J 2009 Multi-walled carbon nanotubes as adsorbents for the removal of parts per billion levels of hexavalent chromium from aqueous solution *J. Hazard. Mater.* **166** 1067–75
[26] Metcalf & Eddy Inc. 2003 *Wastewater engineering: treatment and resource recovery* (New Delhi: Tata McGraw-Hill Education) 5th
[27] Yang Y, Li X, Gu Y, Lin H, Jie B, Zhang Q and Zhang X 2022 Adsorption property of fluoride in water by metal organic framework: optimization of the process by response surface methodology technique *Surf. Interfaces* **28** 101649
[28] Nguyen V T, Nguyen T B, Dat N D, Huu B T, Nguyen X C, Tran T, Bui M H, Dong C D and Bui X T 2022 Adsorption of norfloxacin from aqueous solution on biochar derived from spent coffee ground: master variables and response surface method optimized adsorption process *Chemosphere* **288** 132577
[29] Mourabet M, El Rhilassi A, El Boujaady H, Bennani-Ziatni M and Taitai A 2017 Use of response surface methodology for optimization of fluoride adsorption in an aqueous solution by Brushite *Arabian J. Chem.* **10** S3292–302
[30] Iqbal M, Iqbal N, Bhatti I A, Ahmad N and Zahid M 2016 Response surface methodology application in optimization of cadmium adsorption by shoe waste: a good option of waste mitigation by waste *Ecol. Eng.* **88** 265–75
[31] Murthy Z V and Chaudhari L B 2009 Separation of binary heavy metals from aqueous solutions by nanofiltration and characterization of the membrane using Spiegler–Kedem model *Chem. Eng. J.* **150** 181–7
[32] Nguyen C M, Bang S, Cho J and Kim K W 2009 Performance and mechanism of arsenic removal from water by a nanofiltration membrane *Desalination* **245** 82–94
[33] Ahmad A L and Ooi B S 2010 A study on acid reclamation and copper recovery using low pressure nanofiltration membrane *Chem. Eng. J.* **156** 257–63
[34] Muthukrishnan M and Guha B K 2008 Effect of pH on rejection of hexavalent chromium by nanofiltration *Desalination* **219** 171–8
[35] Yang Z, Zhou Y, Feng Z, Rui X, Zhang T and Zhang Z 2019 A review on reverse osmosis and nanofiltration membranes for water purification *Polymers* **11** 1252
[36] Waqas S, Harun N Y, Bilad M R, Samsuri T, Nordin N A, Shamsuddin N, Nandiyanto A B, Huda N and Roslan J 2022 Response surface methodology for optimization of rotating

biological contactor combined with external membrane filtration for wastewater treatment *Membranes* **12** 271
[37] Wang M, Liu C, Fan M, Liu M and Shen S 2022 Optimization of nanofiltration hollow fiber membrane fabrication process based on response surface method *Membranes* **12** 374
[38] Maher A, Sadeghi M and Moheb A 2014 Heavy metal elimination from drinking water using nanofiltration membrane technology and process optimization using response surface methodology *Desalination* **352** 166–73
[39] Goyal R K, Jayakumar N S and Hashim M A 2011 A comparative study of experimental optimization and response surface optimization of Cr removal by emulsion ionic liquid membrane *J. Hazard. Mater.* **195** 383–90
[40] Zhou W, Wu P, Zhang L, Zhu D, Zhao X and Cai Y 2022 Heavy metal ions and particulate pollutants can be effectively removed by a gravity-driven ceramic foam filter optimized by carbon nanotube implantation *J. Hazard. Mater.* **421** 126721
[41] Saja S, Bouazizi A, Achiou B, Ouammou M, Albizane A, Bennazha J and Younssi S A 2018 Elaboration and characterization of low-cost ceramic membrane made from natural Moroccan perlite for treatment of industrial wastewater *J. Environ. Chem. Eng.* **6** 451–8
[42] Ohji T and Fukushima M 2012 Macro-porous ceramics: processing and properties *Int. Mater. Rev.* **57** 115–31
[43] Twigg M V and Richardson J T 2007 Fundamentals and applications of structured ceramic foam catalysts *Ind. Eng. Chem. Res.* **46** 4166–77
[44] Obada D O *et al* 2021 Fabrication of novel kaolin-reinforced hydroxyapatite scaffolds with robust compressive strengths for bone regeneration *Appl. Clay Sci.* **215** 106298
[45] Obada D O *et al* 2021 Solution combustion synthesis of strontium-doped hydroxyapatite: effect of sintering and low compaction pressure on the mechanical properties and physiological stability *Mater. Lett.* **304** 130613
[46] Salami K A, Obada D O, Alabi A A, Csaki S and Oyedeji A N 2022 Utilization of sawdust as a pore-former in the fabrication of ceramic adsorbents for water purification *J. Aust. Ceram. Soc.* **58** 1297–309
[47] Gregorová E, Pabst W, Sofer Z, Jankovský O and Matějíček J 2012 Porous alumina and zirconia ceramics with tailored thermal conductivity *J. Phys.: Conf. Ser.* **395** 012022
[48] Darus F, Isa R M, Mamat N and Jaafar M 2018 Techniques for fabrication and construction of three-dimensional bioceramic scaffolds: effect on pores size, porosity and compressive strength *Ceram. Int.* **44** 18400–7
[49] Jean G, Sciamanna V, Demuynck M, Cambier F and Gonon M 2014 Macroporous ceramics: novel route using partial sintering of alumina-powder agglomerates obtained by spray-drying *Ceram. Int.* **40** 10197–203
[50] Guo Q, Xiang H, Sun X, Wang X and Zhou Y 2015 Preparation of porous YB4 ceramics using a combination of *in situ* borothermal reaction and high temperature partial sintering *J. Eur. Ceram. Soc.* **35** 3411–8
[51] Kim M, Franco R A and Lee B T 2011 Synthesis of functional gradient BCP/ZrO_2 bone substitutes using ZrO_2 and BCP nanopowders *J. Eur. Ceram. Soc.* **31** 1541–8
[52] Gómez S Y, Alvarez O A, Escobar J A, Rodrigues Neto J B, Rambo C R and Hotza D 2012 Relationship between Rheological Behaviour and Final Structure of Al ð O ð and YSZ Foams Produced by Replica *Adv. Mater. Sci. Eng.* **2012** 549508

[53] Bucharsky E C, Schell K G, Oberacker R and Hoffmann M J 2010 Preparation of transparent glass sponges via replica method using high-purity silica *J. Am. Ceram. Soc.* **93** 111–4
[54] Sarhadi F, Shafiee Afarani M, Mohebbi-Kalhori D and Shayesteh M 2016 Fabrication of alumina porous scaffolds with aligned oriented pores for bone tissue engineering applications *Appl. Phys.* A **122** 1–8
[55] Studart A R, Gonzenbach U T, Tervoort E and Gauckler L J 2006 Processing routes to macroporous ceramics: a review *J. Am. Ceram. Soc.* **89** 1771–89
[56] Petit C, Meille S, Maire E, Tadier S and Adrien J 2016 Mechanical behaviour of a β-TCP ceramic with a random porosity: study of the fracture path with x-ray tomography *J. Eur. Ceram. Soc.* **36** 3225–33
[57] McNamara S L, Rnjak-Kovacina J, Schmidt D F, Lo T J and Kaplan D L 2014 Silk as a biocohesive sacrificial binder in the fabrication of hydroxyapatite load bearing scaffolds *Biomaterials* **35** 6941–53
[58] Zhang L, Mao Z, Thomason J D, Wang S and Huang K 2012 Synthesis of a homogeneously porous solid oxide matrix with tunable porosity and pore size *J. Am. Ceram. Soc.* **95** 1832–7
[59] Konegger T, Williams L F and Bordia R K 2015 Planar, polysilazane-derived porous ceramic supports for membrane and catalysis applications *J. Am. Ceram. Soc.* **98** 3047–53
[60] Pokhrel A, Seo D N, Lee S T and Kim I J 2013 Processing of porous ceramics by direct foaming: a review *J. Korean Ceram. Soc* **50** 93–102
[61] Barg S, Soltmann C, Andrade M, Koch D and Grathwohl G 2008 Cellular ceramics by direct foaming of emulsified ceramic powder suspensions *J. Am. Ceram. Soc.* **91** 2823–9
[62] Zhang L Y, Zhou D L, Chen Y, Liang B and Zhou J B 2014 Preparation of high open porosity ceramic foams via direct foaming molded and dried at room temperature *J. Eur. Ceram. Soc.* **34** 2443–52
[63] Butscher A, Bohner M, Hofmann S, Gauckler L and Müller R 2011 Structural and material approaches to bone tissue engineering in powder-based three-dimensional printing *Acta Biomater.* **7** 907–20
[64] Bose S and Das C 2013 Preparation and characterization of low cost tubular ceramic support membranes using sawdust as a pore-former *Mater. Lett.* **110** 152–5
[65] Abdullayev A, Bekheet M F, Hanaor D A and Gurlo A 2019 Materials and applications for low-cost ceramic membranes *Membranes* **9** 105
[66] Rhodes D 2015 *Clay and Glazes for the Potter.* (Ravenio Books)
[67] Jana S, Purkait M K and Mohanty K 2010 Preparation and characterization of low-cost ceramic microfiltration membranes for the removal of chromate from aqueous solutions *Appl. Clay Sci.* **47** 317–24
[68] Zhou J, Zhang X, Wang Y, Larbot A and Hu X 2010 Elaboration and characterization of tubular macroporous ceramic support for membranes from kaolin and dolomite *J. Porous Mater.* **17** 1–9
[69] Ali M B, Hamdi N, Rodriguez M A, Mahmoudi K and Srasra E 2018 Preparation and characterization of new ceramic membranes for ultrafiltration *Ceram. Int.* **44** 2328–35
[70] Kalló D 2001 Applications of natural zeolites in water and wastewater treatment *Rev. Mineral. Geochem.* **45** 519–0
[71] Hay R L 1986 Geologic occurrence of zeolites and some associated minerals *Studies in Surface Science and Catalysis* (Amsterdam: Elsevier)28 35–40

[72] Roque-Malherbe R, Del Valle W, Marquez F, Duconge J and Goosen M F 2006 Synthesis and characterization of zeolite based porous ceramic membranes *Sep. Sci. Technol.* **41** 73–96

[73] Bensalah H, Bekheet M F, Younssi S A, Ouammou M and Gurlo A 2017 Removal of cationic and anionic textile dyes with Moroccan natural phosphate *J. Environ. Chem. Eng.* **5** 2189–99

[74] Bensalah H, Bekheet M F, Younssi S A, Ouammou M and Gurlo A 2018 Hydrothermal synthesis of nanocrystalline hydroxyapatite from phosphogypsum waste *J. Environ. Chem. Eng.* **6** 1347–52

[75] Masmoudi S, Larbot A, El Feki H and Amar R B 2006 Elaboration and properties of new ceramic microfiltration membranes from natural and synthesised apatite *Desalination* **190** 89–103

[76] Masmoudi S, Larbot A, El Feki H and Amar R B 2007 Elaboration and characterisation of apatite based mineral supports for microfiltration and ultrafiltration membranes *Ceram. Int.* **33** 337–44

[77] Brazón E M *et al* 2016 Photodegradation of pharmaceutical persistent pollutants using hydroxyapatite-based materials *J. Environ. Manage.* **182** 486–95

[78] Bates R L and Jackson J A 1987 *Glossary of Geology* (Alexandria, VA: American Geological Institute) 788–865

[79] Foughali L, Barama S, Harabi A, Bouzerara F, Guechi A and Boudaira B 2016 Effect of sodium phosphate addition on mechanical properties of porous Sigue quartz sand *Desalin. Water Treat.* **57** 5286–91

[80] Yao Z T, Ji X S, Sarker P K, Tang J H, Ge L Q, Xia M S and Xi Y Q 2015 A comprehensive review on the applications of coal fly ash *Earth Sci. Rev.* **141** 105–21

[81] Marchetti P, Jimenez Solomon M F, Szekely G and Livingston A G 2014 Molecular separation with organic solvent nanofiltration: a critical review *Chem. Rev.* **114** 10735–806

[82] Dong Y, Ma L, Tang C Y, Yang F, Quan X, Jassby D, Zaworotko M J and Guiver M D 2018 Stable superhydrophobic ceramic-based carbon nanotube composite desalination membranes *Nano Lett.* **18** 5514–21

[83] Zhao J, Wu Q, Tang Y, Zhou J and Guo H 2022 Tannery wastewater treatment: conventional and promising processes, an updated 20-year review *J. Leather Sci. Eng.* **4** 10

[84] Akpan E S, Dauda M, Kuburi M, Obada L S and D O 2021 Box–Behnken experimental design for the process optimization of catfish bones derived hydroxyapatite: a pedagogical approach *Mater. Chem. Phys.* **272** 124916

[85] Akpan E S, Dauda M, Kuburi L S, Obada D O and Dodoo-Arhin D 2020 A comparative study of the mechanical integrity of natural hydroxyapatite scaffolds prepared from two biogenic sources using a low compaction pressure method *Results Phys.* **17** 103051

[86] Akpan E S, Dauda M, Kuburi L S and Obada D O 2020 A facile synthesis method and fracture toughness evaluation of catfish bones-derived hydroxyapatite *MRS Adv.* **5** 1357–66

[87] Akpan E S, Dauda M, Kuburi L S, Obada D O, Bansod N D and Dodoo-Arhin D 2021 Hydroxyapatite ceramics prepared from two natural sources by direct thermal conversion: from material processing to mechanical measurements *Mater. Today Proc.* **38** 2291–4

[88] Obada D O *et al* 2023 Hydroxyapatite materials-synthesis routes, mechanical behavior, theoretical insights, and artificial intelligence models: a review *J. Aust. Ceram. Soc.* **59** 565–96

[89] Obada D O, Dauda E T, Abifarin J K, Dodoo-Arhin D and Bansod N D 2020 Mechanical properties of natural hydroxyapatite using low cold compaction pressure: effect of sintering temperature *Mater. Chem. Phys.* **239** 122099

[90] Osuchukwu O A, Salihi A, Abdullahi I and Obada D O 2022 Taguchi grey relational optimization of sol–gel derived hydroxyapatite from a novel mix of two natural biowastes for biomedical applications *Sci. Rep.* **12** p 17968

[91] Osuchukwu O A, Salihi A, Abdullahi I, Etinosa P O and Obada D O 2023 Weibull modulus of a novel mixture of natural hydroxyapatite materials produced from biowastes *Res. Mater.* **18** p 100394

[92] Osuchukwu O A, Salihi A, Abdullahi I, Etinosa P O and Obada D O 2023 A comparative study of the mechanical properties of sol–gel derived hydroxyapatite produced from a novel mixture of two natural biowastes for biomedical applications *Mater. Chem. Phys.* **297** p 127434

[93] Osuchukwu O A, Salihi A, Abdullahi I and Obada D O 2022 Experimental data on the characterization of hydroxyapatite produced from a novel mixture of biowastes *Data in Brief* **42** p 108305

[94] Osuchukwu O A, Salihi A, Abdullahi I and Obada D O 2022 Synthesis and characterization of sol–gel derived hydroxyapatite from a novel mix of two natural biowastes and their potentials for biomedical applications *Mater. Today Proc.* **62** pp 4182–7

[95] Obada D O *et al* 2022 Trends in the development of hydroxyapatite from natural sources for biomedical applications *WIT Trans. Ecol. Environ.* **257** 199–210

[96] Akpan E S, Dauda M, Kuburi L S and Obada D O 2023 Bio-inspired synthesis of hydroxyapatite materials from two natural sources: a crack behavior and biological insight *J. Aust. Ceram. Soc.* **59** 511–9

[97] Osuchukwu O A, Salihi A, Abdullahi I, Abdulkareem B and Nwannenna C S 2021 Synthesis techniques, characterization and mechanical properties of natural derived hydroxyapatite scaffolds for bone implants: a review *SN Appl. Sci.* **3** pp 1–23

[98] Abifarin J K, Obada D O, Dauda E T and Dodoo-Arhin D 2019 Experimental data on the characterization of hydroxyapatite synthesized from biowastes *Data Brief* **26** p 104485

[99] Obada D O, Dauda E T, Abifarin J K, Bansod N D and Dodoo-Arhin D 2021 Mechanical measurements of pure and kaolin reinforced hydroxyapatite-derived scaffolds: a comparative study *Mater. Today Proc.* **38** pp 2295–300

Chapter 14

Cost and environmental evaluations and comparisons of commonly used sorbents

Kingsley Obodo and Uyiosa Osagie Aigbe

This chapter examines the costs and environmental implications of utilizing commonly available sorbents. Sorbents are solid materials capable of selectively capturing and reversibly releasing specific gaseous or liquid substances. They can be categorized as absorbents or adsorbents based on the occurrence of structural changes or surface adhesion. Sorbents possess high surface-area-to-volume ratios, elastic porous structures, and fluid extraction capabilities that use various mechanisms. Overall, this chapter recognizes the challenges of comparing and evaluating sorbent materials in terms of cost due to variations in sorption capability testing methods, material sources, and environmental impacts.

14.1 Introduction

The cost implications of deploying sorbents—also sometimes referred to as 'adsorbents'—for reasons of environmental sustainability in the cleaning up of oil spills to prevent ecological disaster have been explored. Strictly speaking, 'sorbent' refers to a solid material that can selectively capture as well as reversibly capture and release specific gaseous or liquid substances. Broadly, sorbents can be categorized into two types, namely 'ab-sorbent' and 'ad-sorbent' depending on whether they undergo structural changes when materials are captured on them or whether materials adhere to the surface of the sorbent material [1]. Sorbent materials have a large surface-area-to-volume ratio; they can have elastic porous structures and also possess the ability to extract fluids through suction, mechanical deformation, surface attraction, etc. The application of sorbents in different environmental situations is a topic that keeps attracting attention. Commonly used sorbents as well as various sophisticated sorbents, such as nanomaterials, functionalized surfaces, etc. are important in different applications, such as the oil industry, air cleaning, water cleaning, etc. but cost and environmental impact are paramount considerations [2–6].

doi:10.1088/978-0-7503-5598-8ch14

Patalano *et al* [1] showed that the different characteristics of crude oil and nonpolar materials such as density, weathering, and viscosity make the practical deployment of sorbents challenging due to their cost, unlike deployment at the laboratory scale, where they are attractive. Their study highlighted that the precursor cost, manufacturing cost, reusability, and disposal make the application of sorbents impractical in oil spillage cleanup, but they stated that further study should examine cost reduction and ease of deployment at oil spills. A study by Yadav *et al* [7] assessed the environmental impact of producing polymeric membranes; they found that the negative environmental consequences associated with membrane production primarily stem from the use of toxic solvents such as N-methyl-2-pyrrolidone (NMP), which accounts for approximately 40%–60% of the total environmental impact. The use of environmentally friendly solvents such as ethylene carbonate (EC) leads to significant ways to reduce membrane cost and environmental impact, potentially achieving a reduction of up to 35% in both. A study by Lawler *et al* [8] highlighted the benefits of sustainability in reverse osmosis (RO) membrane disposal within the desalination industry, taking into consideration the reuse and recycling of these membranes. Several options were highlighted, such as the available affordable second-hand ultrafiltration (UF) and/or nanofiltration (NF) membranes that can be applied in the water treatment sector, in which such membranes can be used as low-cost pretreatment RO solutions in humanitarian water treatment for temporary or remote water treatment operations. They proposed that the membranes should be disposed of via extracting energy from the old RO modules, which offers the opportunity to transform an expensive waste into an asset. These suggested options would lead to carbon footprint reduction and improve the cradle-to-grave life span.

Zhao *et al* [9] reviewed several polymeric nanocomposites and showed that introducing different nanoparticles, such as carbon nanotubes, carbon nanofibers, and activated carbon (AC), can improve the gas separation efficiency of membranes. Different diffusion mechanisms should be considered, such as the Knudsen diffusion mechanism, as well as the application of artificial intelligence models to search for new and improved membranes with improved sorbent properties as well as increased efficacy at temperatures above 100 °C. Godage *et al* [10] showed that substantial advancements in environmentally friendly sample preparation techniques were achieved by downsizing the sampling devices and reducing the required sorptive-phase quantity, which led to the effective removal of the targeted compounds. The use of natural sorbents also presents an advantageous and convenient approach for the preparation of green samples, which offer additional benefits such as their ready availability from renewable sources, biodegradability, and minimal toxicity. Leonzio *et al* [6] showed that the equilibrium loading of sorbent was found to be a critical parameter that determines the cost, energy consumption, and adsorption capacity according to a comparative analysis of various sorbents used for the direct air capture of CO_2. They considered metal–organic framework (MOF) processes that showed that the improved loading led to a reduction in the thermal energy cost.

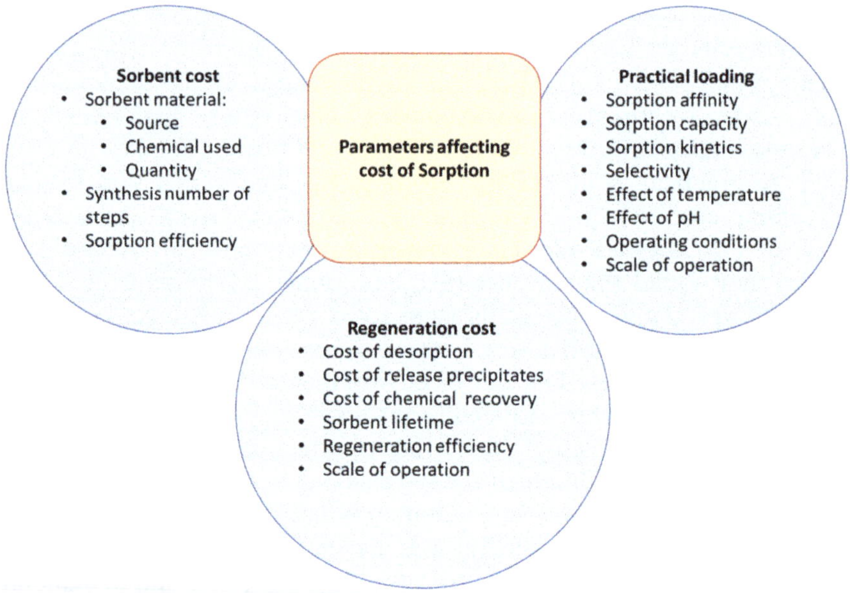

Figure 14.1. General parameters affecting the cost of commonly used sorbents.

Overall, cost comparisons and evaluations of different sorbent materials pose a significant challenge due to the variations in the sorption capability testing methods, the use of different materials for the sorption, environmental impacts, etc. which makes effective comparisons of materials difficult. In section 14.2, the general parameters affecting the cost of sorbents are discussed in detail.

14.2 General parameters affecting the cost of sorbents

The cost of commonly used sorbent materials, such as carbon-based materials, clay, silica, zeolites, and MOFs can be affected by several parameters. This section discusses some of the key factors. Figure 14.1 illustrates typical parameters that influence the cost [11, 12], which are discussed in the following subsections.

14.2.1 Sorbent material

The cost of the adsorbent material is itself a significant factor. Different sorbents have varying costs, and the selection of an adsorbent with a higher cost can increase the overall expense. In order to understand the various costs associated with sorbent materials, it is necessary to consider the precursor cost, manufacturing cost, reusability, and finally the disposal cost. These associated costs for the sorbent material depend on the area of application, such as gas separation and cleaning (membrane materials can be applied, which can be expensive), heavy metal extraction (clay-based sorbent materials, which are relatively cheaper), etc. The relationship between the sorbent material and cost is a multivariate dependence, as discussed above.

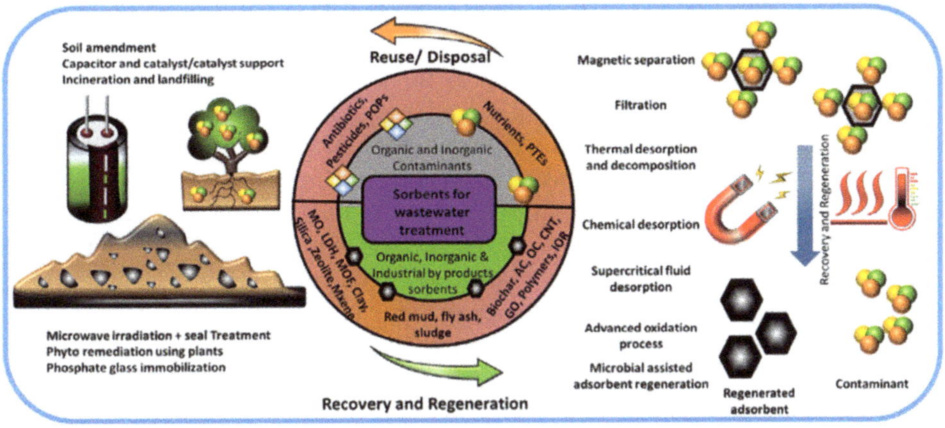

Figure 14.2. A schematic of recovery and regeneration as well as the reuse/disposal of spent sorbents. Reprinted from [15], Copyright (2022), with permission from Elsevier.

14.2.2 Sorbent quantity

The quantity of adsorbent used can vary for different adsorption processes, and this eventually affects the cost. The larger the quantity of adsorbent used for a specific adsorption process, the higher the cost, and the higher the environmental impact due to the increased energy and resources required to manufacture a greater quantity. The sorbent quantity and the cost of the specific quantity have a direct relationship.

14.2.3 Sorbent regeneration

The ability to regenerate and reuse a specific sorbent material affects its cost and environmental impact. Regeneration entails the recovery or rapid recycling of spent sorbents via scientifically and economically practical means. Sorbent regeneration can occur via magnetic separation, filtration, thermal desorption, chemical desorption (which can consist of supercritical fluid desorption, advanced oxidation processes, microbe-assisted processes, etc.) [13, 14] and a combination of methods such as advanced oxidation processes, electrochemical, thermal, and KOH activation, hydrothermal carbonization, thermochemical treatments, water washing and vacuum filtration, solvent washing (ethanol and water, methanol wash), sonication, alkaline regeneration, or microwave or ultraviolet irradiation [15], as presented in figure 14.2. The sorbent replacement rate is significantly reduced if there is more frequent regeneration and vice versa. The cost and environmental impact associated with adsorbent regeneration have an inverse relationship.

14.2.4 Operating conditions

Operational conditions such as temperature and pressure play an important role in the sorption processes of various commonly used sorbents; these have associated costs as well as environmental implications. Optimal operational conditions have been shown to yield better sorptive capacity [16]. Certain polymeric membranes have specific ranges of operational temperatures and pressures; if these are not adhered to, rubberization and polymeric matrix degradation start to occur. Under specific operating conditions, for instance, in the presence of polar or nonpolar sorbates, Woo *et al* [17] found that conventional isotherm models produce notable inconsistencies in the estimates of anomalous adsorption isotherms. Thus, they applied a hybrid isotherm model consisting of Langmuir and Sips isotherms and demonstrated optimal compatibility for the adsorption isotherms of four different commercial silica gels [17]. Generally, the specificity of operational conditions with narrow tolerances results in increased energy consumption and operational expense as well as higher environmental impact; there are, however, certain exceptions, such as sorbents that operate at room temperature and pressure.

14.2.5 Sorption efficiency

The efficiency of the adsorption process influences the cost. Higher efficiency means that fewer adsorbents are needed, leading to eventual cost savings. In wastewater treatment, the interactions between various metal ions in aqueous solutions play a crucial role in the sorption efficiencies of the different ions. In a batch experiment, the effects of several parameters, such as the pH, particle size, contact time, temperature, ion concentration, dose, etc. can affect the sorption efficiency [18]. The parameter under consideration varies depending on the model and the scale of the experiment under consideration. The parameters that affect the sorption efficiency can be evaluated by varying one parameter and keeping the others constant [19] or by carrying out a multivariate analysis.

14.2.6 Contaminant concentration

The initial concentration of contaminants in the feed stream affects the cost. Higher concentrations may require more adsorbents or longer contact times, increasing the overall expense. One way to address the issues associated with pollution resulting from heavy metals is by implementing techniques such as precipitation, ultra-filtration, and electrodeposition. These techniques are costly and ineffective when dealing with low concentrations of heavy metals. Adsorption has a high potential to eliminate, recover, and recycle heavy metals, even at low concentrations [20–22]. It has emerged as a viable technique because of its effectiveness, affordability, and environmentally friendly nature at different contaminant concentrations.

14.2.7 Sorption capacity

The adsorption capacity of the adsorbent is a crucial parameter. Adsorbents with higher capacities can treat larger volumes of contaminants, resulting in lower costs. Tamjidi *et al* [23] showed that the application of algae as a bioadsorbent is promising because it is low cost, environmentally friendly, easily accessible, and, in particular, it has a high sorption capacity. In addition, other materials such as biopolymer-clay-based nanocomposites show excellent sorption capacity as a result of their increased surface area and active sites [24]. The sorption/adsorption capacity can be described from the sorption isotherms using the first-order adsorption rate; this behavior is described by Lagergren's equation (which considers liquid and solid phases) as well as other first-order equations such as Elovich's equation, Ritchie's equation, etc. The second-order rate equations (also referred to as the pseudo second-order equations) include the Langmuir isotherm, the Temkin isotherm, the Freundlich isotherm, the Dubinin–Radushkevich isotherm, and the Sips isotherm [25]; these provide details of the chemisorption of gases by considering ion exchange reactions [26]. Functionalization, magnetism [27], and increasing temperature have been reported to result in a higher sorption capacity for Ni [28]. Thus, different sorbent materials act differently based on the operational conditions,HHH resulting in different costs and environmental impacts.

14.2.8 Sorption kinetics

The sorption kinetics can impact the cost of sorbent materials. If a sorbent has fast kinetics, the contact time can be reduced, leading to cost savings. Qasem *et al* [20] showed that an MOF's stability and sorption kinetics can be enhanced using different functionalization techniques. Different kinetic models have been developed based on the reactant concentration and the adsorption, including pseudo first-order kinetics, the intraparticle diffusion model, and the second-order kinetics double exponential model [25].

14.2.9 Regeneration efficiency

The efficiency of regenerating the spent adsorbent can affect the cost. The implication of higher regeneration efficiency is that less waste is generated and the replacement costs are reduced. The regeneration efficiency achieved using waste heat and solar energy can result in cost savings as well as the beneficial environmental impacts of heat transfer and energy storage [29]. Some of the important factors that affect regeneration efficiency include the pH, the regeneration temperature, and proper time management for the desorption of the pollutants [30].

14.2.10 The scale of operation

The scale of the sorption process can influence the cost. Generally, large-scale operations benefit from economies of scale, thus reducing overall expenses [31].

14.2.11 Other parameters

The life cycle assessment (LCA) is another parameter that plays a crucial role in evaluating the cost and environmental impact of adsorbents. A recent study evaluated different sorbents used in direct air capture [32], with an emphasis on their environmental performance in capturing CO_2 from ambient air. In the LCA, accuracy and precision are essential in order to assess the cost and environmental impact throughout the life cycle stages of the sorbent. The different stages generally include the source of the raw material, raw material acquisition, adsorbent manufacture, and the various relevant process. A comprehensive LCA and an impact assessment provide information that is valuable when choosing a sorbent as well as the associated costs and environmental impacts.

It is important to note that the relative importance of these parameters varies depending on the specific application and context. Thus, further analysis such as life cycle analysis can be carried out for various sorbent materials.

14.3 The costs and environmental evaluations of commonly used sorbents

14.3.1 Overview

Sorbents are materials that are used to capture or remove pollutants, contaminants, or other substances from air, water, or other solutions. The costs and environmental evaluations of commonly used sorbents mainly depend on several factors, such as the sorbent type, production technique, resource availability, disposal considerations, etc. to name a few. It is also important to note that the costs and environmental impacts of sorbents can vary based on factors such as regional availability, scale of production, and waste management practices. Thus, it is important to understand the costs and environmental impacts as well as to compare commonly used sorbents, as presented in table 14.1. This is because it drives their applicability for different kinds of pollutant removal. In addition, the performance and suitability

Table 14.1. Commonly used sorbents along with their costs and environmental evaluations.

Sorbent	Cost	Environmental impact
AC	• Widely used and relatively affordable. • The source of raw material used, the activation process, and the application affect its cost. • Cost-effective option in the long run.	• The source can be an organic material such as coconut shell, wood, or coal. • Production uses high-temperature treatment, which requires energy. • Disposal can be a concern, but regeneration and reuse in certain applications can mitigate the environmental impact.
Clay	• Relatively inexpensive compared to other sorbents. • Readily available in large quantities.	• Extraction and processing could have a negative impact on the environment, such as deforestation and soil erosion. • Clay minerals are biodegradable. • Minimal long-term environmental impact.
Zeolites	• Crystalline aluminosilicate minerals. • The zeolite type and its source affect cost; e.g. synthetic zeolites	• Zeolites are naturally occurring minerals and can be mined.

(*Continued*)

Table 14.1. (*Continued*)

Sorbent	Cost	Environmental impact
	are generally more expensive than natural zeolites.	• The mining process used to extract zeolites can have environmental implications. • High capacity for adsorption and can be regenerated and reused, which can reduce the overall environmental impact.
Silica gel	• A commonly used desiccant. • A widely used adsorbent material. • Low cost and readily available	• Synthesized from sodium silicate, a compound derived from sand. • The production process involves high-temperature treatment and requires a lot of energy. • Can be regenerated and reused, reducing waste generation.
Molecular Sieves	• Synthetic materials with highly ordered porous structures. • Cost varies based on the type and application. • Some molecular sieves are more expensive due to their specialized properties.	• The production of molecular sieves involves energy-intensive processes. • Their environmental impact varies based on the raw materials and synthesis methods. • Can be regenerated and reused in some applications, minimizing waste generation.
Polymeric sorbents	• Cost varies significantly, depending on the specific type, manufacturing process, and application.	• Disposal methods depend on the material used, which determines the environmental impact. • Release toxic by-products during production or degradation. • Can be recycled or incinerated with energy recovery.
MOFs	• A relatively new class of sorbents, which can be more expensive compared to traditional sorbents. • Costs arise from complex synthesis processes and the use of specific metal components.	• The environmental impact of MOFs is attributed to the particular metal components used and the manufacturing process. • Metals and solvents used in MOF synthesis can raise environmental concerns. • MOFs can offer high sorption capacity, tunable properties, etc. • The regeneration potential of the MOF can contribute to its sustainability.

Chitosan	• Biopolymer is derived from chitin, which is found in the shells of crustaceans. • Cost varies based on the source and purification process. • Generally more expensive than traditional sorbents.	• Biodegradable and obtained from a renewable resource. • The extraction and purification processes can have environmental impacts. • Modified for enhanced sorption properties and can be biodegraded or disposed of without significant environmental impact.

of a sorbent for a specific application should also be considered alongside cost and environmental evaluations.

It is important to note that the costs and environmental evaluations mentioned above and presented in table 14.1 are general considerations and vary based on the applications, regional availability, and market dynamics. In the selection of a sorbent material, it is crucial to consider factors and parameters that influence the cost (as discussed in section 14.3.2), the application, environmental aspects, and compatibility with the intended application.

14.3.2 Recent interest in common sorbents for environmental remediation

Interest in the utilization of common sorbents for environmental remediation is continuously increasing. As discussed above, sorbents are materials that can selectively adsorb or absorb pollutants from air, water, or soil, aiding in the mitigation of pollution and contamination. Sorbents are either natural or synthetic and have properties that make them valuable tools in various remediation applications. Some of the recent interest in the use of sorbents for environmental remediation is described in the following sections.

14.3.2.1 Water remediation

Commonly used sorbents play an important role in the removal of pollutants from water sources. For example, AC and clay are often applied in the sorption of organic pollutants, heavy metals, and developing contaminants such as medicines and personal care products in water sources [33]. Innovative sorbents for improved water treatment efficiency, such as carbon-based systems (e.g. graphene-based materials, MXenes, nanomaterials, etc.) and modified clays for improved water treatment efficiency, are under continuous investigation and have the potential to reduce sorbent costs and environmental impacts.

14.3.2.2 Air pollution control

The removal of toxic gases and particulate matter found in the air using sorbents is vital in air pollution control technologies. Traditional sorbents such as AC, zeolites, and MOFs are being applied to the sorption of volatile organic compounds (VOCs) and odorous chemicals. Novel sorbent materials such as nanomaterials, transition-based carbides, and nitrides [2] are also being investigated for their potential to sorb VOCs, store carbon dioxide (CO_2) emissions, etc. Such advances in sorbent materials can contribute to cost reduction and environmental remediation.

14.3.2.3 Soil remediation

Through the sorbing process and/or contaminant immobilization, sorbents aid the remediation of contaminated soils [34]. This is attributed to their capacity to retain heavy metals and organic molecules. Clay-based sorbent minerals such as bentonite and kaolinite have been extensively applied in soil remediation [35]. The use of biochar, a carbon-rich substance derived from biomass, has shown promise in the adsorption of organic contaminants and the improvement of soil quality. Other modified commonly used sorbents such as zeolites are essential in soil remediation.

14.3.2.4 Oil spill remediation

In oil spill remediation, the use of sorbents is effective in absorbing and recovering spilled oil from water surfaces. Polymeric materials such as polypropylene fibers, cellulose-based materials, and natural sorbents such as peat moss and sawdust can be modified to improve their sorptive capacity as oil sorbents. Recent advancements include the development of superhydrophobic [1] and magnetically responsive sorbents for efficient oil spill remediation [36].

14.3.2.5 Emerging contaminant removal

The increased awareness of emerging contaminants such as per- and polyfluoroalkyl substances (PFASs) and microplastics [37, 38] has led to a renewed focus on developing sorbents and nanosorbents specifically designed to target these pollutants. Continuous research into the development of novel and modified clays, polymeric materials, combinations of sorbents, carbon-based materials such as nanotubes, nanowires, etc. with enhanced affinity for the abstraction of these emerging contaminants from different environmental matrices is important from both cost and environmental perspectives.

14.4 Conclusions

The choice of a sorbent material is generally influenced by its cost and its specific application, as discussed and highlighted in the current chapter. However, the choice of a sorbent should also be guided by its environmental impact. While commonly used sorbents such as AC, clay, polymeric materials, silica gel, zeolites, MOFs, molecular sieves, and chitosan are generally cost-effective and readily available, their production and disposal processes can have negative consequences for the environment.

The chapter has emphasized the need to move beyond a focus on cost-effectiveness alone and consider the broader environmental implications of sorbent selection. It is crucial to adopt sustainable practices that minimize resource consumption, pollution, and waste generation throughout the sorbent life cycle. This can include utilizing renewable resources, implementing the principles of green chemistry, and promoting recycling and reusability.

A comprehensive approach to sorbent selection involves the use of LCAs to evaluate the overall environmental impact. LCAs provide valuable insights into the carbon footprints, energy consumption, water usage, and waste generation associated with sorbent materials. By incorporating LCA results into decision-making processes, stakeholders can make informed choices that are aligned with the principles of a circular economy and minimize environmental harm.

However, this chapter also acknowledges the challenges of comparing and evaluating sorbent materials in terms of their costs and environmental impacts. Variations in sorption capability testing methods, material sources, and the lack of standardized metrics make it challenging to perform direct cost comparisons. In addition, assessing the environmental impact requires a comprehensive understanding of the entire life cycle, from production to disposal.

In conclusion, the choice of sorbent material should consider both cost and environmental impact. By recognizing the challenges and addressing them through sustainable practices and LCAs, stakeholders can make more informed decisions and thereby select effective and environmentally responsible sorbents. This chapter provides valuable insights into the parameters influencing sorbent cost and aims to contribute to the advancement of sustainable environmental remediation practices.

References

[1] Patalano A *et al* 2019 Scaling sorbent materials for real oil-sorbing applications and environmental disasters *MRS Energy Sustain* **6** 1–13

[2] Obodo K O *et al* 2021 Sn_3C_2 monolayer with transition metal adatom for gas sensing: a density functional theory studies *Nanotechnology* **32** 355502

[3] Unuabonah E I, Günter C, Weber J, Lubahn S and Taubert A 2013 Hybrid clay: a new highly efficient adsorbent for water treatment *ACS Sustain. Chem. Eng.* **1** 966–73

[4] Shukla S K, Al Mushaiqri N R S, Al Subhi H M, Yoo K and Al Sadeq H 2020 Low-cost activated carbon production from organic waste and its utilization for wastewater treatment *Appl. Water Sci.* **10** 1–9

[5] Kausar A *et al* 2018 Dyes adsorption using clay and modified clay: a review *J. Mol. Liq.* **256** 395–407

[6] Leonzio G, Fennell P S and Shah N 2022 A Comparative study of different sorbents in the context of direct air capture (DAC): evaluation of key performance indicators and comparisons *Appl. Sci.* **12** 2618

[7] Yadav P *et al* 2021 Assessment of the environmental impact of polymeric membrane production *J. Memb. Sci.* **622** 118987

[8] Lawler W *et al* 2012 Towards new opportunities for reuse, recycling and disposal of used reverse osmosis membranes *Desalination* **299** 103–12

[9] Zhao C *et al* 2022 Polymeric nanocomposite membranes for gas separation: performance, applications, restrictions and future perspectives *Case Stud. Therm. Eng.* **38** 102323

[10] Godage N H and Gionfriddo E 2020 Use of natural sorbents as alternative and green extractive materials: a critical review *Anal. Chim. Acta* **1125** 187–200

[11] Kumar P S, Korving L, van Loosdrecht M C M and Witkamp G J 2019 Adsorption as a technology to achieve ultra-low concentrations of phosphate: research gaps and economic analysis *Water Res.* X **4** 100029

[12] Gkika D A, Mitropoulos A C and Kyzas G Z 2022 Why reuse spent adsorbents? The latest challenges and limitations *Sci. Total Environ.* **822** 153612

[13] Alsawy T, Rashad E, El-Qelish M and Mohammed R H 2022 A comprehensive review on the chemical regeneration of biochar adsorbent for sustainable wastewater treatment *npj Clean Water* **5** 29

[14] Momina M, Shahadat M and Isamil S 2018 Regeneration performance of clay-based adsorbents for the removal of industrial dyes: a review *RSC Adv.* **8** 24571–87

[15] Baskar A V *et al* 2022 Recovery, regeneration and sustainable management of spent adsorbents from wastewater treatment streams: a review *Sci. Total Environ.* **822** 153555

[16] Aghel B, Mohadesi M, Gouran A and Razmegir M H 2020 Use of modified Iranian clinoptilolite zeolite for cadmium and lead removal from oil refinery wastewater *Int. J. Environ. Sci. Technol.* **17** 1239–50

[17] Woo S Y, Lee H S, Ji H and Kim Y D 2021 Adsorption isotherm model for analyzing the adsorption characteristics of water vapor to commercially available silica gel adsorbents for adsorption desalination applications *J. Chem. Eng. Data* **66** 1144–56

[18] Zhao G 2011 Sorption of heavy metal ions from aqueous solutions: a review *Open Colloid Sci.* J **4** 19–31

[19] Elkhaleefa A *et al* 2021 Evaluation of the adsorption efficiency on the removal of lead(II) ions from aqueous solutions using Azadirachta indica leaves as an adsorbent *Processes* **9** 559

[20] Qasem N A A, Mohammed R H and Lawal D U 2021 Removal of heavy metal ions from wastewater: a comprehensive and critical review *npj Clean Water* **4** 1–15

[21] Gil A *et al* 2021 A review of organic-inorganic hybrid clay based adsorbents for contaminants removal: synthesis, perspectives and applications *J. Environ. Chem. Eng.* **9** 105808

[22] Jha S, Gaur R, Shahabuddin S and Tyagi I 2023 Biochar as sustainable alternative and green adsorbent for the remediation of noxious pollutants: a comprehensive review *Toxics* **11** 1–23

[23] Tamjidi S and Ameri A 2020 A review of the application of sea material shells as low cost and effective bio-adsorbent for removal of heavy metals from wastewater *Environ. Sci. Pollut. Res.* **27** 31105–19

[24] Orta M and del M *et al* 2020 Biopolymer-clay nanocomposites as novel and ecofriendly adsorbents for environmental remediation *Appl. Clay Sci.* **198** 105838

[25] Singh S, Kapoor D, Khasnabis S, Singh J and Ramamurthy P C 2021 Mechanism and kinetics of adsorption and removal of heavy metals from wastewater using nanomaterials *Environ. Chem. Lett.* **19** 2351–81

[26] Ho Y S 2006 Review of second-order models for adsorption systems *J. Hazard. Mater.* **B136** 681–9

[27] Fan L, Luo C, Sun M, Li X and Qiu H 2013 Highly selective adsorption of lead ions by water-dispersible magnetic chitosan/graphene oxide composites *Colloids Surf. B Biointerfaces* **103** 523–9

[28] Velarde L, Nabavi M S, Escalera E, Antti M L and Akhtar F 2023 Adsorption of heavy metals on natural zeolites: a review *Chemosphere* **328** 138508

[29] Feng C *et al* 2021 Key technology and application analysis of zeolite adsorption for energy storage and heat-mass transfer process: a review *Renew. Sustain. Energy Rev.* **144** 110954

[30] Jain A, Kumari S, Agarwal S and Khan S 2021 Water purification via novel nano-adsorbents and their regeneration strategies *Process Saf. Environ. Prot.* **152** 441–54
[31] Luo J *et al* 2021 Critical review of advances in engineering nanomaterial adsorbents for metal removal and recovery from water: mechanism identification and engineering design *Environ. Sci. Technol.* **55** 4287–304
[32] Leonzio G, Fennell P S and Shah N 2022 A comparative study of different sorbents in the context of direct air capture (DAC): evaluation of key performance indicators and comparisons *Appl. Sci.* **12** 2618
[33] Pourhakkak P, Taghizadeh M, Taghizadeh A and Ghaedi M 2021 Fundamentals of adsorption technology *Adsorption: Fundamental Processes and Applications* ed M Ghaedi (Amsterdam: Elsevier Interface Science and Technology)Vol 33 1–70
[34] Das P, Arias V A, Kambala V, Mallavarapu M and Naidu R 2013 Remediation of perfluorooctane sulfonate in contaminated soils by modified clay adsorbent: a risk-based approach *Water Air Soil Pollut.* **224** 1714
[35] Otunola B O and Ololade O O 2020 A review on the application of clay minerals as heavy metal adsorbents for remediation purposes *Environ. Technol. Innov.* **18** 100692
[36] Abujaber F *et al* 2018 Magnetic cellulose nanoparticles coated with ionic liquid as a new material for the simple and fast monitoring of emerging pollutants in waters by magnetic solid phase extraction *Microchem. J.* **137** 490–5
[37] Chen Y J *et al* 2020 Metal-organic framework-based foams for efficient microplastics removal *J. Mater. Chem.* A **8** 14644–52
[38] Goh P S *et al* 2022 Nanomaterials for microplastic remediation from aquatic environment: why nano matters? *Chemosphere* **299** 134418

Chapter 15

The challenges of, and perspectives on adsorption applications for environmental sustainability

Kingsley Eghonghon Ukhurebor, Uyiosa Osagie Aigbe, Robert Birundu Onyancha, Kenneth Kennedy Adama, Osikemekha Anthony Anani, Joseph Onyeka Emegha, Benedict Okundaye, Rout George Kerry, Atala Bihari Jena, Francis Jesmar P Montalbo, Olusoji Anthony Ayeleso, Grace Jokthan and Aizebeoje Balogun Vincent

Biosorption (adsorption or sorption) is a crucial procedure for the treatment of wastewater that is highly effective, cost-effective, and environmentally sustainable. This technique mainly uses the adsorption (or biosorption or sorption) mechanism and procedure to remove toxicants from wastewater. Hence, this chapter covers the main challenges and the probable pollutant removal mechanisms of sorbents based on agricultural and nonagricultural materials. This also helps to clarify how modifying or treating these sorbents with other composites can improve their efficiency. The major research gaps as well as potential future research areas (perspectives) are also highlighted in this chapter.

List of unusual acronyms used

Acronym	Definition
ESS	Environmental safety and sustainability
eqb	Equilibrium
exp	Experimental
HMs	Heavy metals
ISM	Isotherm
max	Maximum
NMs	Nanomaterials
NTech	Nanotechnology
PNEs	Potentially noxious elements

doi:10.1088/978-0-7503-5598-8ch15

15.1 Introduction

Water is one of the most essential natural resources that influence the natural environment [1–4]; consequently, its conservation and preservation should be taken seriously in order to support environmental safety and sustainability (ESS) [5–7]. As water represents a significant environment for living organisms (both plants and animals), it is essential to avert contamination by both organic (biological) and inorganic (mineral) contaminants. Nevertheless, some technologies employed for this purpose emit secondary impurities or by-products, which additionally adulterate the ecosystem [8–10].

In order to battle hunger, poverty, and malnutrition, there is an urgent need for enough food to be produced year-round. This makes it imperative to reuse wastewater for irrigation [11, 12]. The last few decades have seen increasing interest among academics in industrial wastewater recycling as a means of reusing water. Because wastewater recycling improves agricultural output and has numerous benefits for farming operations, wastewater should be treated carefully before being applied to crops [12, 13]. This will protect the environment, save money, and increase agricultural productivity [14, 15].

Wastewater treatment presents a much bigger problem than it first seems. The contaminants in wastewater can be categorized according to two major sources [1, 16]; (i) organic (natural) contaminants, which are due to, for example, volcanicity, erosion of land (soil), and rock weathering; and (ii) inorganic (mineral) contaminants, which are due to contaminant dispersion via human events, such as waste dumps, urban/municipal runoff, mining, the production of printed circuit boards, agronomic (farming) events, the purification (treatment) and electroplating of metal surfaces, the burning of fuels, fabric or textile dyeing, and the production of semiconductors. Heavy metals (HMs), dyes, excessive nutrients, and other noxious inorganic and organic contaminants present significant environmental and health risks when they are present in the wastewater produced by agriculture, industry, and households, among other sources [1, 17, 18].

Potentially noxious elements (PNEs) are HMs and metalloids that fall into the category of trace elements with a mass greater than 4.10 g cm^{-3}. They include Al, As, Cd, Cr, Cu, Co, Fe, Pb, Mn, Hg, Ni, Ag, Sn, and Zn. The maximum (max) permitted contaminant level guidelines for the riskiest HMs, along with their respective associated negative impacts on the environment and human health, are listed in table 15.1 [1, 17]. These PNEs are present in wastewater/effluents that are not adequately treated before being discharged into freshwater resources, where they have a variety of negative impacts on the environment and human health [19, 20]. These PNEs also adversely affect human health because they are absorbed by aquatic creatures, crops, and other plant species before entering the human food chain [1, 17, 21, 22].

Environmental toxicity is caused by a variety of chemical pollutants (including HMs and dyes) that are released into the environment through spills, leakages, emissions, effluent, and/or solid refuse, and their management at excess amounts and/or due to overexposure is challenging [23, 24]. These chemical pollutants are

Table 15.1. The max permitted contaminant level guidelines for the riskiest HMs along with their respective associated negative impacts on the environment and human health.

PNE	Max permitted contaminant level (mg l^{-1})	Associated negative impacts on human health
As	0.050	Visceral cancers, vascular illness/diseases, and skin symptoms/ manifestations
Cd	0.010	Damage to the kidneys, renal diseases, and human cancer
Cr	0.050	Headache, sickness, vomiting, diarrhea, and carcinogens
Cu	0.250	Wilson's disease, insomnia, and liver damage
Pb	0.006	Disorders of the kidneys, heart, blood vessels, and nervous system, as well as injury to the fetal brain
Hg	0.00003	Disorders of the kidneys, heart, and nervous system, as well as rheumatoid arthritis
Ni	0.200	Human carcinogen, dermatitis, nausea, persistent asthma, coughing
Zn	0.800	Lethargy, depression, increased thirst, and neurological symptoms

introduced to the environment (air, soil, and water), where they find their way into the food chain and have a detrimental effect on the health and survival of plants and animals [25–27].

There are two main categories of wastewater or effluent pollutants, viz organic (biological) and inorganic (mineral) pollutants, in addition to natural and human-induced causes. Dyes, biphenyls, fats, herbicides, hydrocarbons, oils, pesticides, pharmaceuticals, phenols, proteins, and starches are examples of organic pollutants, whereas chemical fertilizers, PNEs, and an abundance of minerals are examples of inorganic pollutants. These pollutants lead to a decline in water quality and significant environmental issues [1, 17, 28, 29]. Numerous methods or technologies are employed to reduce the risks that wastewater or effluent presents to the environment and human health. These methods or technologies include adsorption (biosorption or sorption), chemical precipitation, coagulation/flocculation, electrodialysis, flotation, ion exchange, membrane filtration, solvent extraction, oxidation, and reverse osmosis [1, 17, 30, 31]. In comparison to the other methods or technologies mentioned above, sorption (biosorption or adsorption) involves chemical renewal requirements, plant fouling, adsorbent corrosion, and exhaust air discharge [1, 17, 32–34]. Adsorption (also known as biosorption or sorption) utilizes various adsorbent constituents and is considered to be easier to carry out and also to be more affordable [1, 17, 32–34]. In addition to its main advantages, the generation of by-products during this procedure does not result in secondary (subordinate) pollution [1, 17]. Different adsorption (biosorption or sorption) processes exist for removing different pollutants from wastewater, depending on the organic (biological) and inorganic (mineral) substances that are frequently utilized as adsorbents (such as activated clay inorganic minerals, agricultural wastes, biofuels, carbon, industrial by-products, polymer inorganic minerals, zeolite, etc) [1, 17, 18, 33]. To intensify the capability of adsorbent constituents to capture or confiscate pollutants (effluents), some researchers have combined them with other chemicals and substances to create hybrids. Virtually every domain of human interest has recently witnessed the utilization of nanotechnology (NTech), including wastewater (effluent) purification or treatment, soil (land) and natural resource management, and ecological clean-up [1, 17, 35]. In several cases, nanoadsorbents have been even more efficient than conventional adsorbents at confiscating organic (biological) and inorganic (mineral) contaminants from wastewater for ESS. Carbon nanotubes, metal-oxide-based nanomaterials (NMs), graphite, plant/agro-based nanocomposites, and other materials have been used to make these adsorbents [1, 17, 20, 36].

However, there is a pressing need for wastewater (effluent) treatment that employs sorption (biosorption or adsorption) methods or technologies based on various adsorbents or nanoadsorbent (sorbent) materials that are more effective and affordable. Therefore, this chapter discusses the main challenges and possible methods or technologies for removing pollutants from sorbents (adsorbents or nanoadsorbents) made from both agricultural and nonagricultural materials. This chapter also clarifies how these sorbents (adsorbents or nanoadsorbents) can be modified or treated with other composites to increase their effectiveness for ESS. This chapter also highlights the major study gaps and potential future research areas (perspectives) for ESS.

15.2 The effects of wastewater pollution on the environment

Protecting the environment from PNEs has been the subject of numerous studies and efforts. However, the 2030 Sustainable Development Goals (SDGs) of the United Nations also place a strong emphasis on advancing our understanding of resource conservation and promoting overall health and wellbeing (i.e. terrestrial/soil, aquatic/water, and aerial/atmospheric environments) [1, 17]. An excellent example of this is the utilization of wastewater or effluent for irrigation in areas without access to high-quality water [1, 17, 37]. However, frequent wastewater irrigation can lead to a buildup of PNEs in the soil and plants, which can have negative impacts on human health [1, 17]. Mineral nutrients such as nitrogen (N) and phosphorus (P), PNEs, organic (biological) materials, and microbes are the primary pollutants in wastewater [20, 38].

Several pathogenic microorganisms, including bacteria, fungi, protozoa, and viruses that instigate waterborne illnesses, are primarily found in organic material in wastewater or effluent [39]. In addition, wastewater or effluent may have a large salt concentration that builds up in the root zone of the soil, impacting its productivity, quality, and properties. The continual application of effluent or wastewater that contains a lot of salt damages the structure of the soil and reduces soil yield [40]. The salinization of soil caused by effluent irrigation has been widely documented lately [41]. In addition, it has been claimed that agronomic regions in Mexico and the US that have been irrigated by effluents/wastewater for over 50 years have meager soil health and a constrained capacity to sustain the productivity of crops [42–44]. Another reason that untreated effluent/wastewater is utilized for irrigation in developing nations is the lack of appropriate equipment, facilities, and accessibility [45]. This has an impact on about 2.45×10^3 km^2 of aquatic/marine environments, piscaries (fisheries), and constituents that eventually reach the food web [46]. According to a 2012 World Health Organization (WHO) report, inadequate sanitation, water pollution, and a lack of vital nutrients were to blame for about 842 000 mortalities in low- and middle-income nations. These diseases are typically seen in infants/babies or young children under the age of five [47]. As a result, wastewater/effluent treatment possibilities, food security, and hazard evaluation have received a lot of global attention [1, 17]. In most societies, people now use wastewater treatment rather than untreated use due to improved knowledge of the preclusion of waterborne diseases. This is because effluent/wastewater treatment plays a substantial role in averting and spreading waterborne as well as other water-related illnesses, ecological disruptions, eutrophication, waterlogging, the emergence of salt in agronomic land/soils, as well as other adverse effects on human health caused by the consumption of wastewater-irrigated food crops [48–50].

15.3 Treatment methods or technologies for wastewater

As previously mentioned, various methods are employed to eliminate pollutants from wastewater. Physical, pharmacological, and biological methods of treatment are available [36, 51–53]. Treatment procedures are crucial for the purification and reuse of wastewater. Primary, secondary, and tertiary treatments are a few of these therapies. Initial screening and the sedimentation of larger waste particles from household, municipal, and commercial sources are examples of primary treatments. Wastewater is handled aerobically and anaerobically in lagoons or ponds in the second stage of treatment. This stage includes the processing of treatable wastes through microbial action, biodegradation, and bioremediation. Chemical treatment techniques such as flocculation, coagulation, ion exchange, membrane filtering, etc. are included in tertiary treatment. Multiple treatment modalities are, however, occasionally used in conjunction with other forms of remediation. Figure 15.1 shows most of the treatment methods or technologies for wastewater and their possible weaknesses. However, among the various new water treatment technologies for confiscating pollutants from effluent, wastewater, or other water, adsorption (biosorption or sorption) continues to be the most efficient approach due to its affordability and simplicity of operation and design [1, 54].

15.4 Challenges of adsorption applications for environmental safety and sustainability

It has been shown that in the current water treatment scenarios based on adsorption (biosorption or sorption) used for ESS, we need to focus more on basic factors such as the type of adsorbent, the operational approach, the modeling, and the regeneration (recycling), as well as the application in actual water models or samples

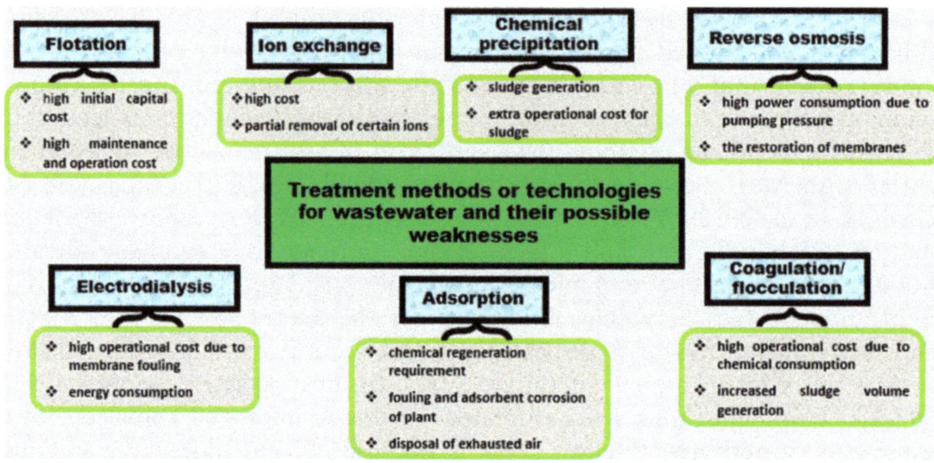

Figure 15.1. Most of the treatment methods or technologies for wastewater and their possible weaknesses.

[1, 16, 17, 55, 56]. Considering this evidence, the academic and industrial benefits, and our own experience, we believe that the primary difficulties (challenges) in water management and purification (treatment) by biosorption (adsorption or sorption) for ESS are as follows [1, 17, 56]:

- Of course, an important consideration for the end-user is the total financial implication of a biosorption (adsorption or sorption) treatment or purification plant. The prices of the adsorbent itself, operating expenses, and regeneration costs are all significant. About 70% of the overall cost is spent on the adsorbent. In this manner, a sustainable response can be provided and the advantages of utilizing a low-cost adsorbent can be realized by the creation of novel and affordable adsorbent constituents from wastes/effluents and biomass-based precursors.
- Although industry has improved waste/effluent purification or treatment and reduced the concentrations of pollutants in its discharges in response to the growing environmental regulations, these levels could possibly be further reduced and should be further reduced. However, researchers conduct our biosorption (adsorption or sorption) research studies of novel bio-adsorbents/adsorbents with an emphasis on: (i) identifying the max or highest sorption (adsorption or biosorption) capacity of a process or system; (ii) conducting equilibrium (eqb) isotherm (ISM) studies employing experimental (exp) data that are based on the monolayer capacity or the max or highest capacity at the topmost/highest eqb figures/values, and (iii) getting or obtaining the whole or complete ranges of ISTs utilizing error analysis procedures that are weighted in the direction of the topmost/highest adsorbate eqb concentrations. Nevertheless, because the industrial effluent concentrations of the pollutant might only be a small fraction of the max or highest eqb saturation capacity (say, only 50%), the adsorbent's uptake capacity will reach its max

or highest value. The number at the vertical operating or effective line that corresponds to the concentration of the fluid eqb stage or phase is the optimal adsorbent capacity value. This represents the upper limit for fixed-bed adsorption (biosorption or sorption) devices. The shape or outline of the graph, the quantity of exp points, and the uptake or acceptance curve or arc analysis, therefore, turn out to be more important for systems operating at considerably lower uptake or acceptance capacities than the maximal ISM capacities. It is entirely possible that an adsorbent with a reduced max or topmost capacity will be able to hold more of the sorbate at lower solution eqb values.

- Almost all academic research studies of adsorption (biosorption or sorption) use eqb ISM analysis, and only about half of them record any kind of kinetic or diffusion rate experiments or modeling research. However, less than one-tenth (10%) of the published research studies include exp and modeling work based on fixed-bed columns. Due to the fact that all medium/average- and large-scale/comprehensive industrial applications rely on fixed-bed sorption (adsorption or biosorption) processes or systems, this presents a significant problem. Fixed-bed adsorbers work differently from highly agitated, top-agitated, or shaken batch adsorbers.
- The slope/gradient ($-V\ m^{-1}$) in the adsorber mass balance equation controls the operating line for agitated batch adsorbers, which limits the effluent discharge that can be accomplished as well as limiting the effective or operational application or utilization of the bio-adsorbent/adsorbent to this design limit or restraining value. In order to improve adsorption (biosorption or sorption) utilization and achieve a reduced discharge concentration, the use of multibatch biosorption (adsorption or sorption) reactors is of interest. When two or more adsorbers are installed, a thorough optimization analysis is required to determine their economics and operational strategy. In addition, adsorption (biosorption or sorption) in water treatment is inadequately understood for operational modes other than columns. The fluidized bed, the virtual moving bed, the spouted bed, and the continuously stirred tank reactor all need more research.
- In order to model biosorption (adsorption or sorption), it is necessary to form ISM models for multicomponent processes or systems and techniques for the precise determination of the thermodynamic variables/parameters of biosorption. In addition, it is essential to create and use diffusional models for multicomponent processes or systems and to develop correlations to calculate the surface diffusion coefficient. Research studies of composite models that couple ISM and kinetic models with response surface methodology (RSM), artificial neural networks (ANNs), factory information systems (FISs), and adaptive neuro-fuzzy inference systems or adaptive network-based fuzzy inference systems (ANFISs) are to be encouraged.
- In the era of sustainable development, developing efficient regeneration processes or systems must be a priority. These systems must allow for both the regeneration or recycling of the adsorbent for subsequent utilization as

well as the capability to recover, recuperate, reuse, or use the adsorbate. In order to reduce waste production and costs, enable adsorbate recovery, and increase the number/quantity of cycles the adsorbent can be employed in, it is necessary to generate alternative and environmentally friendly regeneration methods.

- Adsorption (biosorption or sorption) needs to be optimized under regulated circumstances before it is used in practical situations. The ideal or model capacity of a single-component biosorption (adsorption or sorption) device is reduced by the virtually constant multicomponent nature of real waters and effluent discharge, which frequently contain competing adsorbates. Then, because selectivity modeling for multicomponent systems is quite analytically challenging, we frequently require empirical parameters, such as diffusivities, transfer of mass coefficients, or/and activity coefficients, that fit the data the best.

15.5 Perspectives on adsorption applications for environmental safety and sustainability

To guarantee or certify high-quality or first-rate water, lessen biological and chemical effluents or contaminants, and improve industrial production or invention processes for ESS, advanced adsorption (biosorption or sorption) technologies are urgently required today [1, 17, 55, 57, 58]. The majority of the research results up to this point have involved small-scale laboratory exp work. The primary drawback of using cost-effective adsorbents in place of activated carbon and other expensive treatment or purification technologies is the lack of knowledge, information, data, or facts about pilot-scale systems [10, 56]. More research is needed before we can encourage the utilization of unconventional or alternative adsorbents/bio- adsorbents on an industrial or large scale. However, one of the best methods for cutting-edge effluent or contaminant treatment is NTech, which involves the adsorption (biosorption or sorption) of contaminants using NMs [6].

Research into effluent and contaminant treatments has been successfully accomplished, and many different NMs have been created. A few of these are worth mentioning: nanoadsorbents (based on the metals Fe, MnO, ZnO, and MgO and carbon nanotubes), photocatalysts, electrocatalysts (Pt and Pd), and nanofilms [1, 17, 27]. In order to increase the effectiveness of the effluent treatment process, these NMs can also added to biological or organic processes (such as algae membranes, anaerobic fermentation, and microbial/microorganism-based fuel cells). Each technology has benefits and varies in how well it confiscates and eliminates contaminants [1, 10, 17]. PNEs such as As, Cr, Cu, Hg, Ni, Pb, and Zn can be separated from effluent using bionanoadsorbents or nanoadsorbents [33, 58]. The alteration of the catalyst constituent opens the possibility of using available areas to absorb radiant energy (sunlight) rather than using pricey synthetic ultraviolet radiation in the treatment of toxic contaminants using NM photocatalysts.

Future research should focus on efficient procedures that only need small concentrations of NMs, because they are still more costly than conventional

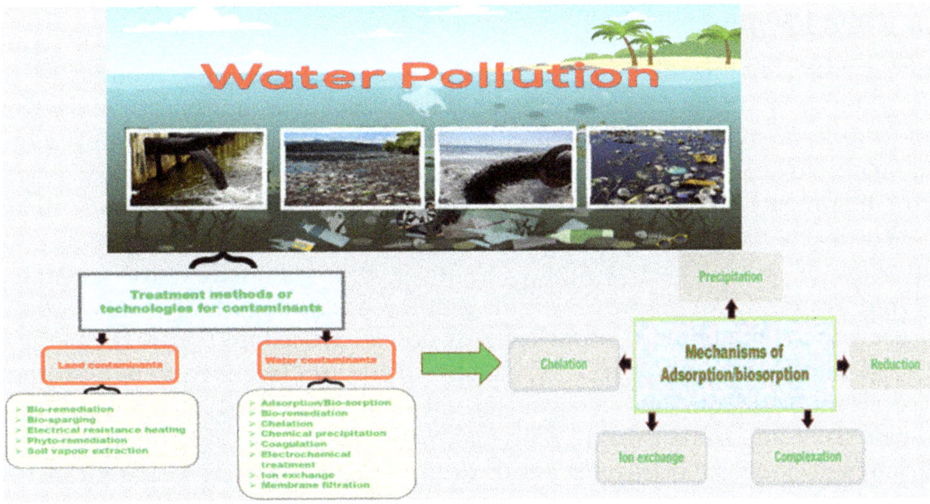

Figure 15.2. A graphical presentation of some of the treatment technologies for contaminants vis-à-vis some aspects or features of the adsorption mechanisms for water purification or treatment. Reprinted by permission from Springer Nature [60], Copyright (2023).

materials (such as activated carbon, etc) [6, 24, 32, 35, 36, 59]. However, the different aspects of adsorption for water purification or treatment, some of the primary problems, and the possibilities of adsorption for treating water were strongly emphasized in a recent review of studies by Ukhurebor *et al* [60]. Some of the treatment technologies for contaminants vis-à-vis some aspects of the adsorption mechanisms for water treatment are graphically illustrated in figure 15.2 (adopted from [60]).

In addition, further work is required to create affordable techniques for NM synthesis as well as to evaluate productivity on a large scale for effective field applications. Future studies need to close the following gaps:

- Studies at the cellular level, as well as at the molecular level, are required to comprehend the workings of adsorbents in wastewater purification.
- In addition to laboratory testing, extra work is required on pilot-scale and field-scale tests.
- Before pilot-scale exp work is conducted, the cost versus effectiveness and requirements need to be estimated and computed.
- It is important to investigate the effectiveness of the discussed adsorbents in combination with other treatment modalities, particularly carbon-based adsorbents.
- It is important to investigate microbial interactions in aquatic habitats.
- The influence of geo-environmental variables on adsorption (biosorption or sorption) mechanisms needs to be studied further.

15.6 Conclusions

Due to greater urbanization and industrialization, the generation, production, and purification or treatment of wastewater or effluents are creating concern in the world today (twenty-first century). Several effluent treatment technologies or techniques, including biological, chemical, as well as physical (primary and secondary purification or treatment) methods, are used to mitigate the issue of the rising ecological risks. The use of various cleaning techniques or methods raises the possibility of the production of secondary pollution. The most effective method for treating or purifying wastewater or effluents with the smallest generation of secondary pollution is to use various materials as adsorbents. A crucial procedure that is both efficient and economical is adsorption (biosorption or sorption). The adsorption (biosorption or sorption) procedure and mechanism are mainly utilized in this method to eliminate toxicants from effluent. The main sorbents based on agrarian and non-agrarian materials were briefly discussed, along with any potential pollutant removal methods. Furthermore, this helped to expound how modifying or treating these sorbents with other substances can advance their effectiveness. This was accompanied by a focus on the significant research issues and likely future research directions.

This chapter's core aim was to discuss the challenges and perspectives of adsorption (biosorption or sorption) applications for ESS. The current situation and the primary difficulties with adsorption (biosorption or sorption) applications for ESS vis-à-vis water purification were explicitly discussed in this chapter. This chapter is intended to act as a quick reference guide for new theories and original research in this area. In conclusion, the chapter highlighted the next steps most likely to be needed in order to advance this field.

References

[1] Younas F *et al* 2021 Current and emerging adsorbent technologies for wastewater treatment: trends, limitations, and environmental implications *Water* **13** 215

[2] Aigbe U, Onyancha R, Ukhurebor K and Obodo K 2020 Removal of fluoride ions using polypyrrole magnetic nanocomposite influenced by rotating magnetic field *RSC Adv.* **10** 595–609

[3] Zhang X, Gu P and Liu Y 2019 Decontamination of radioactive wastewater: state of the art and challenges forward *Chemosphere* **215** 543–53

[4] Alssgeer H, Gasim M, Hanafiah M, Abdulhadi E and Azid A 2019 GIS-based analysis of water quality deterioration in the Nerus River, Kuala Terengganu, Malaysia *Desalin. Water Treat.* **112** 334–43

[5] Inyinbor A, Bello O, Oluyori A, Inyinbor H and Fadiji A 2019 Wastewater conservation and reuse in quality vegetable cultivation: overview, challenges and future prospects *Food Control* **98** 489–500

[6] 2022 Current existing techniques for environmental monitoring *Nanobiosensors for Environmental Monitoring: Fundamental and Applications* ed R Singh, K Ukhurebor, J Singh, C Adetunji and K Singh (Berlin: Springer Nature) 239–62 pp

[7] Aigbe U, Ukhurebor K, Onyancha R, Ama M, Osibote O, Kusuma H, Okanigbuan P, Azi S and Osifo P 2022 Dendrimers for environmental remediation *Nanotechnology for Environmental Remediation* ed T Sabu, P L.A and M Thomas (Berlin: Wiley)

[8] Ganjegunte G, Ulery A, Niu G and Wu Y 2018 Organic carbon, nutrient, and salt dynamics in saline soil and switchgrass (*Panicum virgatum* L.) irrigated with treated municipal wastewater *Land Degrad. Dev.* **29** 80–90

[9] Ukhurebor K, Aigbe U, Onyancha R, Nwankwo W, Osibote O, Paumo H, Ama O, Adetunji C and Siloko I 2021 Effect of hexavalent chromium on the environment and removal techniques: a review *J. Environ. Manage.* **280** 111809

[10] Aigbe U, Ukhurebor K, Onyancha R, Okundaye B and Osibote O 2023 Green Nanomaterials for Wastewater Treatment Analysis *Green Nanoarchitectonics* ed K Pal (Singapore: Jenny Stanford Publishing)

[11] Bartolomeu M, Neves M, Faustino M and Almeida A 2018 Wastewater chemical contaminants: remediation by advanced oxidation processes *Photochem. Photobiol. Sci.* **17** 1573–98

[12] Onyancha R, Aigbe U, Ukhurebor K, Kusuma H, Darmokoesoemo H, Osibote O and Pal K 2022 Influence of magnetism-mediated potentialities of recyclable adsorbents of heavy metal ions from aqueous solutions—an organized review *Res. Chem.* **4** 100452

[13] Aigbe U, Ukhurebor K, Onyancha R, Okundaye B, Pal K, Osibote O, Esiekpe E, Kusuma H and Darmokoesoemo H 2022 A facile review on the sorption of heavy metals and dyes using bionanocomposites *Adsorpt. Sci. Technol.* **2022** 1–36

[14] Sudarni D, Aigbe U, Ukhurebor K, Onyancha R, Kusuma H, Darmokoesoemo H, Osibote O, Balogun V and Widyaningrum B 2021 Malachite green removal by activated potassium hydroxide clove leaves agro-waste biosorbent: characterization, kinetic, isotherm, and thermodynamic studies *Adsorpt. Sci. Technol.* **2021** 1–15

[15] Souza R, Seibert D, Quesada H, Bassetti F, Fagundes-Klen M and Bergamasco R 2020 Occurrence, impacts and general aspects of pesticides in surface water: a review *Proc. Saf. Environ. Prot.* **135** 22–37

[16] Aigbe U, Ukhurebor K, Onyancha R, Osibote O, Darmokoesoemo H and Kusuma H 2021 Fly ash-based adsorbent for adsorption of heavy metals and dyes from aqueous solution: a review *J. Mater. Res. Technol.* **14** 2751–74

[17] Dotto G and McKay G 2020 Current scenario and challenges in adsorption for water treatment *J. Environ. Chem. Eng.* **8** 1–6

[18] El-Nemr M, Aigbe U, Ukhurebor K, Onyancha R, El Nemr A, Ragab S, Osibote O and Hassaan M 2022 Adsorption of Cr^{6+} ion using activated pisum sativum peels-triethylenetetramine *Environ. Sci. Pollut. Res.* 1–25 2022

[19] Eleryan A *et al* 2022 Copper (II) ion removal by chemically and physically modified sawdust biochar *Biomass Convers. Biorefin.* **2022** 1–38

[20] Chowdhary P, Raj A and Bharagava R 2018 Environmental pollution and health hazards from distillery wastewater and treatment approaches to combat the environmental threats: a review *Chemosphere* **194** 229–46

[21] Hussain M, Wang J, Bibi I, Shahid M, Niazi N, Iqbal J, Mian I, Shaheen S, Bashir S and Shah N 2020 Arsenic speciation and biotransformation pathways in the aquatic ecosystem: the significance of algae *J. Hazard. Mater.* **403** 124027

[22] Ahmed M and Hameed B 2018 Removal of emerging pharmaceutical contaminants by adsorption in a fixed-bed column: a review *Ecotoxicol. Environ. Saf.* **149** 257–66

[23] Saravanan A, Kumar S, Jeevanantham S, Karishma S, Tajsabreen B, Yaashika P and Reshma B 2023 Effective water/wastewater treatment methodologies for toxic pollutants removal: processes and applications towards sustainable development *Chemosphere* **280** 130595

[24] Cherkasov N 2020 Liquid-phase adsorption: common problems and how we could do better *J. Mol. Liq.* **301** 112378

[25] Neolaka Y, Riwu A, Aigbe U, Ukhurebor K, Onyancha R, Darmokoesoemo H and Kusuma H 2023 Potential of activated carbon from various sources as a low-cost adsorbent to remove heavy metals and synthetic dyes *Results Chem.* **5** 100711

[26] Salgot M and Folch M 2018 Wastewater treatment and water reuse *Curr. Opin. Environ. Sci. Health* **2** 64–74

[27] Escudero L, Quintas P, Wuilloud R and Dotto G 2019 Recent advances on elemental biosorption *Environ. Chem. Lett.* **17** 409–27

[28] Eldeeb T, Aigbe U, Ukhurebor K, Onyancha R, El-Nemr M, Hassaan M, Ragab S, Osibote O and El Nemr A 2022 Adsorption of methylene blue (MB) dye on ozone, purified and sonicated sawdust biochars *Biomass Convers. Biorefin.* 1–23

[29] Eldeeb T and Nemr *et al* 2022 Biosorption of acid brown 14 dye to mandarin-biochar-CO-TETA derived from mandarin peels *Biomass Convers. Biorefin.* 1–21

[30] Kusuma H, Aigbe U, Ukhurebor K, Onyancha R, Okundaye B, Ama O, Darmokoesoemo H, Widyaningrum B, Osibote O and Balogun V 2023 Biosorption of methylene blue using clove leaves waste modified with sodium hydroxide *Results Chem.* **5** 100778

[31] Eleryan A, Hassaan M, Aigbe U, Ukhurebor K, Onyancha R, El Nemr M, Ragab S, Hossain I and El Nemr A 2023 Kinetic and isotherm studies of acid orange 7 dye absorption using sulphonated mandarin biochar treated with TETA *Biomass Convers Biorefin.* 1–12 2023

[32] Pauletto P, Dotto G and Salau N 2020 Diffusion mechanisms and effect of adsorbent geometry on heavy metal adsorption *Chem. Eng. Res. Des.* **157** 182–94

[33] Alyasi H, Mackey H, Loganathan K and McKay G 2020 Adsorbent minimisation in a two-stage batch adsorber for cadmium removal *J. Ind. Environ. Chem.* **81** 153–60

[34] Salau N 2020 Diffusion mechanisms and effect of adsorbent geometry on heavy metal adsorption *Chem. Eng. Res. Des.* **157** 182–94

[35] Vargas V, Paveglio R, Pauletto P, Salau N and Dotto G 2020 Sisal fiber as an alternative and cost-effective adsorbent for the removal of methylene blue and reactive black 5 dyes from aqueous solutions *Chem. Eng. Commun.* **207** 523–36

[36] Rossatto D, Netto M, Jahn S, Mallmann E, Dotto G and Foletto E 2020 Highly efficient adsorption performance of a novel magnetic geopolymer/Fe_3O_4 composite towards removal of aqueous acid green 16 dye *J. Environ. Chem. Eng.* **8** 103804

[37] Caicedo C, Rosenwinkel K H, Exner M, Verstraete W, Suchenwirth R, Hartemann P and Nogueira R 2019 Legionella occurrence in municipal and industrial wastewater treatment plants and risks of reclaimed wastewater reuse *Water Res.* **149** 21–34

[38] Akpor O, Otohinoyi D, Olaolu D and Aderiye B 2014 Pollutants in wastewater effluents: impacts and remediation processes *Int. J. Environ. Res. Earth Sci.* **3** 50–9

[39] Jegatheesan V, Weragoda S and Visvanathan C 2008 Rapid water quality characterization for chlorine demand and THM formation in drinking waters *J. Water Supply Res. Technol. AQUA* **57** 259–72

[40] Khalid S, Shahid M, Bibi I, Sarwar T, Shah A and Niazi N 2018 A review of environmental contamination and health risk assessment of wastewater use for crop irrigation with a focus on low and high-income countries *Int. J. Environ. Res. Public Health* **15** 895

[41] Morugán-Coronado A, García-Orenes F, Mataix-Solera J, Arcenegui V and Mataix-Beneyto J 2011 Short-term effects of treated wastewater irrigation on Mediterranean calcareous soil *Soil Tillage Res.* **112** 18–26

[42] Maass O and Grundmann P 2016 Added-value from linking the value chains of wastewater treatment, crop production and bioenergy production: a case study on reusing wastewater and sludge in crop production in Braunschweig (Germany) *Resour. Conserv. Recycl.* **107** 195–211

[43] Cantu J, Beruvides M and Fedler C 2017 An economic framework for adaptive wastewater reuse for future crop production sustainability *67th Annual Conference and Expo of the Institute of Industrial Engineers 2017 (Pittsburgh, PA, USA)* (Peachtree Corners, GA: Institute of Industrial Engineers) 1566–70

[44] Magwaza S, Magwaza L, Odindo A and Mditshwa A 2020 Hydroponic technology as decentralised system for domestic wastewater treatment and vegetable production in urban agriculture: a review *Sci. Total Environ.* **698** 134154

[45] Murtaza G, Ghafoor A, Qadir M, Owens G, Aziz M and Zia M 2010 Disposal and use of sewage on agricultural lands in Pakistan: a review *Pedosphere* **20** 23–34

[46] Corcoran E 2010 *Sick water? The central role of wastewater management in sustainable development: a rapid response assessment* (Nairobi, Kenya: UNEP Publications) https://wedocs.unep.org/20.500.11822/9156

[47] Wolf J, Prüss-Ustün A, Cumming O, Bartram J, Bonjour S, Cairncross S, Clasen T, Colford J, Curtis V and De France J 2014 Systematic review: assessing the impact of drinking water and sanitation on diarrhoeal disease in low-and middle-income settings: systematic review and meta-regression *Trop. Med. Int. Health* **19** 928–42

[48] Qureshi A, Hussain M, Ismail S and Khan Q 2016 Evaluating heavy metal accumulation and potential health risks in vegetables irrigated with treated wastewater *Chemosphere* **163** 54–61

[49] Sharma A, Katnoria J and Nagpal A 2016 Heavy metals in vegetables: screening health risks involved in cultivation along wastewater drain and irrigating with wastewater *SpringerPlus* **5** 1–16

[50] Libutti A, Gatta G, Gagliardi A, Vergine P, Pollice A, Beneduce L, Disciglio G and Tarantino E 2018 Agro-industrial wastewater reuse for irrigation of a vegetable crop succession under Mediterranean conditions *Agric. Water Manag.* **196** 1–14

[51] Rubalcaba A, Suárez-Ojeda M, Stüber F, Fortuny A, Bengoa C, Metcalfe I, Font J, Carrera J and Fabregat A 2007 Phenol wastewater remediation: advanced oxidation processes coupled to a biological treatment *Water Sci. Technol.* **55** 221–7

[52] Joseph L, Jun B, Flora J, Park C and Yoon Y 2019 Removal of heavy metals from water sources in the developing world using low-cost materials: a review *Chemosphere* **229** 142–59

[53] Carmalin S and Lima E 2018 Removal of emerging contaminants from the environment by adsorption *Ecotoxicol. Environ. Saf.* **150** 1–17

[54] Chang H, Quan X, Zhong N, Zhang Z, Lu C, Li G, Cheng Z and Yang L 2018 High-efficiency nutrients reclamation from landfill leachate by microalgae Chlorella vulgaris in membrane photobioreactor for bio-lipid production *Bioresour. Technol.* **266** 374–81

[55] Diraki A, Abdala A, McKay G and Mackey H 2019 Removal of emulsified and dissolved diesel oil from high salinity wastewater by adsorption onto graphene oxide *J. Environ. Chem. Eng.* **7** 103106
[56] Gkika D, Mitropoulos A and Kyzas G 2022 Why reuse spent adsorbents? The latest challenges and limitations *Sci. Total Environ.* **822** 153612
[57] Adetunji C, Olaniyan O, Anani O, Inobeme A, Ukhurebor K, Bodunrinde R, Adetunji J, Singh K, Nayak V P W and Singh R 2021 Bionanomaterials for green bionanotechnology *Bionanomaterials* ed R Singh and K Singh (Bristol: IOP Publishing) 1–25 pp
[58] Aigbe U, Ukhurebor K, Onyancha R, Ama M, Okundaye B, Esiekpe E, Osibote O, Kusuma H and Osifo P 2021 Utility of bionanocomposites for wastewater treatment *Bionanomaterials for Environmental and Agricultural Applications* ed R Singh and K Singh (London: IOP Publishing) 1–25 pp
[59] Wong J, Tan H, Lau S, Yap P S and Danquah M 2019 Potential and challenges of enzyme incorporated nanotechnology in dye wastewater treatment: a review *J. Environ. Chem. Eng.* **7** 103261
[60] Ukhurebor K, Hossain I, Pal K, Jokthan G, Osang F, Ebrima F and Katal D 2023 Applications and contemporary issues with adsorption for water monitoring and remediation: a facile review *Top. Catal.* 1–16

www.ingramcontent.com/pod-product-compliance
Ingram Content Group UK Ltd.
Pitfield, Milton Keynes, MK11 3LW, UK
UKHW051903100926
8556IPUK00005B/746

9 780750 355964